KT-385-432

TIME-RESOLVED VIBRATIONAL SPECTROSCOPY

Academic Press Rapid Manuscript Reproduction

Based on the Proceedings of the International Conference on TRVS
Held in Lake Placid, New York, on August 16–20, 1982.

TIME-RESOLVED VIBRATIONAL SPECTROSCOPY

Edited by

GEORGE H. ATKINSON

Department of Chemistry
Syracuse University
Syracuse, New York

1983

ACADEMIC PRESS
A Subsidiary of Harcourt Brace Jovanovich, Publishers
NEW YORK LONDON
PARIS SAN DIEGO SAN FRANCISCO SÃO PAULO SYDNEY TOKYO TORONTO

ACADEMIC PRESS, INC.
111 Fifth Avenue, New York, New York 10003

United Kingdom Edition published by
ACADEMIC PRESS, INC. (LONDON) LTD.
24/28 Oval Road, London NW1 7DX

Library of Congress Cataloging in Publication Data
Main entry under title:

Time–resolved vibrational spectroscopy.

Papers presented at the First International Conference on Time–Resolved Vibrational Spectroscopy, Lake Placid, N.Y., Aug. 16–20, 1982.

1. Vibrational spectra––Congresses. 2. Raman spectroscopy--Congresses. I. Atkinson, George H. II. International Conference on Time–Resolved Vibrational Spectroscopy (1st : 1982 : Lake Placid, N.Y.)
QD96.V53T55 1983 543'.0858 83–9928
ISBN -0–12–066280–9

PRINTED IN THE UNITED STATES OF AMERICA

83 84 85 86 9 8 7 6 5 4 3 2 1

CONTENTS

Contributors *xi*
Foreword *xvii*
Preface *xix*

SECTION I THEORETICAL AND COMPUTATIONAL DEVELOPMENTS

Some Theoretical Considerations Concerning Raman Scattering 1
A. C. Albrecht

Intermediate Gauge Formulation of Two-Photon Spectroscopies 11
G. W. Robinson and C. W. Brown

The Local Mode Picture of Molecular Vibrations 23
Ole Sonnich Mortensen

Advantages of the Time Correlator Theory of Resonance Raman Scattering 31
C. K. Chan, J. B. Page, and D. L. Tonks

On the Interaction Operator in the Optical Spectroscopies: Position vs. Momentum 37
Duckhwan Lee and A. C. Albrecht

Semiclassical Simulation of Vibronic Processes 41
Arieh Warshel, P. Stern, and S. Mukamel

Normal Coordinate Analysis of Polyene Chains for Interpreting the Vibrational Spectra of Transient and Less Stable Species 53
Mitsuo Tasumi

Spectra from Molecular Dynamics 59
Peter H. Berens, John P. Bergsma, and Kent R. Wilson

SECTION II INSTRUMENTATION AND TECHNIQUES

Stimulated Raman Techniques for Time-Resolved Spectroscopy of Gas-Phase Molecules 63
P. Esherick, A. J. Grimley, and A. Owyoung

Time-Resolved Study of Subpicosecond Orientational Relaxation in Molecular Liquids 73
C. L. Tang and J-M. Halbout

Dynamic Measurements of Gas Properties by Vibrational Raman Scattering: Applications to Flames 83
Michael C. Drake, Marshall Lapp, Robert W. Pitz, and C. Murray Penney

Time-Resolved Photoacoustic Detection of Collisional Relaxation of Vibrationally Excited HD Molecules 91
Jack Gelfand, Richard B. Miles, Eric Rohlfing, and Herschel Rabitz

Stroboscopic Interferometry: A Means to Follow Broadband Infrared Phenomena Using an FT-IR Spectrometer 97
James L. Chao and T. Kumar

Infrared Multiphoton Excited Visible Luminescence from Hexafluorobenzene 105
Michael T. Duignan and Shammai Speiser

Difference Raman Spectroscopy with a Stopped-Flow Device 113
Yoshifumi Nishimura, Tadayuki Uno, and Masamichi Tsuboi

High Overtone Vibrational Spectrum of SF_6: Evidence for Localized Vibrations 117
Harold B. Levene and David S. Perry

Vibrational Energy Transfer Rate Constants from the Thermal Lens 121
R. T. Bailey, F. R. Cruickshank, R. Guthrie, D. Pugh, and I. J. M. Weir

Slowing of the Vibrational Dephasing Time by Hydrogen Bonding in Glycerol 127
R. Dorsinville, W. M. Franklin, N. Ockman, and R. R. Alfano

Time-Resolved V–T Energy Transfer Measurements in Iodine Monofluoride 135
P. J. Wolf, R. F. Shea, and S. J. Davis

SECTION III CHEMICAL SYSTEMS: GROUND AND EXCITED STATES

Raman Scattering from Excited States with Lifetimes into the Femtosecond Range 139
M. Asano, L. V. Haley, and J. A. Koningstein

Resonance Raman Spectroscopy of the Electronically Excited States of Transition Metal Bipyridine Complexes 147
W. H. Woodruff, R. F. Dallinger, M. Z. Hoffman, P. G. Bradley, D. Presser, V. Malueg, R. J. Kessler, and K. A. Norton

Spatial and Time-Resolved Resonance Raman Spectra of Excited Electronic States of the Uranyl Ion 157
M. Asano and J. A. Koningstein

Time-Resolved Resonance Raman Spectra of Excited $^3B_2$ Phenanthrene 161
D. A. Gilmore and G. H. Atkinson

Time-Resolved Resonance Raman Spectra of the Lowest Triplet State of *N,N,N′,N′*-Tetramethyl-*P*-Phenylenediamine in Solution 167
Kenji Yokoyama

Oxidation Reaction of Excited Triplet *p*-Benzoquinone 173
S. M. Beck and L. E. Brus

Resonance Raman Spectra and Vibrational Analyses of Reaction Intermediates 177
Ronald E. Hester

Photochemically Induced Electron Transfer Reactions of *trans*-Stilbene 183
W. Hub, S. Schneider, F. Dorr, J. T. Simpson, J. D. Oxman, and F. D. Lewis

Electron Transfer Reactions in Stilbene Isomers 191
H. Shindo and G. H. Atkinson

Time-Resolved Raman Studies of Radiation-Produced Radicals 199
G. N. R. Tripathi and R. H. Schuler

Raman Spectra of Phenoxy Radicals 207
Hitoshi Shindo

The Electronic and Resonance Raman Spectroscopy of the Inorganic Transients S_2^-,S_3^-, and Se_2^- Stabilized in Ultramarine 213
Robin J. H. Clark, David P. Fairclough, and Mohamedally Kurmoo

SECTION IV BIOCHEMICAL SYSTEMS

Time-Resolved Resonance Raman Spectroscopy of the K_{610} and O_{640} Photointermediates of Bacteriorhodopsin 219
Steven O. Smith, Mark Braiman, and Richard Mathies

Time-Resolved Resonance Raman Studies of the Photochemical Cycle of Bacteriorhodopsin 231
Thomas Alshuth, Iris Grieger, and Manfred Stockburger

Resonance Coherent Anti-Stokes Raman Spectroscopy of the K Intermediate in the Bacteriorhodopsin Photocycle 239
Aaron Lewis

Resonance Raman Spectra of Photochemical Picosecond Transients: Method and Application to Study Bacteriorhodopsin Primary Processes 251
M. A. El-Sayed, Chung-lu Hsieh, and Malcolm Nicol

Picosecond Events in Biological Processes 263
E. F. Hilinski and P. M. Rentzepis

Applications of Time-Resolved Resonance and Raman Spectroscopy in Radiation Chemistry and Photobiology: Structure and Chemistry of Carotenoids in the Excited Triplet State 273
Robert Wilbrandt and Niels-Henrick Jensen

Influence of Excitation Wavelength on the Time-Resolved Resonance Raman Spectroscopy of Deoxy Heme Proteins 287
M. J. Irwin and G. H. Atkinson

Structure and Dynamics in the Photolysis of CO- and O_2-Hemoglobin, Monitored by Time-Resolved Resonance Raman Spectroscopy 297
Thomas G. Spiro and James Terner

Structure–Function Relationships in Hemoglobin: Transient Raman Studies 307
J. M. Friedman

Resonance Raman Spectra of Photodissociated Carbonmonoxy Hemoglobin (Hb*): Room Temperature Transients and Cryogenically Trapped Intermediates 317
D. L. Rousseau, M. R. Ondrias, and J. M. Friedman

Time-Resolution by Transient Chromophore Creation: Defining the Scissile Bonds in an Enzyme-Substrate Complex to 0.03 Å 325
P. R. Carey, C. P. Huber, H. Lee, Y. Ozaki, and A. C. Storer

Time-Resolved Fluorescence Spectra of Chlorophyll-*A* Dimers and Aggregates Using Selective Fluorescence Quenching 331
A. C. de Wilton, J. A. Koningstein, and L. V. Haley

SECTION V NONLINEAR RAMAN SPECTROSCOPY

Excited State Population Effects on Coherent Raman Spectroscopy 335
R. Bozio, P. L. Decola, and R. M. Hochstrasser

New Results on Vibrational Dephasing and Population Relaxation from Picosecond Techniques 345
H. R. Telle, H. Graener, and A. Laubereau

Raman Spectra of Short-Living Molecular States by CARS 353
A. Lau, H-J. Weigmann, W. Werncke, K. Lenz, and M. Pfeiffer

Resonance CARS of Excited Molecules 361
Shiro Maeda, Haruhiko Kataoka, Toshio Kamisuki, and Yukio Adachi

SECTION VI SURFACED - ENHANCED AND COLLOIDAL RAMAN SCATTERING

Surface Enhancement Mechanisms 369
Richard K. Chang

Time-Resolved Studies of Electrochemical SERS: The Pyridine / Cl^- / Ag Model System 377
A. M. Stacey and R. P. Van Duyne

Combined Surface-Enhanced and Resonance Raman Scattering by Dabsyl Aspartate on Colloidal Silver 387
O. Siiman, L. A. Bumm, R. Callaghan, and M. Kerker

Transient Raman Scattering Studies of Chemical Kinetics in Aqueous Micellar and Semiconductor Colloidal Solutions 391
S. M. Beck and L. E. Brus

CONTRIBUTORS

Numbers in parentheses indicate the pages on which the authors' contributions begin.

Yukio Adachi (361), *Research Laboratory of Resources Utilization, Tokyo Institute of Technology, Nagatsuta, Yokohama, Japan*

A. C. Albrect (1, 37), *Department of Chemistry, Cornell University, Ithaca New York*

R. R. Alfano (127), *Ultrafast Spectroscopy & Laser Laboratory, The City College of New York, New York, New York*

Thomas Alshuth (231), *Max-Planck-Institut für Biophysikalische Chemie, Göttingen, Germany*

M. Asano (139, 157), *Department of Chemistry, Carleton University, Ottawa, Ontario, Canada*

G. H. Atkinson (161, 191, 287), *Department of Chemistry, Syracuse University, Syracuse, New York*

R. T. Bailey (121), *Department of Pure and Applied Chemistry, University of Strathclyde, Glasgow, Scotland*

S. M. Beck (173, 391), *Bell Laboratories, Murray Hill, New Jersey*

Peter H. Berens (59), *Department of Chemistry, University of California, San Diego, La Jolla, California*

John P. Bergsma (59), *Department of Chemistry, University of California, San Diego, La Jolla, California*

R. Bozio (335), *Department of Chemistry and Laboratory for Research on the Structure of Matter, University of Pennsylvania, Philadelphia, Pennsylvania*

P. G. Bradley (147), *Department of Chemistry, University of Texas at Austin, Austin, Texas*

Mark Braiman (219), *Department of Chemistry and Biology, Massachusetts Institute of Technology, Cambridge, Massachusetts*

C. W. Brown (11), *Texas Tech University, Picosecond & Quantum Radiation Laboratory, Lubbock, Texas*

L. E. Brus (173, 391), *Bell Laboratories, Murray Hill, New Jersey*

L. A. Bumm (377), *Department of Chemistry, Clarkson College of Technology, Potsdam, New York*

R. Callaghan (387), *Department of Chemistry, Clarkson College of Technology, Potsdam, New York*

P. R. Carey (325), *Division of Biological Sciences, National Research Council of Canada, Ottawa, Canada*

C. K. Chan (31), *Department of Physics, Arizona State University, Tempe, Arizona*

Richard K. Chang (369), *Section of Applied Physics & Center for Laser Diagnostics, Yale University, New Haven, Connecticut*

James L. Chao (97), *IBM Instruments, Inc., P.O. Box 332, Danbury, Connecticut*

Robin J. H. Clark (213), *Christopher Ingold Laboratory, University College London, London, United Kingdom*

F. R. Cruickshank (121), *Department of Pure and Applied Chemistry, University of Strathclyde, Glasgow, Scotland*

R. F. Dallinger (147), *Department of Chemistry, University of Texas at Austin, Austin, Texas*

S. J. Davis (135), *Air Force Weapons Laboratory, Kirtland AFB, New Mexico*

P. L. Decola (335), *Department of Chemistry and Laboratory for Research on the Structure of Matter, University of Pennsylvania, Philadelphia, Pennsylvania*

F. Dorr (183), *Institut für Physikalische und Theoretische Chemie, Technische Universität, München, Garching, Germany*

R. Dorsinville (127), *Ultrafast Spectroscopy & Laser Laboratory, Physics Department, The City College of New York, New York, New York*

Michael C. Drake (83), *GE Research & Development Center, Schenectady, New York*

Michael T. Duignan (105), *Naval Research Laboratory, Washington, D.C.*

M. A. El-Sayed (251), *Department of Chemistry, University of California, Los Angeles, Los Angeles, California*

P. Esherick (63), *Sandia National Laboratories, Albuquerque, New Mexico*

David P. Fairclough (213), *Christopher Ingold Laboratory, University College London, London, United Kingdom*

W. M. Franklin (127), *Ultrafast Spectroscopy & Laser Laboratory, Physics Department, The City College of New York, New York, New York*

J. M. Friedman (307, 317), *Bell Laboratories, Murray Hill, New Jersey*

Jack Gelfand (91), *Department of Mechanical & Aerospace Engineering, Princeton University, Princeton, New Jersey*

D. A. Gilmore (161), *Department of Chemistry, Syracuse University, Syracuse, New York*

H. Graener (345), *Physikalisches Institut, Universität Bayreuth, Bayreuth, West Germany*

Iris Grieger (231), *Max-Planck-Institut für Biophysikalische Chemie, Göttingen, Germany*

A. J. Grimley (63), *Sandia National Laboratories, Albuquerque, New Mexico*

R. Guthrie (121), *Department of Pure and Applied Chemistry, University of Strathclyde, Glasgow, Scotland*

J-M. Halbout (73), *Bell Laboratories, Murray Hill, New Jersey*

L. V. Haley (139, 331), *Department of Chemistry, Carleton University, Ottawa, Ontario, Canada*

Ronald E. Hester (177), *Department of Chemistry, University of York, York, England*

E. F. Hilinski (263), *Bell Laboratories, Murray Hill, New Jersey*

R. M. Hochstrasser (335), *Department of Chemistry and Laboratory for Research on the Structure of Matter, University of Pennsylvania, Philadelphia, Pennsylvania*

M. Z. Hoffman (147), *Department of Chemistry, University of Texas at Austin, Austin, Texas*

Chung-lu Hsieh (251), *Technical Center, The Clorox Company, Pleasonton, California*

W. Hub (183), *Institut für Physikalische und Theoretische Chemie, Technische Universität, München, Garching, Germany*

C. P. Huber (325), *Division of Biological Sciences, National Research Council of Canada, Ottawa, Canada*

M. J. Irwin (287), *Department of Chemistry, Syracuse University, Syracuse, New York*

Niels-Henrick Jensen (273), *Risø National Laboratory, Roskilde, Denmark*

Toshio Kamisuki (361), *Research Laboratory of Resources Utilization, Tokyo Institute of Technology, Nagatsuta, Yokohama, Japan*

Haruhiko Kataoka (361), *Research Laboratory of Resources Utilization, Tokyo Institute of Technology, Nagatsuta, Yokohama, Japan*

M. Kerker (387), *Department of Chemistry, Clarkson College of Technology, Potsdam, New York*

R. J. Kessler (147), *Department of Chemistry, University of Texas at Austin, Austin, Texas*

J. A. Koningstein (139, 157, 331), *Department of Chemistry, Carleton University, Ottawa, Ontario, Canada*

T. Kumar (97), *Department of Chemistry, Kansas State University, Manhattan, Kansas*

Mohamedally Kurmoo (213), *Christopher Ingold Laboratory, University College London, London, United Kingdom*

Marshall Lapp (83), *GE Research & Development Center, Schenectady, New York*

A. Lau (353), *Central Institute of Optics and Spectroscopy, Academy of Sciences of GDR, 1199 Berlin-Adlershof, Rudower Chaussee 5, German Democratic Republic*

A. Laubereau (345), *Physikalisches Institut, Universität Bayreuth, Bayreuth, West Germany*

Duckhwan Lee (37), *Department of Chemistry, Cornell University, Ithaca, New York*

H. Lee (325), *Division of Biological Sciences, National Research Council of Canada, Ottawa, Canada*

K. Lenz (353), *Central Institute of Optics and Spectroscopy, Academy of Sciences of GDR, 1199 Berlin-Adlershof, Rudower Chaussee 5, German Democratic Republic*

Harold B. Levene (117), *Department of Chemistry, University of Rochester, Rochester, New York*

Aaron Lewis (239), *School of Applied and Engineering Physics, Cornell University, Ithaca, New York*

F. D. Lewis (183), *Department of Chemistry, Northwestern University, Evanston, Illinois*

Shiro Maeda (361), *Research Laboratory of Resources Utilization, Tokyo Institute of Technology, Nagatsuta, Yokohama, Japan*

V. Malueg (147), *Department of Chemistry, University of Texas at Austin, Austin, Texas*

Richard Mathies (219), *Department of Chemistry, University of California, Berkeley, California*

Richard B. Miles (91), *Department of Mechanical & Aerospace Engineering, Princeton University, Princeton, New Jersey*

Ole Sonnich Mortensen (23), *Odense University, Fysisk Institut, DK 5230 Odense M, Denmark*

S. Mukamel (41), *Department of Chemical Physics, The Weizmann Institute of Science, Rehovot, Israel*

Malcolm Nicol (251), *Department of Chemistry, University of California at Los Angeles, Los Angeles, California*

Yoshifumi Nishimura (113), *Faculty of Pharmaceutical Science, University of Tokyo, Hongo, Bunkyo-ku, Tokyo, Japan*

K. A. Norton (147), *Department of Chemistry, University of Texas at Austin, Austin, Texas*

N. Ockman (127), *Ultrafast Spectroscopy & Laser Laboratory, The City College of New York, New York, New York*

M. R. Ondrias (317), *University of New Mexico, Albuquerque, New Mexico*

A. Owyoung (63), *Sandia National Laboratories, Albuquerque, New Mexico*

J. D. Oxman (183), *Department of Chemistry, Northwestern University, Evanston, Illinois*

Y. Ozaki (325), *Division of Biological Sciences, National Research Council of Canada, Ottawa, Canada*

J. B. Page (31), *Department of Physics, Arizona State University, Tempe, Arizona*

C. Murray Penney (83), *GE Research & Development Center, Schenectady, New York*

David S. Perry (117), *Department of Chemistry, University of Rochester, Rochester, New York*

M. Pfeiffer (353), *Central Institute of Optics and Spectroscopy, Academy of Sciences of GDR, 1199 Berlin-Adlershof, Rudower Chaussee 5, German Democratic Republic*

Robert W. Pitz (83), *GE Research & Development Center, Schenectady, New York*

D. Presser (147), *Department of Chemistry, University of Texas at Austin, Austin, Texas*

D. Pugh (121), *Department of Pure and Applied Chemistry, University of Strathclyde, Glasgow, Scotland*

Herschel Rabitz (91), *Department of Mechanical & Aerospace Engineering, Princeton University, Princeton, New Jersey*

P. M. Rentzepis (263), *Bell Laboratories, Murray Hill, New Jersey*

G. W. Robinson (11), *Texas Tech University, Picosecond & Quantum Radiation Laboratory, Lubbock, Texas*

Eric Rohlfing (91), *Exxon Research & Engineering, Linden, New Jersey*

D. L. Rousseau (317), *Bell Laboratories, Murray Hill, New Jersey*

S. Schneider (183), *Institut für Physikalische und Theoretische Chemie, Technische Universität, München, Garching, Germany*

R. H. Schuler (199), *Radiation Laboratory, University of Notre Dame, Notre Dame. Indiana*

R. F. Shea (135), *Air Force Weapons Laboratory, Kirtland AFB, New Mexico*

Hitoshi Shindo (191, 207), *National Chemical Laboratory, Yatabe, Ibaraki 305, Japan*

O. Siiman (387), *Department of Chemistry, Clarkson College of Technology, Potsdam, New York*

J. T. Simpson (183), *Department of Chemistry, Northwestern University, Evanston, Illinois*

Steven O. Smith (219), *Department of Chemistry, University of California, Berkeley, California*

Shammai Speiser (105), *Department of Chemistry, Technion-Israel Institute of Technology, Haifa, Israel*

Thomas G. Spiro (299), *Department of Chemistry, Princeton University, Princeton, New Jersey*

A. M. Stacey (377), *Department of Chemistry, Northwestern University, Evanston, Illinois*

P. Stern (41), *Department of Chemical Physics, The Weizmann Institute of Science, Rehovot, Israel*

Manfred Stockburger (231), *Max-Planck-Institut für Biophysikalische Chemie, Göttingen, Germany*

A. C. Storer (325), *Division of Biological Sciences, National Research Council of Canada, Ottawa, Canada*

C. L. Tang (73), *Department of Engineering, Cornell University, Ithaca, New York*

Mitsuo Tasumi (53), *Department of Chemistry, Faculty of Science, University of Tokyo, Bunkyo-ku, Tokyo, Japan*

H. R. Telle (345), *Physikalisches Institut, Universität Bayreuth, Bayreuth, West Germany*

James Terner (299), *Department of Chemistry, Virginia Commonwealth University, Richmond, Virginia*

D. L. Tonks (31), *Group X-7, LANL, Los Alamos, New Mexico*

G. N. R. Tripathi (199), *Radiation Laboratory, University of Notre Dame, Notre Dame, Indiana*

Masamichi Tsuboi (113), *Faculty of Pharmaceutical Science, University of Tokyo, Hongo, Bunkyo-ku, Tokyo, Japan*

Tadayuki Uno (113), *Faculty of Pharmaceutical Science, University of Tokyo, Hongo, Bunkyo-ku, Tokyo, Japan*

R. P. Van Duyne (377), *Department of Chemistry, Northwestern University, Evanston, Illinois*

Arieh Warshel (41), *Department of Chemistry, University of Southern California, Los Angeles, California*

H-J. Weigmann (353), *Central Institute of Optics and Spectroscopy, Academy of Sciences of GDR, 1199 Berlin-Adlershof, Rudower Chaussee 5, German Democratic Republic*

I. J. M. Weir (121), *Department of Pure and Applied Chemistry, University of Strathclyde, Glasgow, Scotland*

W. Werncke (353), *Central Institute of Optics and Spectroscopy, Academy of Sciences of GDR, 1199 Berlin-Adlershof, Rudower Chaussee 5, German Democratic Republic*

Robert Wilbrandt (273), *Risø National Laboratory, Roskilde, Denmark*

Kent R. Wilson (59), *Department of Chemistry, University of California, San Diego, La Jolla, California*

A. C. de Wilton (331), *Department of Chemistry, Carleton University, Ottawa, Ontario, Canada*

P. J. Wolf (135), *Air Force Weapons Laboratory, Kirtland AFB, New Mexico*

W. H. Woodruff (147), *Department of Chemistry, University of Texas at Austin, Austin, Texas*

Kenji Yokoyama (167), *Institute of Physical and Chemistry Research, Wako, Saitama, Japan*

FOREWORD

During the end of the 1970s, it became increasingly clear that the element of time was finding its way into vibrational spectroscopy. It is perhaps interesting to point out that the new developments did not take place in that part of the spectrum where direct transitions between vibrational levels take place, but rather in the methods of light scattering spectroscopy. Thus, time-dependent aspects of the indirect (i.e., Raman) transitions appear to be of more practical use. This, of course, is due to the fact that tunable, pulsed laser sources are now available, which operate in the ultraviolet, visible, and near-infrared regions of the spectrum with a finesse not yet to be found in the infrared. In addition, the detection of fast laser pulses in the former spectral regions is more easily achieved than in the infrared.

The event of the first International Conference on Time-Resolved Vibrational Spectroscopy in Lake Placid, August 16–20, 1982, was particularly timely as judged by the truly international character of the meeting, and thus may be viewed as a forerunner of many things to come. Time-resolved vibrational spectroscopy of gasses, liquids, and solids may experience a renaissance similar to that of laser Raman spectroscopy in the decade of the 1960s.

Professor A. Koningstein
Ottawa, 1982

PREFACE

The wide-ranging contributions of vibrational spectroscopy to the study of molecular structure have been significantly expanded in recent years through the development of time-resolved techniques for rapidly recording vibrational spectra. Generally termed time-resolved vibrational spectroscopy (TRVS), these methods have provided information on both the vibrational structure and the dynamical properties of reactive species with lifetimes as short as picoseconds. The versatility of TRVS, as well as the breadth of information it provides, has fostered extremely diverse applications involving chemical and biochemical reactions, physical relaxation phenomena, surface and colloidal reactions, combustion diagnostics, analytical microanalysis, and electrochemistry.

Research in TRVS had been discussed as part of general meetings on spectroscopy, photochemistry, and photobiology, but it had not previously formed the central topic for a conference until the International Conference on TRVS was held in Lake Placid, New York, on August 16–20, 1982. This conference provided a forum in which discussions of TRVS were undertaken by researchers already active in the field and by scientists interested in initiating work in TRVS. The proceedings of that conference have been adapted for presentation in this publication. The large number of outstanding papers appearing here clearly document the rapid growth of TRVS as a field and suggest that such research will be sustained long into the future.

I would like to recognize several sources of assistance in the preparation of this book. Foremost, I am indebted to Mrs. Ann Dlugozima whose untiring efforts have been invaluable. I would like to gratefully acknowledge the Office of Naval Research, U.S. Army Research Office, Air Force Office of Scientific Research, and Syracuse University for financially supporting the International Conference on TRVS from which the publication is derived. Finally, I wish to express appreciation to my many colleagues who assisted me in the organization of the International Conference on TRVS and in the preparation of this publication.

Professor George H. Atkinson
Syracuse, New York
October 1982

SOME THEORETICAL CONSIDERATIONS CONCERNING RAMAN SCATTERING[1]

A. C. Albrecht

Department of Chemistry
Cornell University
Ithaca, New York

I. INTRODUCTION

Resonance Raman scattering (RRS) is one of the several spectroscopies that can be found at the third order level of the electrical susceptibility of a molecule. In fact the density matrix description of the electrical susceptibility offers a powerful means for describing the interaction of light with matter at various orders of interaction, particularly when damping effects play an important role (1). Parametric spectroscopies, in which matter acts only catalytically to alter states of the radiation field, appear at all orders. Spectroscopies which bring about transitions between distinctly different states of the material system appear in odd orders only. One might term these "passive" and "active" spectroscopies, respectively, since only in the latter is there a significant transfer of work between the radiation field and the material sample. The calculation of the rate of work transferred in any given process leads to an expression for the cross-section for the spectroscopy. At

[1] *This work is supported by grants from the National Science Foundation (CHE 80-16526) and the Materials Science Center of Cornell University*

ISBN 0-12-066280-9

first order one encounters one-photon absorption, or emission, as the active spectroscopy, while linear wave propagation or ordinary dispersion serves as the passive event. Second harmonic frequency generation is one familiar example of the passive processes found at second order. As mentioned, at the $\chi^{(3)}$ level many spectroscopies appear. These include CARS and CSRS as passive or parametric events, and two-photon absorption, two-photon emission, or one-photon absorption-one-photon emission as active processes. The overall event consisting of one-photon absorption, one-photon emission has been called resonant secondary radiation, or resonance luminescence (RL). The phenomenon of RRS is somehow a part of this process. The object of this paper is to briefly summarize what density-matrix theory has to say about RRS and to indicate under what conditions the familiar resonant term of the Kramers-Heisenberg (KH) expression for RRS is recovered. The second part of this paper reviews recent theoretical developments which provide analytical expressions for relating an absorption band to the resonance Raman excitation profile (REP), the Raman scattering cross-section for a given line, as a function of excitation frequency. These so-called transform methods are based on the resonant KH expression for the RL event. This short review cannot be complete, but a full exposure to the literature may be found within the few references listed here.

II. THE DENSITY MATRIX APPROACH TO RESONANCE LUMINESCENCE (2)

The work performed (negative or positive) by incident electromagnetic fields (and the black-body field) upon a material system is related to its electrical susceptibility. (The work is just the time average of the scalar product of the net induced field and the time derivative of the induced polarization.) The work performed in an sth order process is then proportional to the imaginary part of $\chi^{(s)}$. The density matrix treatment of the induced polarization in a molecular eigenstate basis leads to quantum-mechanical expressions for the tensor elements of each order of the electrical susceptibility. At s^{th} order a given tensor element has $(2^s s!)/(s_1!....s_c!)$ distinct terms, each with s+1 transition dipole matrix elements in the numerator and s (complex) energy factors in the denominator. The $s_j!$ (j=1,...,c) represent possible color degeneracies in the incident beam. (In a true multicolor experiment all s_j=1.) It turns out that diagramatic techniques of Yee and Gustafson (3) are extremely useful for writing down the various terms existing at any given order. The damping factors, which appear in the energy denominators, represent dephasing or coherence loss rate constants for a given pair of optically coupled states in the basis set. Dephasing occurs by both population decay mechanism ($1/T_1$, inelastic, or longitudinal relaxation) and pure-dephasing processes ($1/T_2'$, quasielastic, or transverse relaxation). At the $\chi^{(3)}$ level (s=3) there are at most forty-eight terms to consider. For those one-photon absorption-one-photon emission processes which find no resonance at the intermediate state, the forty-eight terms reduce to just four. These are exactly those given by the modulus square of the full KH expression for Raman scattering, $|KH^- + KH^+|^2$, where KH^- and KH^+ are the familiar resonant and non-resonant KH terms, respectively. In addition, a simple Lorentzian line-shape function is obtained with a damping parameter representing the coherence loss rate constant between the initial and the final state of the system.

When an intermediate state resonance exists, one encounters a complex luminescence best termed resonance luminescence (RL). Shen (4) finds that the cross-section for this RL can be written as a sum of two terms, each sharing the same modulus square of the absorption transition moment times the emission transition moment. One term carries three complex energy denominator factors including coherence-loss between the initial and final state. Nowhere in its time evolution description do both bra and ket of the intermediate state coexist. It can be said that an intermediate state population is not achieved. On the other hand the second term, for which both the bra and ket of the intermediate level co-exist for some period during the time evolution of the system, evidently is based on a population of the intermediate state. This term is exactly the product of a Lorentzian absorption and a Lorentzian emission. Emission due to the second term might be identified as resonance fluorescence (RF), while that from the first term as RRS. Alternate characterizations of RF and RRS exist. A popular one requires that RF exist only if pure-dephasing is present (5). The issue of nomenclature aside, it is a fact that the density matrix expressions for RL reduce to $|KH^-|^2$ only when pure-dephasing is absent. At the same time the line-shape of the RL phenomenon, which is not simple when the pure-dephasing component of coherence loss is important, reduces to a simple Lorentzian in which the coherence-loss between the initial and final state must be governed only by inelastic processes. The transform methods to be discussed next have been developed only in the context of the $|KH^-|^2$ expression and therefore do not formally apply to RL. The question of whether so-called experimental resonance-Raman spectra are actually RL and not pure $|KH^-|^2$ emission requires serious attention. Still it must be that the sharp features in an experimental spectrum are principally $|KH^-|^2$ since the dephasing between the initial and final vibrational states in the ground electronic state is much slower than inter-electronic state dephasing, which is responsible for line-shapes found in the RF term. However, if the initial and final vibrational states belong to an excited electronic state (or any short lived transient), then the corresponding RL ("excited state RRS") will not exhibit sharp spectra.

III. TRANSFORM METHODS

In the absence of pure dephasing, or, more approximately, when dealing only with the sharp features of RL, the $|KH^-|^2$ description ought to properly account for what has been called the Raman excitation profile (REP). The problem of course centers upon the explicit sum-over-states present in $\overline{\mathrm{KH}}$, since the resonance region in general is encompassed by one or more relatively broad, often featureless, absorption bands. Even when only one electronic transition is in resonance Franck-Condon (FC) and even Herzberg-Teller (or non-Condon (NC)) vibronic activity of a multitude of vibrational states can bring structure into the electronic band , and in some cases simply cause broadening to an extent where discrete structure is absent. When the resonant band thus consists of a superposition of a very large number (possibly narrow) individual molecular transitions, one encounters the multimode problem in RRS. The point is that the formally correct treatment requires the full sum over resonant or nearly resonant transitions in $\overline{\mathrm{KH}}$, at each incident ν, in order to predict the REP(ν). It is incorrect (though simplifying) to treat the problem using a simple three level model in which the damping factor is used to account for the spectral width. The correct sum-over-transitions is difficult, if not impossible, to achieve particularly if one hopes to assign an individual damping factor to each molecular transition, as theory requires. Analytical methods have recently appeared which tell us how to transform the resonant absorption band, ABS, into the corresponding REP. Such transform methods implicitly recognize that the multimode complexity is already present in ABS, and that most of it carries directly over into the REP. A review of both the multimode problem, as well as transform methods, has just been completed (6). A complete up-to-date literature can be found there. The transform story began in a paper in the late sixties from an Estonian school of theorists (7). One object of this early work was to treat simultaneously RRS and RF (called hot-luminescence (HL) by them) as distinct components of RL (called resonant secondary radiation by them). But a transform relationship between ABS and REP also appeared. It was developed under the constraints of a constant linewidth parameter over the resonant region, FC activity only (no NC effects), and the absence of quadratic coupling effects in the resonant electronic transition. In subsequent papers some of these constraints were removed. For example quadratic coupling effects were included when changes in force constants were allowed in the upper electronic state, as well as mode-mixing (Duschinsky rotation). Much more recently Page and Tonks have elaborated upon and greatly clarified these developments in a series of papers only the first of which (8) and the most recent (9) are mentioned here. While these studies proceeded using the time-correlator approach and a similar time domain treatment has also appeared (10), the same transform law can be obtained in the frequency domain as well (11,12).

The simplest form of the transform prescription in effect states that the REP(ν) is just proportional to the square modulus of the difference between the molecular polarizability at the incident frequency, ν, and the molecular polarizability at the scattered frequency, ν_s. As is well known, the molecular polarizability at a given frequency is a complex function whose pure real part is the Kramers-Kronig transform of ABS and whose pure imaginary part is just ABS itself. Thus REP = TR(ABS) is determined.

Very recently we have explored the transform method within a framework in which the adiabatic approximation is applied only to the scattered mode (12). Vibronic (NC) effects are included to lowest order and the transform laws are given for each of the three rotationally invariant polarizability tensors. The limited use of the adiabatic approximation relieves to a considerable extent the need for a fixed damping constant over the resonant region. Including NC terms points to a method for not only determining FC coupling constants (potential-energy curve shifts along the coordinate of the scattered mode upon electronic excitation), already available at the simplest level, but actually provides a measure of vibronic coupling matrix elements, once the absolute REP is measured for a given vibration. Furthermore, given the explicit transform expressions for each of the three polarizability tensors (12), one has the proper transform laws for depolarization excitation profiles, as well as reversal excitation profiles.

Since the transform law, in its simplest form, involves the destructive interference of the molecular polarizability at two neighboring frequencies, the incident and the scattered, it is evident that as these frequencies approach each other, the REP cross-section must greatly weaken. This can be avoided only if the FC coupling constant, which scales the transform expression, correspondingly increases. This means that in order for low frequency vibrations to exhibit RRS they must possess particularly strong FC coupling in the absorption band. If they are NC active, however, such destructive interference is absent (12).

The transform method at the FC and NC levels has been applied to the absolute RRS in the Soret region of the 1362 cm^{-1} oxidation marker band in both ferrous and ferric cytochrome-c (13). The fit of theory to experiment is excellent and is achieved in the former case by introducing a vibronically induced (NC) transition moment into the Soret absorption band that is about 3% of the allowed component. The FC parameters obtained from the fit show how the principal source of broadening in the Soret band is the multimode phenomenon.

Finally, in this brief survey, we turn to the concept of the inverse transform. It turns out that in principle an observed REP can be transformed into the ABS that is responsible for it [ABS = TR^{-1}(REP)] (14). The inverse transform can be carried out with and without NC contributions with similar ease. It has already been applied to the Soret REP in cytochrome-c, where it is seen how the Soret ABS is recovered only if a small NC contribution is present, as we already know (12). Another application is to the REP observed for 351 cm^{-1} metal-sulfur vibration in cytochrome P-450 to which a camphor substrate is bound. The inverse transform shows that the resonantly active ABS is a small absorption band underlying the blue wing of the Soret band (14). In principle the inverse transform offers a new method for resolving complex absorption spectra since the REP's of different vibrations might well respond to different electronic spectral components of the overall resonant region.

In conclusion it is important to emphasize that the transform methods are based on the $|KH^-|^2$ expression for resonance scattering. Nevertheless they do offer a means to extract information about the excited state potential energy surface, provided absolute cross-section measurements are available for the resonantly scattered vibrations. In any case any multimode modelling of ABS and the corresponding REP's must be consistent with the relationship ABS = TR(REP) (or the inverse) to constitute a valid model (15). In the end we must determine the damping constants for the individual molecular transitions to complete the story. Line shape analyses of RL and transform techniques certainly go part way in that direction.

ACKNOWLEDGMENTS

This brief survey has incorporated results from the density matrix studies by Mr. Duckhwan Lee, multimode contributions of Dr. P. M. Champion, and the transform research by Dr. B. R. Stallard, Professor Patrik Callis, and Dr. Paul Champion. The significant contributions of these colleagues is gratefully acknowledged.

REFERENCES

1. The work of N. Bloembergen and his school is well known. Reference can be made to Bloembergen, N., in "Nonlinear Optics", W. A. Benjamin, Reading, MA, (1977); as well as to many papers before and since.

2. The present material is a synopis of a more extensive examination of this topic in Lee, D. and Albrecht, A. C., manuscript in preparation for *J. Phys. Chem.*
3. Yee, T. K. and Gustafson, T. K., *Phys. Rev.* *A18*, 1597 (1978).
4. Shen, Y. R., *Phys. Rev*, *B9*, 622 (1974).
5. Rousseau, D. L., Patterson, G. D., and Williams, P. F., *Phys. Rev. Lett.* *34*, 1306 (1975).
6. Champion, P. M. and Albrecht, A. C. *Ann. Rev. Phys. Chem.* *33*, 353 (1982).
7. Rebane, K., Hizhnyakov, V. and Tehver, I., *ENSV TA Toimet. Fuus. Matem.* *16*, 202 (1967). (Proceedings of the Academy of Sciences of the Estonian SSR, Physics and Mathematics.)
8. Tonks, D. L. and Page, J. B., *Chem. Phys. Lett.* *66*, 449 (1979).
9. Tonks, D. L. and Page, J. B., *J. Chem. Phys.* *76*, 5820 (1982).
10. Lee, S.-Y. and Heller, E. J., *J. Chem. Phys.* *71*, 4777 (1979).
11. Hassing, S. and Mortensen, O., *J. Chem. Phys* *73*, 1078 (1980).
12. Stallard, B. R., Champion, P. M., Callis, P. R., and Albrecht, A. C., submitted to *J. Chem. Phys.*
13. Stallard, B. R., Callis, P. R., Champion, P. M., and Albrecht, submitted to *J. Chem. Phys.*
14. Stallard, B. R., Ph.D. Thesis, Cornel University, August (1982). See also, Stallard, B. R., Champion, P. M., and Albrecht, A. C., submitted to *J. Chem. Phys.*
15. Champion, P. M. and Albrecht, A. C., *Chem. Phys. Lett.* *82*, 410 (1981).

INTERMEDIATE GAUGE FORMULATION OF TWO-PHOTON SPECTROSCOPIES

G. W. Robinson[1]
C. W. Brown[2]

Picosecond and Quantum Radiation Laboratory
Texas Tech University
Lubbock, Texas

I. PREVIEW

In 1941 Pauli (1) distinguished two classes of gauge transformations. The more familiar type, which he termed <u>gauge transformations of the second kind</u> (GT2), concerns real fields, such as electromagnetic fields. The transformation from Coulomb gauge to Lorentz gauge is an example of GT2. Transformations of complex fields, which refer to particles, were called <u>gauge transformations of the first kind</u> (GT1). An example of GT1 is the transformation (2,3) whereby the matter/radiation Hamiltonian containing $\vec{A}\cdot\vec{p}$ and A^2 terms (velocity-form) is changed to one containing $\vec{E}\cdot\vec{r}$ (length-form).

The question about gauge is not an idle one. The length- and velocity-forms of the interaction matrix elements can give results differing by an order of magnitude or more in the case where inexact eigenfunctions (an incomplete set in the sum-over-

[1]Robert A. Welch Professor. Research supported by the Robert A. Welch Foundation, Grant D-950.
[2]Robert A. Welch Postdoctoral Fellow.

ISBN 0-12-066280-9

states or other approximate forms) are used. A result that is not gauge invariant may incur errors that are as large as those associated with switching between the two gauge-forms.

II. INTERMEDIATE GAUGE FORMS

The field and matter variables within the full gauge-invariant matter/radiation Hamiltonian can be intermixed in an infinite variety of ways. One class of intermediate gauge transformations can provide an infinite set of matter/radiation Hamiltonians each containing a different mixture of $\vec{A}\cdot\vec{p}$ and $\vec{E}\cdot\vec{r}$ interaction terms. This can be demonstrated both by classical and quantum field methods (4). In the classical approach, field (particle) coordinates as well as particle (field) velocities occur in the canonical momentum for each particle (field mode). The degree of this mixing is measured by a gauge parameter $0 \leq \xi \leq 1$. When exact eigenfunctions are used, any one of the infinite set of these Hamiltonians gives precisely the same physical result.

The practical importance of the intermediate gauge is that in principle an inexact, but convenient, matter basis can be used in conjunction with a gauge-optimized total Hamiltonian to obtain more accurate cross-sections than those conventionally obtained using a fixed gauge and a more exact matter basis. In the case of Raman scattering or two-photon absorption, for example, one would like to replace the full sum (integral) over bound (continuous) electronic states by an inexact set containing only a limited number of well characterized electronic states.

III. OFF-RESONANCE RAMAN AND TWO-PHOTON SPECTROSCOPIES

A. Amplitude Functions

Energy conservation dictates that $\omega(f)-\omega(i)-\omega_1 \pm \omega_2 \doteq 0$, where the upper sign refers to Raman and the lower sign to two-photon absorption; i and f label initial and final matter states; and ω_1 and ω_2 are the frequencies of the two photons. For Raman, we have taken ω_1 to be the frequency of the incident photon and ω_2 that of the scattered photon. A relative amplitude function is first calculated. It is then squared to obtain a relative cross-section. To obtain the absolute cross-section, this relative cross-section is multiplied by physical constants appropriate to the experiment.

If the starting point is chosen as the length-form of the dipole interaction, the relative amplitude function is (5),

$$S[L] = \sum_n (1+P_{12}^{\pm}) \frac{\vec{e}_2 \cdot \langle f|\Sigma_\alpha \vec{r}_\alpha|n\rangle\langle n|\Sigma_\alpha \vec{r}_\alpha|i\rangle \cdot \vec{e}_1}{\omega(n)-\omega(i)-\omega_1} , \quad [1]$$

for $\omega_1 \neq [\omega(n)-\omega(i)]$. Equation [1] is the Kramers-Heisenberg equation: Σ_n represents a sum over quantum numbers for the discrete states and integration over those for the continuous spectrum, and Σ_α is a sum over the system electrons. The operator P_{12} interchanges the frequencies ω_1, ω_2 and the polarizations $\vec{e}_1, \vec{e}_2$ of the two photons. The symbol $\pm$ on the P indicates that the sign of ω_2 is (+) for Raman scattering but (-) for two-photon absorption. Combining the contributions $(1+P_{12}^{\pm})$ in Eq. [1] yields a compact form of the Kramers-Heisenberg equation (6),

$$S[L] = +\frac{1}{\omega_1\omega_2} \sum_n F_n C_n \quad . \quad [2]$$

Expression [2] proves to be very convenient for mathematical manipulations. The factors in [2] are defined,

$$F_n = \frac{\omega_1\omega_2}{[\omega(n)-\omega(i)\pm\omega_2][\omega(n)-\omega(i)-\omega_1]} , \quad [3]$$

and,

$$C_n = (1+P_{12}^{\pm})[\vec{e}_1 \cdot \langle f|\Sigma_\alpha \vec{r}_\alpha|n\rangle\langle n|\Sigma_\alpha \vec{r}_\alpha|i\rangle \cdot \vec{e}_2][\omega(n)-\omega(i)-\omega_1] . \quad [4]$$

The dimensionless frequency factor F_n is characteristic of the length-form. If the starting point were instead the velocity-form, Eq. [2] is altered,

$$S[V] = \pm \frac{1}{\omega_1 \omega_2} \sum_n F'_n C_n \quad , \qquad [5]$$

where S[V] is the amplitude function in the velocity-form, and,

$$F'_n = \frac{[\omega(n)-\omega(f)][\omega(n)-\omega(i)]}{[\omega(n)-\omega(i)\pm\omega_2][\omega(n)-\omega(i)-\omega_1]} \quad . \qquad [6]$$

Using the energy conservation condition, it can be seen that the two frequency factors are simply related,

$$F'_n = 1 \pm F_n \quad . \qquad [7]$$

When the set n is a complete set of exact eigenfunctions, Eqs. [2] and [5] both yield the exact amplitude function.

B. The Sum-Rule

Because of the equivalence of the length- and velocity-forms of the interaction, one can equate S[L] and S[V] from Eqs. [2] and [5]. Then, using relationship [7], a sum-rule,

$$\sum_n C_n = 0 \quad , \qquad [8]$$

can be derived (6). This sum-rule of course holds only for a complete set of intermediate states n. It can also be derived in a more conventional manner (7). Commonly used second-order amplitude equations, such as a truncated Kramers-Heisenberg sum, violate the sum-rule, are not gauge invariant, and suffer the ambiguity of the gauge question with its attendant error.

C. Inexact Sets

For consideration of inexact sets, the contributions to Eq. [2] can be split into two parts: $S^c[L]$ from "characterized" states and $S^u[L]$ from "uncharacterized" states (6),

$$S(L) = S^c[L] + S^u[L] = + \frac{1}{\omega_1 \omega_2} \left[\sum_n^c F_n C_n + \sum_n^u F_n C_n \right] \quad , \qquad [9]$$

where the superscripts c,u label partial sums over characterized and uncharacterized states. Since F_n is smoothly varying, an approximation to the Σ_n^u terms in Eq. [9] can be made through removal from inside the sum of a factor F_K, where F_K is the same as F_n except that an effective frequency $\omega(K)$ is substituted for $\omega(n)$. This effective frequency approximation is not equivalent to the one that is traditionally employed in Raman or two-photon intensity calculations, where the sum-rule is not utilized, and gauge invariance is in fact violated (6).

Using the sum-rule [8] together with the effective $\omega(K)$ approximation, it is possible to write a modified expression for the amplitude function (6),

$$S = \frac{1}{\omega_1\omega_2}\sum_n^c (F_n - F_K)C_n \quad . \qquad [10]$$

Since from Eq. [7], $\pm F_n = F_n' - 1$ and in addition $\pm F_K = F_K' - 1$, this result is independent of the initial gauge form chosen even though an inexact set of intermediate states n has been employed. A gauge label is therefore unneccessary. For a complete set, S, S[L] and S[V] are all equivalent and are equal to the exact amplitude function.

D. Symmetry Considerations

A separation into symmetry classifications is possible since the sum-rule Eq. [8] is separable according to symmetry. This separation depends upon the factorization (or approximate factorization) of the molecular eigenfunction. A separate sum-rule must therefore apply to each symmetry class. It is not possible, for example, to use a $p\pi$ sum-rule to make up for missing information about contributions from $s\sigma$ states. Explicitly this means that a sum-rule such as Eq. [8] applies independently to every component of the scattering tensor.

IV. APPLICATION TO OFF-RESONANCE RAMAN SCATTERING

A. Intermediate Gauge

The concept of intermediate gauge enters into the calculation of absolute Raman cross-sections in two ways. The sum-rule given by Eq. [8] is a direct consequence of gauge invariance. This sum-rule provides a useful average frequency approximation, Eq. [10], which is not only "closed" with respect to residual ("uncharacterized") electronic states but is also independent of whether the starting point is the length or velocity gauge. Equation [10] is not independent of gauge. In fact, as F_K changes from zero to its maximum possible value, the gauge varies from the pure length-form towards the velocity-form. If $\omega_1 \approx \omega_2 \lesssim \frac{1}{2}[\omega(K)-\omega(i)]$, then the maximum value of F_K is around $\frac{1}{3}$. This small maximum value of F_K confines Raman intensity calculations to the neighborhood of the length-form results. Though our past persuasions (8,9) leaned toward the velocity gauge Hamiltonian for Raman calculations, the present paper shows conclusively that the length-form is the best limiting form to use when the inexact set is composed of low-lying electronic states. However, either "open ended" fixed gauge form contains great intrinsic uncertainty when used with incomplete sets because of the unknown magnitude of residual state contributions.

The second way that the intermediate gauge concept enters Raman intensity calculations is in the assessment of the vibronic matrix elements. The contributions to the state sum depend on how the magnitude of the dipoles are distributed among the vibronic levels, and the relative phases of the contributions depend on the mathematical properties of the full vibronic eigenfunctions. For the description of symmetry allowed transitions (a_{1g} Raman intensity), a paper by Brown (10) using the concept of intermediate gauge is useful. This paper shows that the k,j vibronic matrix elements associated with two electronic states n,m in the state sum of Eq. [1] can be factored into purely vibrational (χ) and electronic (ϕ) parts,

$$|\omega(nk;mj)|^{\xi-1} \times \langle\phi_n(R^0,\vec{r}1,\vec{r}2,\cdots)|\vec{\gamma}(\xi)|\phi_m(R^0,\vec{r}1,\vec{r}2,\cdots)\rangle \times \langle\chi_k^n(\vec{R}_1,\vec{R}_2,\cdots)|\chi_j^m(\vec{R}_1,\vec{R}_2,\cdots)\rangle \ , \quad [11]$$

where ξ, which varies between $\xi=1$ (length) and $\xi=0$ (velocity), is the gauge parameter of Section II, R^0 represents a fixed nuclear configuration, and $\vec{\gamma}(\xi)$ is a generalized electronic operator. This expression allows Franck-Condon factorization in an extended sense, as originally proposed by Robinson and Auerbach (8,9).

B. Raman Cross-Section Calculations

Calculations incorporating the modifications of Eqs. [10] and [11] can begin with the well known expression (randomly oriented sample; polarized laser excitation) employing derivatives of the components of the polarizability (11). The relationship between these components and the dipole matrix elements is used, and the concept of oscillator strength is then introduced (12). The result for the $\rho\sigma$ (ρ,σ=x,y,z) contribution to the absolute total cross-section for a fundamental vibration in a nondegenerate electronic ground state (m=0,j=0,1) is then given by,

$$Q_{\rho\sigma} = Q_0 \frac{\omega_2}{\omega_1} \left| \sum_{n,k} g(n)^{-1} \mathscr{F}(n)(F_{nk} - F_K)\bar{C}_{nk} \right|^2_{\rho\sigma} \ . \quad [12]$$

The sum $\Sigma_{n,k}$ is typically carried out over a limited electronic set n, but over all vibrational levels in each vibronic envelope. The constant Q_0, a "universal scattering constant", has the form $e^4/8\pi\varepsilon_0^2m^2c^4$ (SI units) [$2\pi e^4/m^2c^4$ (CGS units)] whose numerical value is 4.989×10^{-25} cm^2. All other factors in Eq. [12] are dimensionless. The factor g(n) accounts for vibronic degeneracy. Matrix elements of $\vec{\gamma}(\xi)$ in Eq. [11] have been related (10) to the integrated oscillator strength $\mathscr{F}(n)$ appearing in Eq. [12]:

$$\mathscr{F}(n) = \frac{2mg(n)}{3\hbar} |\langle n|\gamma_x|0\rangle|^2 \sum_k \omega(nk;00)^{2\xi-1} |\langle k|0\rangle|^2 \quad . \qquad [13]$$

The sum in Eq. [13] is independent of vibrational state. It can be summed directly. However, a good approximation is $[\omega(n0)]^{2\xi-1}$, where $\omega(n0)$ represents the vibronic absorption maximum. The other dimensionless factors F_{nk} and $\bar{C}_{nk}$ in Eq. [12] are then given by,

$$F_{nk} = \frac{\omega_1\omega_2}{[\omega(nk;00)+\omega_2][\omega(nk;00)-\omega_1]} \quad ,$$

and,

$$\bar{C}_{nk} = \frac{[2\omega(nk;00)-\omega_1+\omega_2]\langle 1|k\rangle\langle k|0\rangle}{[\omega(nk;00)]^{1-\xi}[\omega(nk;01)]^{1-\xi}[\omega(n0)]^{2\xi-1}} \quad .$$

The greatest off-resonance a_{1g} Raman intensity arises from electronic transitions with large 1-0 band strength. Transitions to states exhibiting moderately different vibrational constants contribute more than those undergoing little change (e.g. certain Rydberg states) or a massive change (e.g. dissociative states) in these constants. The absorption profile alone is not a good key to the source of off-resonance Raman intensity.

C. Benzene

Absolute intensities for the C-C 995.4 cm^{-1} a_{1g} vibration in gaseous benzene have been measured using 514.5 nm and 337.1 nm exciting light (13). See also Refs. (14,15). [Under the laser excitation used in Ref. (13), $15(\alpha'^2_{xx}+\alpha'^2_{yy}+\alpha'^2_{zz})$, not that for "cylindrical" excitation (14), is the "scattering activity"]. The solution values of Ziegler and Albrecht (15) might not be directly comparable (14). The depolarization ratios ρ_n under unpolarized exciting light are known near the two relevant wavelengths (16): ρ_n=.05 at 546.1 nm and ρ_n=.07 at 312.6 nm. Using these data (D_{6h} symmetry), experimental values of Q_{xx} may be inferred (14,17). These can be compared with the theoretical results from Eq. [12]. Note that Q_{xx} is multiplied by 2.23

(visible) [2.16 (ultraviolet)] to obtain the overall cross-section ($Q_{xx}+Q_{yy}+Q_{zz}$) under laser excitation.

The information required for the theoretical calculations is conveniently obtained from experimental data. Only the intense ($\mathscr{F}$ = .88) $^1E_{1u}-^1A_{1g}$ transition (18) has been included in the state sum. The full vibronic sum Σ_k over harmonic oscillator levels was carried out (8,9), overlap integrals being evaluated using well known methods (19). The input parameters are the vibrational constants for the ground electronic state together with estimated values for the excited state constants [r_0(CC) $\approx$ .1434 nm; $\bar{\nu} \approx 900$ cm^{-1}] derived from knowledge about isoconfigurational $^1B_{2u}$ and $^1B_{1u}$ states (20). Results were found to be quite insensitive to reasonable variations of these parameters.

D. Conclusions

In past work (8,9) there was misplaced optimism about calculating one-state absolute Raman cross-sections for molecules having the electronic complexity of benzene. This was partly due to a misinterpretation of published experimental data and partly due to realities caused by the sum-rule constraint. Reference (13) stated that <u>differential</u> cross-sections were being reported. These are to be multiplied by a factor $\frac{8}{9}\pi$ for comparision with Eq. [12]. However, upon closer examination [cf. Ref. (14)], it was discovered that the data of Ref. (13) are not differential cross-sections. This introduced an extra factor of three disagreement between theory and experiment.

Depending upon the value of ξ, the $^1E_{1u}$ one-state Raman theory of benzene can account for at most ten per cent of the experimental cross-section. Better results can be obtained if additional electronic transitions in the far ultraviolet are included. This conclusion coincides with views expressed by previous authors based on the observed slope of the Raman excitation spectrum (13,15). The slope of the theoretical curve depends slightly on the value of ξ employed. For $\xi=1$, the theoretical slope using only the $^1E_{1u}$ state is twice that found experimentally, a discrepancy that was first pointed out by Udagawa et al (13). Smaller ξ-values give only a slightly lesser slope, contrasting the findings from "open ended" theories (8).

The DIP spectra found earlier (8) are partly an artifact of the velocity gauge formalism. However, DIPs can occur 1) in higher overtone spectra, 2) if the dipole matrix elements were steep functions of internuclear displacement, or 3) if contributions from deep ultraviolet transitions with opposing phase were dominant off-resonance but were overtaken approaching resonance.

Considering that an oscillator strength of only .88 was utilized out of a total of 42, perhaps it is to be expected that better results were not obtained. See Ref. (21) for an overview of the benzene absorption spectrum to 35 eV. The lack of agreement is caused by the inability of the sum-rule to retrieve information about non-participating electrons in the $^1E_{1u}-^1A_{1g}$ transition. This defect in the theory is transparent for a molecule having "localized" electronic transitions, but is absent for systems having very few electrons (6).

The moral of this story is that one can get something for nothing, but not nearly enough. It is hoped, however, that future Raman intensity and polarization calculations can profit from the lessons learned here about the effect of neglected electronic states, both those that are dealt with by the sum-rule and those that are not.

REFERENCES

1. Pauli, W., Rev. Mod. Phys. 13, 203 (1941).
2. Power, E. A., "Introductory Quantum Electrodynamics", Chap. 8. American Elsevier, New York (1965).
3. Ahranov, Y., and Au, C. K., Phys. Rev. A20, 1553 (1979).

4. Brown, C. W., and Robinson, G. W., to be published.
5. Bassani, F., Forney, J. J., and Quattropani, A., Phys. Rev. Letters 39, 1070 (1977).
6. Robinson, G. W., Phys. Rev. A26, 1482 (1982).
7. Dirac, P. A. M., "The Principles of Quantum Mechanics", pp. 244-248, 3rd ed. Oxford, Clarendon Press (1947).
8. Robinson, G. W., and Auerbach, R. A., Chem. Phys. Letters, 74, 237 (1980).
9. Robinson, G. W., and Auerbach, R. A., J. Chem. Phys. 74, 2083 (1980). Equations in Refs. (8) and (9) use the velocity gauge as starting point.
10. Brown, C. W., J. Chem. Phys., in press.
11. Placzek, G., "The Rayleigh and Raman Scattering", pp. 29-31. UCRL-TRANS-526(L), N.T.I.S., U. S. Dept. of Commerce (1962); Albrecht, A. C., J. Chem. Phys. 34, 1476 (1961).
12. Ford, A. L., and Browne, J. C., Phys. Rev. A7, 418 (1973).
13. Udagawa, Y., Mikami, N., Kaya, K., and Ito, M., J. Raman Spectry. 1, 341 (1973).
14. Schrötter, H. W., and Bernstein, H. J., J. Mol. Spectry. 12, 1 (1964).
15. Ziegler, L., and Albrecht, A. C., J. Chem. Phys. 67, 2753 (1977).
16. For example, Zubov, V. A., Opt. Spectry., 13, 861 (1962).
17. Herzberg, G., "Infrared and Raman Spectra of Polyatomic Molecules", pp. 246-249. D. Van Nostrand Co. Inc., Princeton (1945).
18. Hammond, V. J., and Price, W. C., Trans. Faraday Soc. 51, 605 (1955).
19. Ansbacher, F., Z. Naturforsch. 14a, 889 (1959).
20. Herzberg, G., "Electronic Spectra and Electronic Structure of Polyatomic Molecules", p. 666. D. Van Nostrand Co. Inc., Princeton (1966).
21. Koch, E. E., and Otto, A., Chem. Phys. Letters, 12, 476 (1972).

THE LOCAL MODE PICTURE OF MOLECULAR VIBRATIONS

Ole Sonnich Mortensen

Fysisk Institut
Odense Universitet
Denmark

INTRODUCTION

In quantum mechanics the word "picture" stands for the <u>representation</u> of the physics of a system in a particular basis. The physical results do not depend on the picture in which they are described, although the description may be more or less simple, depending on whether the right choice has been made.

The normal mode picture occupies a canonical position in the theory of vibrational spectroscopy, and so it may perhaps seem heretic to consider any alternative picture. The point of the present paper is nevertheless, to argue that the local mode picture is, for many physical systems, a more appropriate picture for describing molecular vibration than the conventional normal mode picture.

Before discussing the technical details of the local mode picture and its relation to the normal mode picture it seems reasonable to briefly sum up the essential differences between the two pictures. The differences refer to the coordinates used and to the basis functions employed and are summarized in Table I.

ISBN 0-12-066280-9

TABLE I Local Mode and Normal Mode Pictures

	Local	Normal
Coordinates	true internal (bond lengths, angles)	linear combination of cartesian, or linearized internal
Basis Functions	anharmonic oscillator (e.g. Morse)	harmonic oscillator

VIBRATIONAL HAMILTONIANS

Let $\{x_i\}$ be a set of massweighted cartesian coordinates, and $\{q_k\}$ internal coordinates given as (nonlinear) functions of $\{x_i\}$. Then the vibrational Hamiltonian in internal coordinates is;

$$H = \tfrac{1}{2} \sum_{k\ell} p_k \, g_{k\ell} \, p_\ell + \Delta V + V \quad , \tag{1}$$

where p_k is the momentum conjugate to q_k

$$p_k = - i\hbar \frac{\partial}{\partial q_k} \quad , \tag{2}$$

and $g_{k\ell}$ is given by

$$g_{k\ell} = \sum_i \frac{\partial q_k}{\partial x_i} \frac{\partial q_\ell}{\partial x_i} \quad . \tag{3}$$

ΔV is a pseudopotential, a function of the q_k, and is given as a rather complicated function of various commutators involving the $g_{k\ell}$. Since $g_{k\ell}$ is not a constant but a function of $\{q_k\}$ the order of the factors in Eq. (1) is important.

In the normal mode picture one uses linearized internal coordinates $\{q_k^o\}$ related to the cartesian coordinates through the linear transformation:

$$q_k^o = \sum_i B_{ki}^o \, x_i \quad . \tag{4}$$

In terms of these coordinates the vibrational Hamiltonian becomes:

$$H = \tfrac{1}{2} \sum_{k\ell} g_{k\ell}^o \, p_k^o \, p_\ell^o + V \quad , \tag{5}$$

where $g^{o}_{k\ell}$ is the Wilson G-matrix, and is a constant.

It is seen that the vibrational Hamiltonians are formally quite similar in the local - and in the normal mode picture. It should be remembered, however, that the form of the potential energy V is not the same in the two pictures. Indeed it is known that when the potential energy is expressed in the true internal coordinates it is diagonal to a much higher degree than when expressed in the linearized coordinates.

To sum up: in cartesian coordinates T is simple and V is most complicated, for linearized internal coordinates T is relatively simple and V is relatively complicated, while for the true internal coordinates T is relatively complex but V is simple.

THE LOCAL MODE PICTURE

In the local mode picture the vibrational Hamiltonian of Eq. (1) is split into two parts, the first being a sum of pure local mode Hamiltonians, depending on one coordinate only:

$$H = \tfrac{1}{2} \sum_{k}[p_k g_{kk} p_k + V(q_k)] + [\sum_{k<\ell} p_k g_{k\ell} p_\ell + V_{NL}] \quad . \tag{6}$$

V_{NL} is that part of the potential (including the pseudopotential) which depends on more than one coordinate at a time.

The local mode states $\{|v_k\rangle\}$ are now defined as the eigenstates of the local mode Hamiltonians:

$$[\tfrac{1}{2}p_k\ g_{kk}\ p_k + V(q_k)]\,|v_k\rangle = \varepsilon_{v_k}|v_k\rangle \quad . \tag{7}$$

If q_k corresponds to a bondstretch, $g_{k\ell}$ will be a simple constant, and $V(q_k)$ will typically have the form of a Morse function.

In the local mode picture the vibrational problem is then solved by expanding the exact wavefunction in terms of basis states formed as simple products of local mode states. In terms of these basis functions the Hamiltonian may be written:

$$(H-E_o) = \sum_k \omega_k v_k [1-x_k-x_k v_k] + \sum_{k<\ell} \sqrt{\omega_k\omega_\ell} [\gamma_{k\ell} p_k p_\ell + \phi_{k\ell} q_k q_\ell] + V_{NL}^{(3...)} , \quad (8)$$

where E_o is the energy of the ground state, ω_k and x_k the frequency and anharmonicity of the k'th local mode, and $\gamma_{k\ell}$ and $\phi_{k\ell}$ are the kinetic - and the second order potential energy constants respectively. They are given by:

$$\gamma_{k\ell} = - \tfrac{1}{2} g_{k\ell}/\sqrt{g_{kk}g_{\ell\ell}} \; ; \quad \phi_{k\ell} = \tfrac{1}{2} f_{k\ell}/\sqrt{f_{kk}f_{\ell\ell}} , \quad (9)$$

where $f_{k\ell}$ are the usual forceconstants. $V_{NL}^{(3...)}$ is the non-local anharmonic part of the potential and usually very small. Also $\phi_{k\ell}$ will generally be small, since the potential energy is predominantly local in nature. Thus, the most important coupling in the local mode picture will be the second order kinetic energy coupling.

The coupling constant $\gamma_{k\ell}$ is a function of the internal coordinates defining the molecular configuration. If q_k and q_ℓ are two bond stretch coordinates then $\gamma_{k\ell}$ is simple:

$$\gamma_{k\ell} = \tfrac{1}{2} \cos\theta/(1+m_c/m) \quad (10)$$

if the two coordinates have an atom of mass m_c in common, and the angle between them is θ.

From Eq. (10) it is seen that γ ranges from zero for a molecule like SF_6, through 0.01 typical of C-H vibrations, to a value of 0.36 for CS_2, an extreme case of a linear molecule with a light center atom.

Although the local modes are anharmonic oscillators the couplings of Eq. (8) can to a good approximation be calculated as for harmonic oscillators. Then

$$p_k = (a_k^+ - a_k) , \quad q_k = (a_k^+ + a) , \quad (11)$$

where a_k^+ (a_k) are the usual creation (annihilation) operators. To this approximation a pure local mode state, say $|v_k, v_\ell\rangle$, is then only coupled to the states: $|v_k+1, v_\ell+1\rangle$, $|v_k-1, v_\ell-1\rangle$, $|v_k+1, v_\ell-1\rangle$, $|v_k-1, v_\ell+1\rangle$. The first two states are separated by two quanta of oscillation from the state $|v_k, v_\ell\rangle$ and so will only couple very weakly, while the separation of the last two states from the original state is given by the difference: $(\omega_k-\omega_\ell)$ and so will vanish for equivalent oscillators. Thus, to a first, and very good approximation, only equivalent local modes will couple strongly. This coupling can be conveniently handled by symmetry.

SYMMETRY IN THE LOCAL MODE PICTURE

In the normal mode picture symmetry is introduced already at the level of the coordinates, with the normal coordinates formed from linear combinations of symmetry coordinates. In the local mode picture, on the other hand, the symmetrization is formed at the level of the basis states.

Suppose we have just two equivalent local modes with quantum numbers v_k and v_ℓ. Then the unsymmetrized local mode states are $|v_k v_\ell\rangle$ and symmetrization is performed by forming the linear combinations:

$$|v_k v_\ell\rangle_\pm = \frac{1}{\sqrt{2}} [\,|v_k v_\ell\rangle \pm |v_\ell v_k\rangle] \; . \tag{12}$$

The plus and minus states have different symmetry and so cannot mix. Thus the Hamiltonian matrix is split into two submatrices, and the problem of diagonalizing it is greatly reduced.

For three or more equivalent oscillators the symmetrization problem is only slightly more complicated, and has been discussed in detail elsewhere. Here we shall only stress that the symmetrization, because it takes place at the state, rather than at the coordinate level, is similar for all levels of excitation, and in practice much simpler than the symmetrization problem one encounters when considering higher excited states in the normal mode picture.

THE LOCAL MODE VERSUS THE NORMAL MODE PICTURE

Space does not permit us to give a more detailed description of the local mode picture and its connection with and differences from the normal mode picture. Here we shall therefore only briefly highlight some of these differences and connections.

It is seen that the couplings in the local mode picture depend essentially on the ratio of the kinetic energy coupling γ to the local mode anharmonicity x. If this ratio is small we have the local mode limit. The limit is realized in practice by molecules containing CH and OH bonds, giving rise for the higher excited states to the local mode spectra that have attracted so much interest in recent years. In the local mode limit the normal mode picture is hopelessly complicated and can not be used.

When the γ/x ratio is high, on the other hand, we approach the normal mode limit, where the symmetrized local mode states are strongly coupled. Then the normal mode picture might initially seen to offer a better starting point. This is not necessarily so for the following reasons. The potential V is still far more diagonal in the true internal coordinates of the local mode picture than in the normal coordinates. Also, as has been shown in detail elsewhere, it is simple by means of symmetry to express the normal mode limit wavefunctions in terms of the symmetrized local mode states. Thus, for example for two equivalent oscillators, the normal mode limit wavefunctions are simply:

$$
\begin{aligned}
|2_+0_-\rangle &= \tfrac{1}{2}(|20\rangle + |02\rangle) + \frac{1}{\sqrt{2}}|11\rangle \\
|1_+1_-\rangle &= \frac{1}{\sqrt{2}}(|20\rangle - |02\rangle) \qquad\qquad (13) \\
|0_+2_-\rangle &= \tfrac{1}{2}(|20\rangle + |02\rangle) - \frac{1}{\sqrt{2}}|11\rangle \quad ,
\end{aligned}
$$

where the symbolism on the left side refers to the normal modes.

Of course these are not the exact wavefunctions since coupling to other manifolds have been neglected. This coupling is small, however, and so the local mode states above form an excellent approximation to the true eigenstates. As a quantitative example it has been shown, that the overlap between the exact normal mode state and the local mode state at the fundamental level, i.e. $\langle(10)_+|1_+0_-\rangle = 0.998$, for $\gamma = 0.1$. Thus, we may conclude that the local mode picture forms an excellent basis even in the harmonic limit.

The local mode picture is particularly convenient for the treatment of the vibrational structure of electronic transitions. Since the normal coordinates depend on the particular electronic state, normal coordinate rotation will often take place in electronic transitions, and lead to a most complicated vibrational structure problem. In the local mode picture no coordinate rotation can take place, and so the analysis of the vibrational structure entails only the evaluations of simple one-dimensional Franck-Condon factors.

REFERENCES

1. Møller, H. S., and Mortensen, O. Sonnich, Chem. Phys. Lett. 66, 539 (1979).
2. Mortensen, O. Sonnich, Henry, B. R., and Mohammadi, M. A., J. Chem. Phys. 75, 4800 (1981).
3. Mortensen, O. Sonnich, "The Local Mode Picture of Molecular Vibrations", to be published.

ADVANTAGES OF THE TIME CORRELATOR THEORY OF RESONANCE RAMAN SCATTERING[1]

C. K. Chan
J. B. Page

Department of Physics
Arizona State University
Tempe, Arizona

D. L. Tonks

Group X-7
LANL
Los Alamos, New Mexico

We have recently been engaged (1-5) in applying many-body techniques within the time-correlator framework of resonance Raman (RR) scattering (6). The advantages of this approach over more traditional methods separate naturally into three interrelated categories, which were outlined in (7) and will be discussed here with somewhat different emphases.

A. Within well-defined model assumptions, the time-correlator framework lends itself to the use of finite-temperature phonon many-body techniques such as Wick's theorem and the linked cluster theorem, allowing one to derive non-perturbative, nonzero temperature, closed expressions for the optical absorption and RR scattering of multimode systems. For instance, within the adiabatic,

[1]Supported by NSF Grant No. DMR-8007456.

ISBN 0-12-066280-9

Condon, single excited electronic state and harmonic approximations, we have obtained exact and tractable solutions for first-order RR scattering for the linear plus quadratic electron-phonon coupling case of normal mode frequency shifts, normal coordinate changes ("mode mixing") and atomic equilibrium position changes in the resonant electronic state for a general multimode system (2).

Moreover, the time-correlator theory leads naturally to a separation of RR scattering into orders which brings out in a useful way the close connection between the optical absorption and RR excitation profiles, facilitating the development of integral transform relations between the absorption and profile lineshapes (Section B). This separation also leads to very efficient numerical modeling of the absorption and RR profiles of multimode systems at nonzero temperatures via one-dimensional fast Fourier transforms (Section C).

B. For the case of just linear electron-phonon coupling (i.e. atomic equilibrium position shifts, but no normal mode coordinate or frequency changes under electronic excitation), the lineshape $i'_f(\omega_L)$ of the first-order Stokes profile of mode f of a multimode system is given exactly by

$$i'_f(\omega_L) = \omega_L(\omega_L-\omega_f)^3|\Delta(\omega_f)\Phi(\omega_L)|^2, \qquad (1)$$

where ω_L is the incident light frequency, the prime denotes that we are just considering the ω_L -dependent part of the profile, ω_f is the mode frequency, and $\Delta(\omega_f)\Phi(\omega_L) = \Phi(\omega_L)-\Phi(\omega_L-\omega_f)$ (1,3). The complex function $\Phi(\omega_L)$ is proportional to the resonant electronic polarizability and is simply related to the optical absorption through a Kramers-Kronig analysis. Other authors who have recently discussed related absorption→profile transform techniques are referenced in (4). For the more realistic case of small vibrational frequency shifts under electronic excitation, we have recently shown that Eq. (1) holds approximately if $\Delta(\omega_f)$ is replaced by $\Delta(\omega_f^e)$, where ω_f^e is the frequency of mode f in the *excited* electronic state (4). In (1), (4) and (7), we have illustrated Eq. (1) by using measured absorption spectra to generate profile lineshapes.

It is important to note that additional absorption→profile transform relations may be obtained by going beyond the assumptions discussed above in connection with Eq. (1). Recently (8) we have derived *exact* T = 0K profile expressions which include non-Condon terms arising from the linear dependence of the electronic transition dipole matrix element $M(\underline{d})$ on the normal coordinates $\{d_f\}$, i.e. $M(\underline{d}) = M(\underline{0})[1+\Sigma_f m_f (2\omega_f/\hbar)^{\frac{1}{2}} d_f]$. For $\Sigma_f (2m_f\xi_f + m_f^2) < 1$, where ξ_f is the Stokes loss parameter of mode f in the notation of (9), the first-order RR profile lineshape may be written as an infinite series in *explicit* powers of the m_f 's. The first two terms of this series are

$$\begin{aligned} i_f'(\omega_L) = \omega_L(\omega_L-\omega_f)^3 \Big(& |\xi_f[1-\sigma(\omega_f)]\Phi(\omega_L) - m_f[1+\sigma(\omega_f)]\Phi(\omega_L)|^2 \\ & -2\mathrm{Re}\big(\{\xi_f[1-\sigma(\omega_f)]\Phi(\omega_L)\}\{m_f[1+\sigma(\omega_f)] \\ & \times\Sigma_{f'} m_{f'} \xi_{f'} [1-\sigma(\omega_{f'})]\Phi^*(\omega_L)\}\big) + \ldots\Big). \end{aligned} \tag{2}$$

Here $\sigma(\omega_f)\Phi(\omega_L) \equiv \Phi(\omega_L-\omega_f)$, and $\Phi(\omega_L)$ is again proportional to the resonant electronic polarizability. We note that Eq. (2) also contains m_f 's *implicitly*, through $\Phi(\omega_L)$. Neglecting the last term of Eq. (2), we recover Tehver's (9) approximate generalization of Eq. (1). It is interesting that for a "pure" Franck-Condon active mode (i.e. $m_f = 0$), Eq. (2) yields a transform relation formally identical with Eq. (1), even though $\Phi(\omega_L)$ may now contain non-Condon terms ($m_{f'}$'s) from other modes. Indeed, the non-Condon contributions to $\Phi(\omega_L)$ could even dominate the profile lineshape of such a mode.

C. The correlator approach leads to very efficient numerical modeling algorithms in terms of one-dimensional fast Fourier transforms (FFT). For the general case of an n-mode system with linear plus quadratic electron-phonon coupling, nonzero temperatures, and mode mixing involving all modes, the most time-consuming step in the numerical evaluation of our exact multimode expressions is the inversion of an n×n symmetric matrix for each value of the time argument in the FFT. This results in the computing time increasing roughly as n^3 for large n. In contrast, if one uses the conventional method of overlap integrals and sums over just two intermediate vibrational states, omitting temperature effects, the computing time increases with n roughly as $2^n \gg n^3$. With no mode mixing, the matrices in our model expressions become diagonal, and the computing time only increases as $n \lll 2^n$. For the T = 0K case, related computational advantages have also been discussed in (10).

As an illustration, we have calculated the optical absorption and RR profile lineshapes for azulene in CS_2, using a seven mode model which includes equilibrium position shifts and vibrational frequency shifts, but no mode mixing, under electronic excitation. Results for the 1260 cm^{-1} mode profile are shown in Fig. 1, and the agreement with the experimental data (11) is seen to be good. Similar results were obtained for the other totally symmetric modes of this system, as will be detailed elsewhere (12). In this seven-mode model, the FFT computing times for 2100 time points on an IBM 3081 were 1.7 sec. of cpu time for the optical absorption

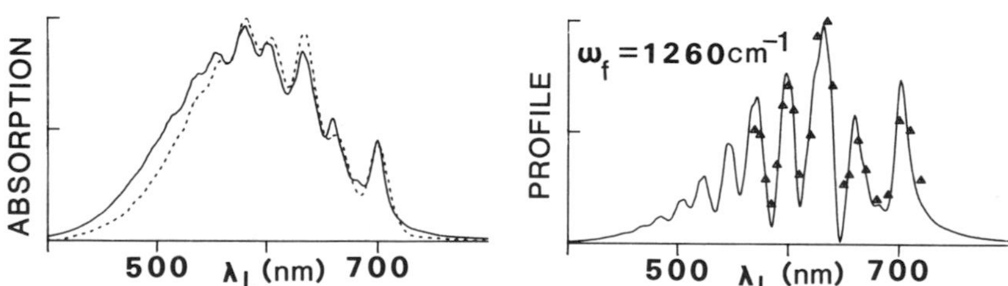

FIGURE 1. Room temperature absorption (dashed line) and profile data (triangles) (11) for azulene in CS_2. The theoretical (solid) curves were calculated via FFT techniques and were scaled via least-squares to the experimental data (12).

and 0.4 sec. for each profile. The profile times are comparatively short because the form of our model expressions is such that large portions of the absorption calculation can be stored and reused in computing the profiles. This is yet another useful result of the time-correlator approach.

REFERENCES

1. D. L. Tonks and J. B. Page, Chem. Phys. Lett. 66, 449 (1979).
2. D. L. Tonks and J. B. Page, Chem. Phys. Lett. 79, 247 (1981).
3. J. B. Page and D. L. Tonks, J. Chem. Phys. 75, 5694 (1981).
4. D. L. Tonks and J. B. Page, J. Chem. Phys. 76, 5820 (1982).
5. C. K. Chan and J. B. Page, Bull. Am. Phys. Soc. 27, 164 (1982), and to be published.
6. V. V. Hizhnyakov and I. Tehver, Phys. Status Solidi 21, 755 (1967).
7. J. B. Page, C. K. Chan, and D. L. Tonks, Proceedings of the International Conference on Raman Scattering VIII, Bordeaux, France, Sept. 1982, in press.
8. C. K. Chan and J. B. Page, to be published.
9. I. J. Tehver, Optics Comm. 38, 279 (1981).
10. D. J. Tannor and E. J. Heller, J. Chem. Phys. 77, 202 (1982).
11. O. Brafman, B. Khodadoost, C. T. Walker, C. K. Chan, and J. B. Page, to be published.
12. C. K. Chan, J. B. Page, O. Brafman, B. Khodadoost, and C. T. Walker, to be published.

ON THE INTERACTION OPERATOR IN THE OPTICAL SPECTROSCOPIES; POSITION vs. MOMENTUM[1]

Duckhwan Lee and A. C. Albrecht

Department of Chemistry
Cornell University
Ithaca, New York 14853

I. INTRODUCTION

The interaction of light with matter is conventionally treated in terms of field-induced transitions among the eigenstates of the field-free Hamiltonian. The minimal-coupling Hamiltonians take on different forms, depending on the gauges chosen for electromagnetic potentials, which are related by local gauge transformations and equivalently describe the evolution of the material in the presence of the radiation field through the time-dependent Schrödinger equation. Although all spectroscopic parameters must be gauge-invariant, Lamb pointed out how the "momentum" (from the Coulomb gauge) and "position" (from Lamb gauge) forms of interaction operator lead to significantly different lineshapes in one-photon absorption at the electric-dipole approximation (EDA) [1].

In this paper, we will show explicitly how the "momentum" form

[1] *supported by a grant from the National Science Foundation(CHE 80-16526) and by the Material Science Center of Cornell University*

ISBN 0-12-066280-9

of interaction operator gives gauge-dependent results as Lamb has mentioned. Then, Yang's gauge-invariant formulation (GIF) will be introduced [2]. All of the "active" and "passive" spectroscopies of any order will be discussed. The detailed treatment will be published elsewhere [3].

II. CONVENTIONAL AND GAUGE INVARIANT FORMULATIONS

The wavefunction, $|\Psi(t)\rangle$, for a molecule subject to time varying radiation fields is a solution of the time-dependent Schrödinger equation with the minimal-coupling Hamiltonian,

$$H(\vec{r},t) = (1/2m)\ \{\vec{p} - (q/c)\vec{A}(\vec{r},t)\}^2 + V(\vec{r}) + qA_o(\vec{r},t)\ . \tag{1}$$

$\vec{A}(\vec{R},t)$ and $A_o(\vec{R},t)$ are the vector and scalar potentials, respectively, which are not uniquely defined. By a local gauge transformation with a gauge function, $\Lambda(\vec{R},t)$,

$$\vec{A}'(\vec{R},t) = \vec{A}(\vec{R},t) + \vec{\nabla}\Lambda(\vec{R},t)\ ;$$
$$A_o'(\vec{R},t) = A_o(\vec{R},t) - (1/c)(\partial/\partial t)\Lambda(\vec{R},t)\,. \tag{2}$$

Any physically meaningful quantity should not be changed under such a transformation. Although the time-dependent Schrödinger equation is form-invariant, the wavefunction in the primed gauge suffers a phase change according to

$$|\Psi'(t)\rangle = \exp\{(iq/c\hbar)\Lambda(\vec{r},t)\}|\Psi(t)\rangle\ , \tag{3}$$

Conventionally (CF), a state vector, $|\Psi(t)\rangle$, is expressed as a linear combination of a complete set of orthonormal dark eigenstates, $\{|\phi_n\rangle\}$, of the dark Hamiltonian, $H_o(\vec{r}) = \vec{p}^2/2m + V(\vec{r})$, with the expansion coefficient $c_n(t) = \langle\phi_n|\Psi(t)\rangle$. The expansion coefficients in different gauges are related according to

$$c_n'(t) = \sum_m c_m(t)\langle\phi_n|\ \exp\{(iq/c\hbar)\Lambda(\vec{r},t)\}|\phi_m\rangle \neq c_n(t)\ . \tag{4}$$

Thus, $\{c_n(t)\}$, depending on gauge, cannot in general be regarded as physically meaningful or directly related to physical observables. In the CF, the Coulomb and Lamb gauges lead to the "momentum" and "position" forms at the EDA, respectively.

In Yang's GIF, instead of $H_o(\vec{r})$, the so-called *energy operator*

$$H_E(\vec{r},t) = (1/2m)\{\vec{p} - (q/c)\vec{A}(\vec{r},t)\}^2 + V(\vec{r}) , \quad (5)$$

is used as the basis defining Hamiltonian [2]. $H_E(\vec{r},t)$ represents the instantaneous energy of the molecule in the presence of radiation fields. Since the eigenstates of $H_E(\vec{r},t)$, $\{|\psi_n(t)\rangle\}$, also change their phase by a gauge transformation, the expansion coefficient, $a_n(t) = \langle\psi_n(t)|\Psi(t)\rangle$, is gauge-invariant. The equation of motion for $\{a_n(t)\}$ at the EDA is always given by

$$i\hbar\partial a_n(t)/\partial t = \varepsilon_n a_n(t) + \sum_m a_m(t) \langle\phi_n| -q\vec{r}\cdot\vec{E}(t) |\phi_m\rangle , \quad (6)$$

in any gauge. It is noted that the correct interaction operator is the manifestly gauge-invariant "position" form.

III. ACTIVE AND PASSIVE SPECTROSCOPIES

In "active" (inelastic) spectroscopic processes, the material suffers a net gain or loss of energy as a result of the interaction with light. On the other hand, in "passive" (parametric or elastic) spectroscopies, the material acts only in a catalytic way to change the state of radiation field. It is possible without the commutation relation to show that the cross-sections for "active" spectroscopies of any order given by the GIF and the CF are equivalent at *exact resonance* between the initial and final states. It can also be shown that the induced polarizations, for "passive" processes, given by the two formulations are equivalent provided there are no intermediate resonances. The equivalency no longer exists if there are any other intermediate resonances.

In the above considerations, damping effects, which become important whenever resonances are approached, are entirely excluded. It is often satisfactory to introduce (radiative and nonradiative) damping term in the density matrix equation. Only the GIF ("position" form) can gaurantee the gauge-invariance of density matrix elements, and can incorporate damping effects in a gauge-invariant fashion. When damping effects are included only the "position" operator is correct.

If an approximate Hamiltonian contains a "nonlocal" (implicily velocity dependent) potential such as in the HF Hamiltonian, the above limited equivalency is valid only after the difference between the potentials in the dark and minimal-coupling HF Hamiltonians is included as an extra interaction operator in the CF.

IV. CONCLUSIONS

The principle of gauge-invariance requires that the "position" formulation of the interaction of light with matter is the only correct one at the EDA. Nevertheless, it is found that the "dark" basis can give correct spectroscopic parameters in any chosen gauge at *exact resonance* between the initial and final states and when *damping effects are entirely ignored*. In practical calculations with approximate eigenfunctions, the CF will always carry an error due to the break-down of the equivalency with the GIF in addition to the error resulting from the use of an approximate basis, the sole source of error in the GIF.

Although the semi-classical approach has been employed here, we expect that on the basis of the correspondence principle the GIF will remain valid even with field quantized. A study of this problem when the field is quantized is now under way.

REFERENCES

1. W. E. Lamb, Jr., Phys. Rev. *85*, 259 (1952)
2. K-H. Yang, Ann. Phys. (NY) *101*, 62 (1976)
3. D. Lee and A. C. Albrecht, J. Chem. Phys. *in press* (1983)

SEMICLASSICAL SIMULATION OF VIBRONIC PROCESSES

Arieh Warshel

Department of Chemistry
University of Southern California
Los Angeles, California

P. Stern and S. Mukamel

Department of Chemical Physics
The Weizmann Institute of Science
Rehovot, Israel

I. INTRODUCTION

Many theories of vibronic processes are based on using experimental information in analyzing experimental results. Such theories are very powerful in extracting the maximum information content from a given experiment, but this information may not be sufficient to analyze uniquely the given experiment. For example, in most cases one cannot determine what mechanism leads to a given vibronic line broadening, using pure experimental information. In this respect there may be a great benefit in computer simulation approaches which evaluate molecular spectra using realistic molecular models.

Methods for calculating vibronic matrix elements were proposed before[1,2] and gave encouraging results. For example, quantum mechanical calculations of vibronic transition intensity gave reasonable agreement with the corresponding experiments.[2,3] However, approaches which are based on evaluating discrete vibronic transitions do not account properly for the observed line width, either because harmonicity effects are neglected or because time dependent effects are not taken into account.[4] In view of the possible complexity of vibronic relaxation processes it may be preferable to exploit semiclassical approaches in studying large molecules.

ISBN 0-12-066280-9

In this work we report a preliminary use of a semiclassical trajectory approach for studying different types of vibronic processes. This includes studies of the absorption and emission of substituted benzene and the radiationless transition of individual vibronic levels of porphines.

II. THEORETICAL APPROACH

Our theoretical approach can be described in different alternative formalisms. The formalism used here is similar to that used in simulating semiclassically cis-trans isomerization reactions[5] and electron transfer processes.[6]

We will use for simplicity a diabatic representation and mention, when needed, the modifications associated with using the adiabatic representation.

Let us consider two diabatic electronic states ϕ_1 and ϕ_2, which satisfy the time dependent Schrödinger equation

$$\begin{aligned} H(\underset{\sim}{r})\phi_1(\underset{\sim}{r}) &= \varepsilon_1(\underset{\sim}{r})\phi_1(\underset{\sim}{r}) + H_{12}(\underset{\sim}{r})\phi_2(\underset{\sim}{r}) \\ H(\underset{\sim}{r})\phi_2(\underset{\sim}{r}) &= \varepsilon_2(\underset{\sim}{r})\phi_2(\underset{\sim}{r}) + H_{12}(\underset{\sim}{r})\phi_1(\underset{\sim}{r}) \end{aligned} \tag{1}$$

where $\varepsilon_1(r)$ and $\varepsilon_2(r)$ are the diabatic energy surfaces of ϕ_1 and ϕ_2. The time dependent wave function of the system can be expressed as:

$$\begin{aligned} \psi(\underset{\sim}{r},t) &= a_1(t)\phi_1(\underset{\sim}{r}) \exp\{-\frac{i}{\hbar}\int^t \varepsilon_1(t')dt'\} \\ &\quad + a_1(t)\phi_2(\underset{\sim}{r}) \exp\{-\frac{i}{\hbar}\int^t \varepsilon_2(t')dt'\} \end{aligned} \tag{2}$$

Substitution of this equation into the time dependent Schrödinger equation

$$H(\underset{\sim}{r})\psi(\underset{\sim}{r},t) = i\hbar\partial\psi(\underset{\sim}{r},t)/\partial t, \tag{3}$$

and neglecting $\partial\phi(r)/\partial t$ gives the coupled equations

$$\begin{aligned} \dot{a}_1 &= \frac{i}{\hbar} H_{12}a_2(t) \exp\{\frac{i}{\hbar}\int^t \Delta\varepsilon_{21}(t')dt'\} \\ \dot{a}_2 &= \frac{i}{\hbar} H_{12}a_1(t) \exp\{-\frac{i}{\hbar}\int^t \Delta\varepsilon_{12}(t')dt'\} \end{aligned} \tag{4}$$

where $\Delta\varepsilon_{21} = \varepsilon_2 - \varepsilon_1$. In the adiabatic representation H_{12} is zero and $\partial\phi(r)/\partial t$ is not neglected. In this representation eq. (4) is changed in such a way that $\langle\phi_1|\partial\phi_2/\partial t\rangle$ is substituted for $(H_{12}/\hbar)$.[6] The semiclassical approximation for eq. (4)

involves substitution of $\Delta\varepsilon_{21}(t)$ by $\Delta\varepsilon_{21}(\underset{\sim}{r}(t))$ where $\underset{\sim}{r}(t)$ is the coordinate vector of the system along the classical trajectory pathway. The overall probability amplitudes $a_1^2(t)$ and $a_2^2(t)$ are obtained by

$$a_i^2(t) = |\langle\int_1^2 \dot{a}_i(t)dt\rangle|^2 \tag{5}$$

where the average $\langle\ \rangle$ is over the initial conditions for the coordinates and momentum of the trajectory. The integration from 1 to 2 indicates that at points of large values of a^2 the trajectory should "hop" from surface 1 to surface 2.[7] However, in this preliminary study we propagate trajectories only on one surface.

Eqs. (4) and (5) can be extended to cases that involve absorption of light by expressing the off diagonal element of the system as the matrix element of the radiation field Hamiltonian.

$$H_{12} = k\mu_{12}e^{-i\omega t} \tag{6}$$

where k is a constant, μ_{12} is the transition diple and ω is the frequency of the radiation field. Eqs. (6), (5), and (4) give, for the initial conditions, $a_1(0) = 1$, $a_2(0) = 0$), the transition probability per unit time

$$I(\omega) = \frac{a_2^2(\tau,\omega)}{\tau} = \frac{1}{\tau}\,|\langle\int^{\tau} k\mu_{12}\exp\{-\frac{i}{\hbar}\int^{t}\Delta\varepsilon_{21}(t')dt'\}$$
$$\times \exp(-i\omega t)dt\rangle|^2 \tag{7}$$

This expression involves the Fourier transform of $a_2(\tau)$ of eq. (5). In fact, one can consider $I(\omega)$ as the crossing probability for the surfaces $\varepsilon_1 + h\omega$ and ε_2.

It is sometimes useful to express eq. (7) by expanding in a Fourier series at discrete frequencies. This gives

$$a_2^2\ (\tau,\omega) \simeq (k\mu)^2|\langle\int \exp\{\sum_s(A_s/(\hbar\omega_s)e^{i\omega_s t}\}\exp(-i\omega t)\rangle|^2 \tag{8}$$

where A_s are the Fourier component of $\Delta\varepsilon(t)$ and μ is approximated by its average value. For small A_s we obtain (by expanding eq. (8) and retaining only terms of first order in A_s) the approximated expression

$$a_2^2\ (\tau,\omega) \propto |\langle\int[1 + \sum_s (A_s/\hbar\omega_s)\exp(i\omega_s t)]\exp(-i\omega t)dt\rangle|^2 \tag{9}$$

Replacing the square integral by a double integral we obtain

$$a_2^2(\tau,\omega) \propto \iint \langle [1+\sum_s (A_s/\hbar\omega_s)\exp(i\omega_s t)][1+\sum_{s'}(A_{s'}/h\omega_{s'})$$

$$\times \exp(i\omega_{s'}t')]\exp(-i\omega(t-t'))dtdt'\rangle \qquad (10)$$

which gives (retaining only A_2^2 terms) the following expression

$$a_2^2(\tau,\omega) \simeq \sum_s (A_s/\hbar\omega_s)^2(\delta(\omega_s-\omega)) + \delta(\omega) \qquad (11)$$

Eq. (9) gives peaks of equal intensities for positive and negative ω_s. This corresponds to the high temperature limit (classical limit).[8] ω_s corresponds to surface hopping from $\varepsilon_1+h\omega_0-k\omega_s$ to ε_2 where ω_0 is the energy gap between the minima of ε_2 and ε_1. This surface hopping does not conserve energy and should give small crossing probability in procedures that integrate the actual hoping step in complex time. Indeed in the quantum mechanical case we may use the fluctuation-dissipation theorem and at T=0 we get:

$$a_2^2(\tau,\omega) \propto \sum_{s'}(A_{s'}/\hbar\omega_{s'})^2(2\delta(\omega_{s'}-\omega))+\delta(\omega) \qquad (12)$$

where s' runs over the positive ω_s.

III. RESULTS AND DISCUSSION

In order to implement eq. (7) in practical calculations of large molecules one needs the proper potential surfaces. The QCFF/PI potential surfaces[2] can be used for this purpose since they are given in an analytical form that allows for efficient trajectory calculations. The calculations should also involve an average over initial conditions. For a molecule with more than 5 degrees of freedom such an average is not practical at the present time. Instead it is possible to propagate trajectories on the harmonic part of the ε_1 potential surface with initial conditions that satisfy the corresponding quantum mechanical harmonic oscillator. That is, for the ground vibrational state the calculation starts with initial kinetic energy of $h\omega/2$ and zero potential energy in all normal modes of the system. Such calculations take into account the anharmonicity in $\Delta\varepsilon_{21}$ but the molecular motion is not affected by the anharmonicity in ε_1 (for jet cooled molecules that absorb at the ground vibrational state the anharmonicities in ε_1 can probably be neglected). Alternatively one can run trajectories on the real anaharmonic potential ε_1 with harmonic initial conditions. This, however, might require extensive averaging over initial conditions for cases with significant energy randomization.

We chose to examine our approach by calculating the vibronic transitions for alkylbenzenes, which were subjected to extensive studies under jet cooled conditions.[9,10] Figure 1 and 2

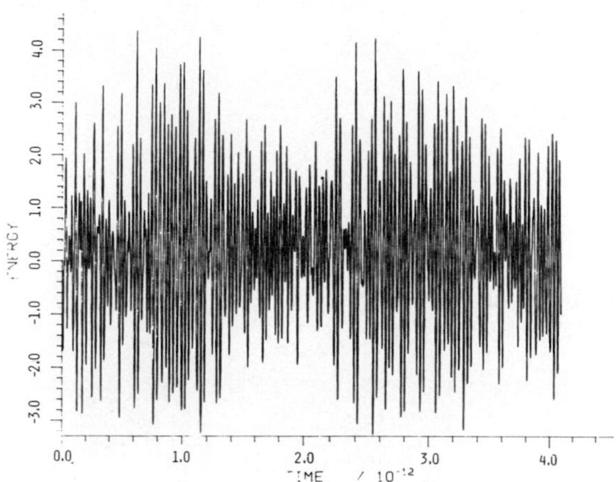

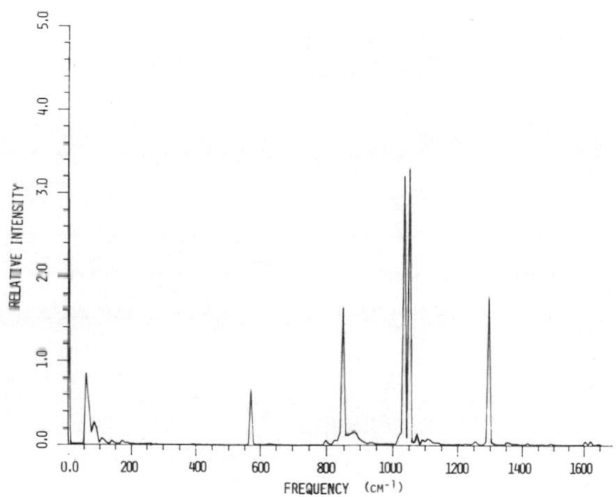

Figure 1. (a) The energy gap, $\Delta\varepsilon_{21}$, for methylbenzene and, (b) the corresponding $(A_s/\omega_s)^2$ of eq. (11) obtained by a Fourier transform of $\Delta\varepsilon_{21}(t)$. The Fourier components at $\omega < 100\ cm^{-1}$ are set to zero (see text). The calculations were done by propagating a trajectory on the anharmonic ground state of the molecule starting with initial momentum that correspond to $h\omega_s/2$ in each of the harmonic normal modes of the system.

describe the result of the calculations for methylbenzene and propylbenzene using the approximated expression of eq. (11), which is valid in the harmonic approximation for vibronic transitions with small origin shifts. The calculations were performed by running trajectories with harmonic initial conditions on the anharmonic ε_1, evaluating the Fourier components of $\Delta\varepsilon_{21}(t)$ by fast Fourier transform and dividing each A_s by the corresponding

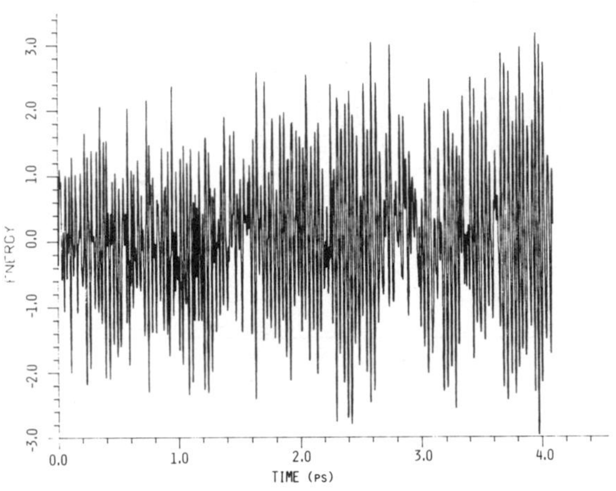

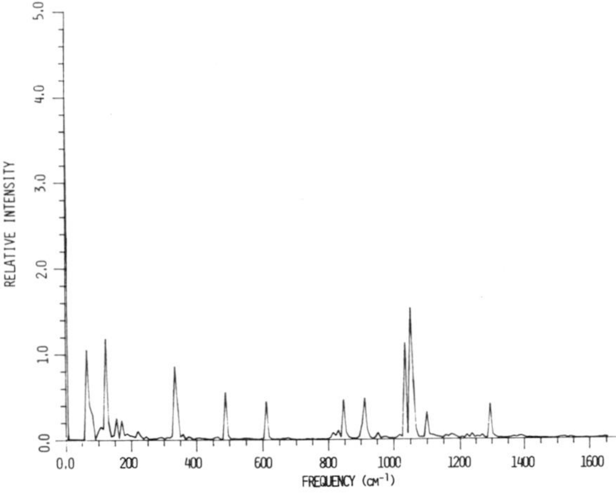

Figure 2. The calculated (A_s/ω_s) for propylbenzene obtained in the same way as in Figure 1.

ω_s. The calculations involved truncation of the Fourier components for $\omega_s < 100\ cm^{-1}$; in this region Eq. (11) gives a large peak (two orders of magnitude larger than the peaks in the figure). This peak is due to the numerical instability of the (A_s/ω_s) function for short time trajectories and to the classical energy transfer between the zero point vibrational levels which will be discussed below.

Figure 3 describes the calculations for methylbenzene obtained by the direct use of eq. (7). Now the calculations do not require any truncation of the peak at $\omega < 100$. However, the lines are much broader than the experimentally observed lines. This non-physical broadening is due to both the numerical instability of the integral of $\Delta\varepsilon(t)$ and to the fundamental problem of semiclassical evaluation of discrete vibronic transitions; the trajectory pathways do not follow a regular orbit (even on the ground vibrational state) but show nonperiodic energy transfer between different normal modes. This point is demonstrated in Figure 4 that describes the time dependent energy redistribution in the ground vibrational state of methylbenzene. The problem of classical preparation of vibrational states might be a major bottleneck in semiclassical calculations of vibronic processes. This problem might be overcome eventually by extensive averaging. At present, however, we recommend performing

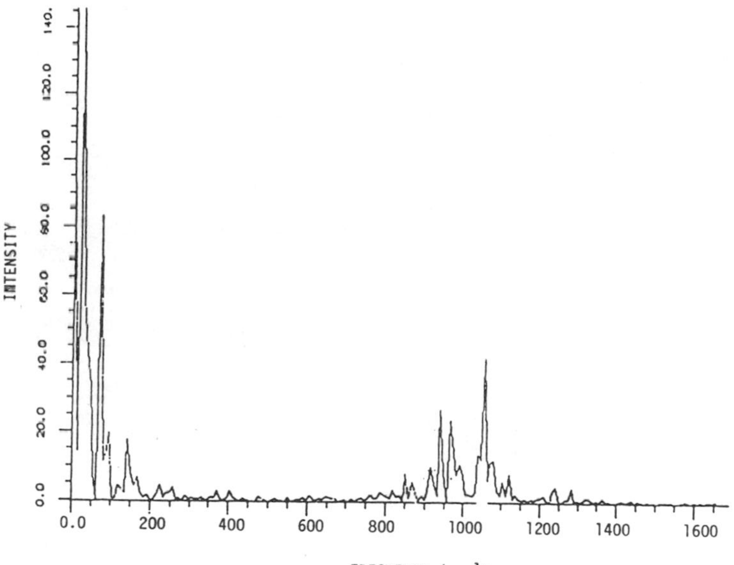

Figure 3. The calculated contribution to the vibronic spectrum of propylbenzene from a single trajectory with the initial conditions given in the caption of Figure 1. The calculations are performed using eq. (7).

absorption calculations propagating trajectories only on the harmonic part of the molecular potential surface or filtering the Fourier components at $\omega_s < 100\ cm^{-1}$.

As was shown above, it is not clear how to prepare a classical vibrational state on an anharmonic potential surface, yet,

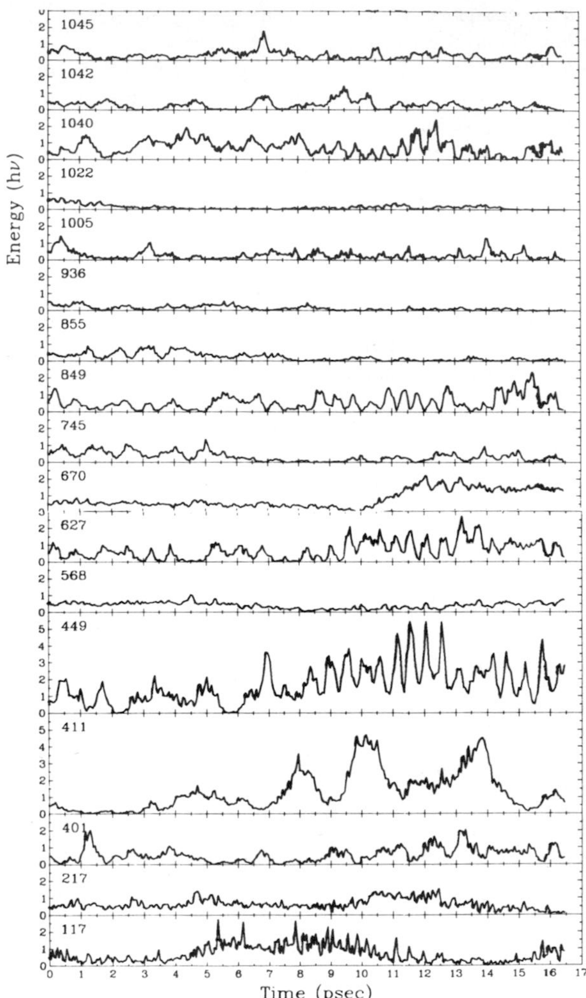

Figure 4. Energy redistribution within the vibrational ground state of methylbenzene. The figure gives the energy content of each of the harmonic normal modes of the molecule along a trajectory that moves on the anharmonic potential surface with the initial conditions of Fig. 1. The figure demonstrates the non periodic nature of the molecular trajectories.

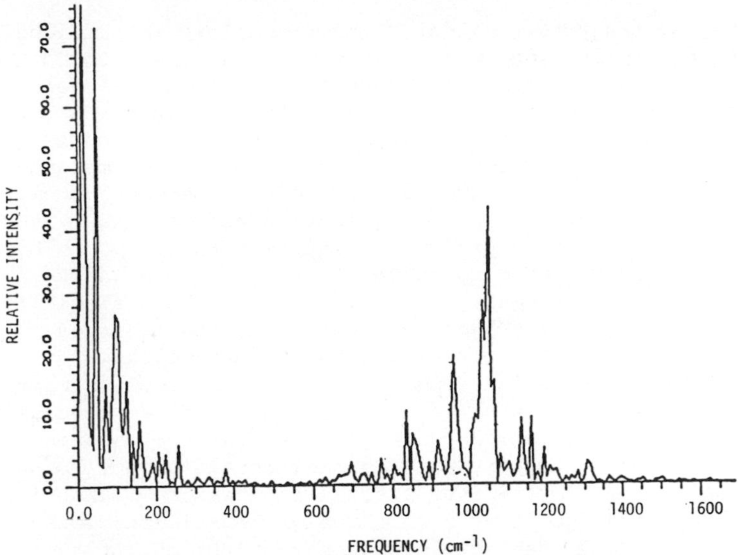

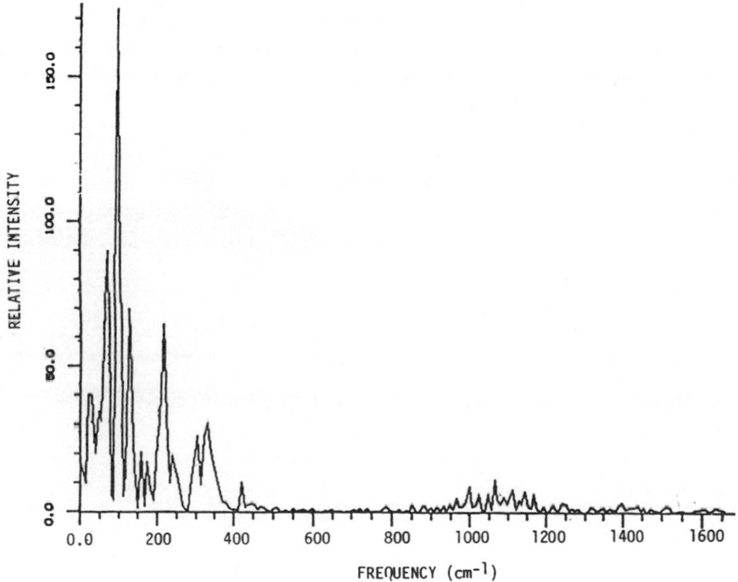

Figure 5. Calculated contribution to the fluorescence of propylbenzene (using eq. (7)) from a single trajectory which starts with $\hbar\omega_s/2$ for all harmonic modes except ω_{12} which was prepared with $3\hbar\omega_s/2$. (a) Calculated for the first 4 picoseconds. (b) Calculated for the time between 12 and 16 picoseconds.

one can use trajectory calculations to explore intramolecular vibrational relaxation in excited electronic states. That is, one can simulate experiments where single vibronic levels are excited by running trajectories with the corresponding harmonic initial conditions on the anharmonic electronic potential surface. Such simulations are expected to provide meaningful results only if the energy transfer between zero point vibrations is much slower than the relaxation process under study. In this work we made a preliminary attempt to study the broadening of the fluorescence spectrum of propylbenzene which is prepared by exciting a single vibronic line.[10] The calculations are summarized in Fig. 5 and indicate that the observed broadening may be due to time dependent intramolecular energy transfer from the excited mode to many other modes. Unfortunately, this result cannot in any way be considered conclusive since the transfer between zero point vibrational states (evaluated for trajectory with initial conditions of $\hbar\omega_s/2$ for all modes) is as fast as the overall relaxation, making it impossible to judge if we have a real relaxation process. Obviously, this fundamental question requires much futher studies.

The semiclassical trajectory approach provides a very powerful tool for studying radiationless transitons. That is, eq. (5) gives the probability for non radiative surface crossing between ε_1 and ε_2. Previous studies with this approach[5,6] did not consider selective preparation in a specific vibronic state. In view of recent advances in preparing isolated excited vibronic states of large molecules and estimating their lifetime we have started preliminary studies of radiationless transitions from excited vibronic states of porphines. This study will be reported elsewhere.[11]

REFERENCES

1. Albrecht, A.C. *J. Phys. Chem. 33,* 169 (1960).
2. Warshel, A. in *Modern Theoretical Chemistry,* Ed. G. Segal, Vol. 7, Plenum Press, New York, 1971.
3. Warshel, A. and Dauber, P. *J. Chem. Phys. 66,* 5477 (1977).
4. Warshel, A. and Karplus, M. *Chem. Phys. Letters, 17,* 7 (1972).
5. Warshel, A. and Karplus, M. *Chem. Phys. Letters, 32,* 11 (1975).
6. Warshel, A. *J. Phys. Chem. 86,* 2218 (1982).
7. Miller, W.H. and George, T.F. *J. Chem. Phys. 56,* 5637 (1972).
8. Mukamel, S. *J. Chem. Phys. 77,* 173 (1982).
9. Hopkins, J.B., Powers, D.E., Mukamel, S. and Smalley, R.E. *J. Chem. Phys. 72,* 5049 (1980).

10. (a) Mukamel, S. and Smalley, R.S. *J. Chem. Phys. 73*, 4156 (1980).
(b) Mukamel, S. in *Relaxation of Elementary Excitation*, Ed. R. Kubo and E. Hunamuro, Springer Verlag Ber. (1980), p. 220.
11. Warshel, A. and Sharon, R., to be published.

NORMAL COORDINATE ANALYSIS OF POLYENE CHAINS FOR INTERPRETING THE VIBRATIONAL SPECTRA OF TRANSIENT AND LESS STABLE SPECIES

Mitsuo Tasumi

Department of Chemistry
Faculty of Science
The University of Tokyo
Bunkyo-ku, Tokyo
Japan

As a result of the recent progress in experimental techniques of time-resolved vibrational spectroscopy, a variety of transient molecular species in either the excited or ground electronic state have been detected, and the experimental data will rapidly increase both in quality and quantity. Under such circumstances analysis of the observed vibrational spectra of transient species has become an important task. It is usually almost impossible to obtain detailed information on the molecular structure of a transient species under study by experimental techniques other than TRVS. This makes the vibrational analysis more difficult but on the other hand more interesting and rewarding.

In this paper the author wishes to describe the recent work of his group on the normal-coordinate analyses of polyene compounds ranging from butadiene to polyacetylene including their less stable isomers. The author has been interested in the resonance Raman studies of all-*trans*-β-carotene (1,2) and its 15,15'-*cis* isomer (2) and intact and doped polyacetylenes (3,4). These studies have naturally led us to the following problems.

(a) Correlation between the force constants and the length of conjugation

(b) Correlation between the Raman intensities and the vibrational modes

Some years ago we calculated the normal frequencies and normal modes of all-*trans* polyacetylene and its deuterated analog using a modified Urey-Bradley-Shimanouchi force field (MUBSFF) (5). In order to gain an insight into the above-mentioned Problem (a), we have studied the force fields of butadiene and hexatriene using both the MUBSFF and the group-coordinate force field (GCFF, formerly called the local-symmetry force field). As described later

ISBN 0-12-066280-9

in more detail, the vibrational analyses of butadiene and hexatriene as well as of β-carotene indicated that the C=C stretching dispersion curve of all-*trans*-polyacetylene given in Ref. 5 should be modified and accordingly some of the force constants should be changed to a certain extent. Therefore, we are now re-examining the force constants of all-*trans*-polyacetylene on the basis of those of butadiene and hexatriene. On the other hand, all-*trans*-β-carotene, all-*trans*-retinal, and their various isomers are potentially useful sources of information but are not as tractable as butadiene and hexatriene, due to the structural complexities. Probably it is not too unreasonable to assume that the ionone-ring vibrations of those compounds do not mix with the chain vibrations. For the models of β-carotene and retinal in which the ionone ring is replaced by a C=C bond and three methyl groups, we are calculating the normal frequencies and modes, starting from the force constants selected from those of butadiene, hexatriene, and 2-butene (which has the methyl groups adjacent to the C=C bond). In short, our approach to the Problem (a) may be illustrated in the following way.

Butadiene → Hexatriene → (Octatetraene) → Polyacetylene
↘ ↙↗
2-Butene → Retinal & β-Carotene

Several years ago Warshel and Dauber calculated the normal frequencies and Raman intensities of all-*trans* forms of butadiene, hexatriene, retinal, and β-carotene (6). Recently Mathies *et al.* calculated the normal frequencies and modes of all-*trans* retinal and its deuterated analogs as well as of demethyl analogs (7,8). Our aim is to extend this kind of calculation to as many related molecules as possible in order to obtain enough information on the force field of polyene compounds. In this way we hope to make a sound basis for interpreting the vibrational spectra of transient species such as β-carotene in an excited triplet state (9-11).

A. 1,3-Butadiene and 1,3,5-Hexatriene

A great number of studies have been done on the vibrational spectra and structures of these molecules, particularly of butadiene. Recently a less stable conformer (either s-*cis* or *gauche*) of butadiene was detected in low-temperature Ar matrices by means of the high-temperature nozzle technique (12,13). We have also observed the infrared spectrum of the less stable conformer of butadiene-d_4 (CD_2=CH-CH=CD_2) using the same technique. We first re-examined the force field of s-*trans*-butadiene using various deuterated analogs. For the in-plane vibrations both the MUBSFF and GCFF gave the calculated frequencies in good agreement with the observed, though some of the MUBS force constants were strongly correlated and consequently had large degrees of uncertainty. The cross term between the adjacent C=C and C-C stretching coordinates and that between the two neighboring C=C stretching coordinates also had large degrees of uncertainty. Therefore, we fixed these two force constants at the values determined by an *ab initio* study (14) and obtained the final set of GCFF force constants. The out-of-plane force constants could be determined without difficulty. Next we calculated the normal frequencies of the planar s-*cis* and *gauche* (rotated by 40° from the planar s-*cis*) forms using the force constants of the s-*trans* form as the initial values. The infrared spectrum of the less stable conformer could be well explained on the basis of the results of calculation for either the planar s-*cis* or *gauche* form, though a few weak bands were better accounted for with the C_2 symmetry of the *gauche* form. This means that it is difficult to distinguish the two forms by means of infrared spectroscopy. From a practical viewpoint the planar s-*cis* form may be assumed in similar cases [for example, hexatriene (see below)] as far as analysis of the infrared spectra is concerned.

1,3,5-Hexatriene has two stable conformers, namely, the *trans* and *cis* isomers with respect to the central C=C bond. If we take into account the conformational isomerism about the two C-C single bond, six conformers are possible as shown in Fig. 1 where the s-*cis* form was assumed. Of these six the tCc and cCc forms are considered to be very unstable due to severe steric hindrance but the tTc and cTc forms may exist in addition to the stable tTt and tCt. In fact, a less stable conformer(s) was produced from tTt using the high-temperature technique. To assign the bands due to the less stable conformer(s) we undertook the normal-coordinate analyses of tTt, tCt, tTc, and cTc. In the first place the force constants of tTt and tCt were obtained by modifying those of butadiene and 2-butene. Then, the force constants

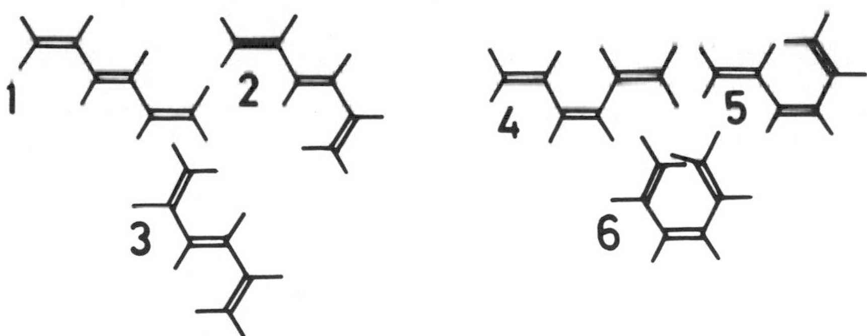

FIGURE 1. Possible conformers of 1,3,5-hexatriene. (1) tTt; (2) tTc; (3) cTc; (4) tCt; (5) tCc; (6) cCc.

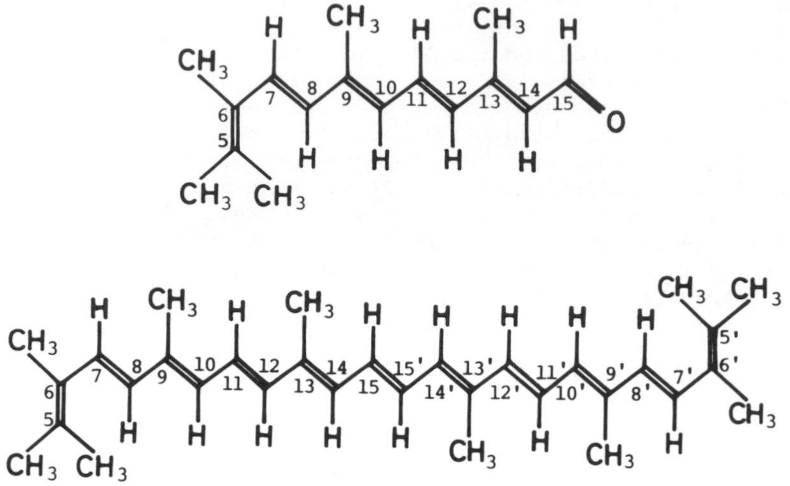

FIGURE 2. *Model structure of retinal (top) and β-carotene (bottom).*

of tTt and those of s-*cis*-butadiene were transferred to tTc and cTc. The results of calculation showed that most of the bands which appeared with the high-temperature nozzle technique were assignable to tTc.

B. *Retinal, β-Carotene, and Polyacetylene*

The purpose of normal-coordinate analyses for longer polyene chains such as retinal, β-carotene, and polyacetylene cannot be the same as those for small molecules. Structural parameters of this class of molecules are not always available and the observed spectral data are not as abundant. In such a case an initial set of force constants should be carefully selected. In this study we made use of the MUBS force constants of butadiene, 2-butene, and polyacetylene (5). We have not yet used the GCFF which requires a greater number of force constants than the MUBSFF. The model structures of retinal and β-carotene are shown in Fig. 2. We are calculating the normal frequencies and modes of various geometric isomers of retinal and β-carotene in order to clarify the vibrational modes of the bands characteristic of each isomer. Recently the Raman spectra of various isomers of β-carotene, namely, 7-*cis*, 9-*cis*, 13-*cis*, 9,13-*dicis*, 9,13'-*dicis*, 13,15-*dicis* were observed by Koyama *et al.* (15). These data are very useful for our purpose.

Although our calculation is still in a preliminary stage,

some interesting features have emerged. For example, the off-diagonal force constant between adjacent C=C and C–C stretchings and that between neighboring C=C stretchings change noticeably as the chain length increases. For s-*trans*-butadiene and tTt-hexatriene totally symmetric C=C stretching frequency (the phase difference δ is close to zero) is higher than the other C=C stretching(s) (δ is close to π or intermediate between 0 and π). However, the C=C stretching dispersion curve for longer polyene chains is close to a bell shape with the maximum frequency at about 1600 cm^{-1}. This was derived from the observation of excitation-wavelength dependence of the C=C stretching region in the Raman spectra of β-carotene (16).

The results of normal-coordinate calculations for all-*trans*- and 15-*cis*-β-carotenes satisfactorily explain the differences between the Raman spectra of these molecules. For example, a characteristic Raman band of 15-*cis* at 1231 cm^{-1} (in the solid state) can be assigned to a calculated mode of 1223 cm^{-1}, which is mainly composed of the CH in-plane bending of the bent part and the central C=C stretching. For the all-*trans* molecule there is no corresponding mode.

Finally we wish to mention the interpretation of the Raman spectra of retinal and β-carotene in the excited triplet state. In the Raman spectrum of all-*trans*-retinal in the excited triplet state the C=C stretching frequency is lower and the putative C–C stretching frequency is higher, respectively, than the corresponding frequencies in the ground state. This is in agreement with the expectation that electron delocalization is increased in going from the ground to the excited triplet state. On the other hand, the putative C–C stretching band of all-*trans*-β-carotene in the excited triplet state is shifted to a higher frequency (as compared with the ground-state frequency), whereas the C=C stretching band is shifted to a lower frequency. This result cannot be explained simply by the increased electron delocalization. We have made normal-coordinate analyses for model structures of all-*trans*-retinal and all-*trans*-β-carotene in the excited triplet state, assuming appropriate changes in bond lengths and force constants from the values in the ground state. The results of such model calculations indicate that the above-mentioned shift of the C–C stretching band of β-carotene is possible due to the rearrangement of the normal mode, even if the C–C stretching force constant is increased in going from the ground to the excited triplet state. In fact, the mode which is usually called the C–C stretching is heaving mixed with the CH in-plane bending. The detailed results of calculation will be published elsewhere.

At the present stage our model calculations would not lead us directly to the conclusion that retinal and β-carotene in the excited triplet state have the all-*trans* conformation. Wilbrandt *et al.* suggested that these molecules in the excited triplet state might have bent structures similar to 9-*cis*-retinal and

15-*cis*-β-carotene, respectively, because some bands of the excited molecules were close in frequency to the bands characteristic of the above-mentioned bent isomers (10,17). However, there is no theoretical basis for such an empirical approach in interpreting the similarity between the Raman spectra observed in resonance with the S_0-S_1 and T-T absorptions. More studies, both experimental and theoretical, are undoubtedly necessary to clarify the molecular structures of retinal and β-carotene in the excited triplet state.

REFERENCES

1. Inagaki, F., Tasumi, M., and Miyazawa, T., *J. Mol. Spectrosc. 50*, 286 (1974).
2. Saito, S., Harada, I., Tasumi, M., and Eugster, C. H., *Chem. Lett.* 1045 (1980).
3. Harada, I., Furukawa, Y., Tasumi, M., Shirakawa, H., and Ikeda, S., *J. Chem. Phys. 73*, 4746 (1980).
4. Furukawa, Y., Harada, I., Tasumi, M., Shirakawa, H., and Ikeda, S., *Chem. Lett.* 1489 (1981).
5. Inagaki, F., Tasumi, M., and Miyazawa, T., *J. Raman Spectrosc. 3*, 335 (1975); Tasumi, M., *J. Raman Spectrosc.* in press (erratum).
6. Warshel, A., and Dauber, P., *J. Chem. Phys. 66*, 5477 (1977).
7. Eyring, G., Curry, B., Mathies, R., Fransen, R., Palings, I., and Lugtenburg, J., *Biochemistry 19*, 2410 (1980).
8. Curry, B., Broek, A., Lugtenburg, J., and Mathies, R., *J. Am. Chem. Soc.* in press.
9. Dallinger, R. F., Guanci, J. J., Jr., Woodruff, W. H., and Rodgers, M. A. J., *J. Am. Chem. Soc. 101*, 1355 (1979).
10. Jensen, N.-H., Wilbrandt, R., Pagsberg, P. B., Sillesen, A. H., and Hansen, K. B., *J. Am. Chem. Soc. 102*, 7441 (1980).
11. Dallinger, R. F., Farquharson, S., Woodruff, W. H., Rodgers, M. A. J., *J. Am. Chem. Soc. 103*, 7433 (1981).
12. Squillacote, M. E., Sheridan, R. S., Chapman, O. L., and Anet, F. A. L., *J. Am. Chem. Soc. 101*, 3657 (1979).
13. Huber-Wälchli, P., and Günthard, Hs. H., *Spectrochim. Acta 37A*, 285 (1981).
14. Bock, C. W., Trachtman, M., and George, P., *J. Mol. Spectrosc. 84*, 243 (1980).
15. Koyama, Y., Kito, M., Takii, T., Saiki, K., Tsukida, K., and Yamashita, J., *Biochim. Biophys. Acta 680*, 109 (1982).
16. Tasumi, M., and Saito, S., *in "Raman Spectroscopy, Linear and Nonlinear"* (J. Lascombe and P. V. Huong, eds.) (*Proc. 8th Int. Conf. Raman Spectrosc., Bordeaux, 1982*), p. 791, John Wiley & Sons, Chichester·New York·Brisbane·Toronto·Singapore, (1982).
17. Wilbrand, R., and Jensen, N. H., *J. Am. Chem. Soc. 103*, 1036 (1981).

SPECTRA FROM MOLECULAR DYNAMICS[1]

Peter H. Berens
John P. Bergsma
Kent R. Wilson

Department of Chemistry
University of California, San Diego
La Jolla, California

Following the work of Roy Gordon(1) we are accustomed to deducing molecular dynamics from spectra. Here we take the inverse approach and compute spectra from molecular dynamics. Specifically, from atomic trajectories we compute infrared, electronic, and nonresonance Raman spectra. Fig. 1 illustrates the general technique. From an initial set of atomic positions and velocities and the forces among the atoms we calculate from Newton's second law the atomic trajectories. For infrared or electronic absorption spectra we connect to the radiation field through the dipole moment $\boldsymbol{\mu}$, and for Raman spectra we connect through the polarizability tensor **P**. The technique can be viewed as an expression of classical linear response theory or as an exercise in classical electromagnetic theory. We use a power spectral approach symbolized as D[ ], taking advantage of fast Fourier analysis with windowing and associated corrections(2, 3). An average, symbolized by <>, of the spectrum is taken over the ensemble appropriate to the experimental conditions (for example a constant temperature and density) and simple quantum corrections to the spectra are applied where needed.

Molecular absorption spectra can be rotational, vibrational, and electronic. All normally involve interaction of nuclear position and radiation field through the dipole moment. Fig. 2 shows an example of a gas phase vibrational-rotational infrared spectrum. Elsewhere(2) we have also computed pure rotational absorption spectra and have extended vibrational and rotational infrared spectral calculations to the liquid phase.

[1] *We thank the National Science Foundation, Chemistry, the Office of Naval Research, Chemistry, the National Aeronautics and Space Administration, Ames Research Center, and the National Institutes of Health, Division of Research Resources for providing the support which has made this work possible.*

ISBN 0-12-066280-9

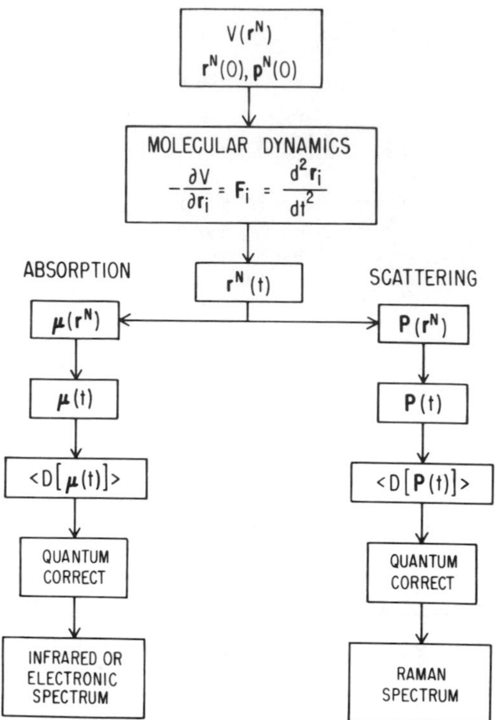

FIGURE 1. Calculation of absorption and scattering spectra from molecular dynamics.

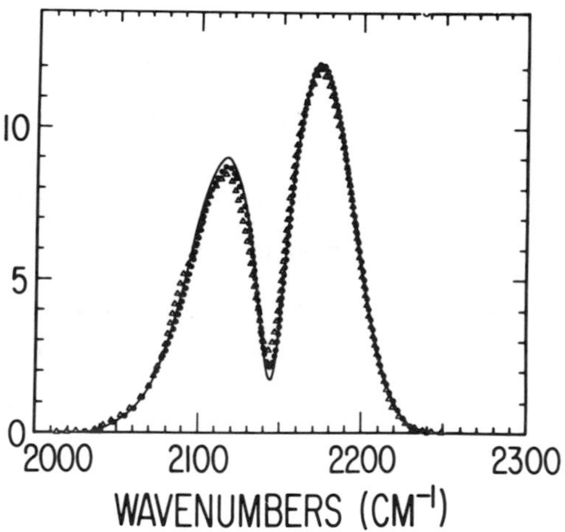

FIGURE 2. Band contours for the infrared gas phase CO fundamental vibrational-rotational band(2). The triangles show our essentially classical molecular dynamics calculation, the solid line shows an accurate quantum calculation and the dots show the experimental measurement(4).

Fig. 3 shows the computed equilibrium and picosecond transient electronic absorption spectra of I_2 photodissociation in liquid Xenon. The connection to the radiation field is through the transition dipole moment, and the Iodine atoms are sufficiently massive that quantum corrections are negligible. In addition to one photon, or absorption spectra, we can also compute two photon, or scattering, processes, as illustrated in Fig. 3 for a liquid solution Raman spectrum.

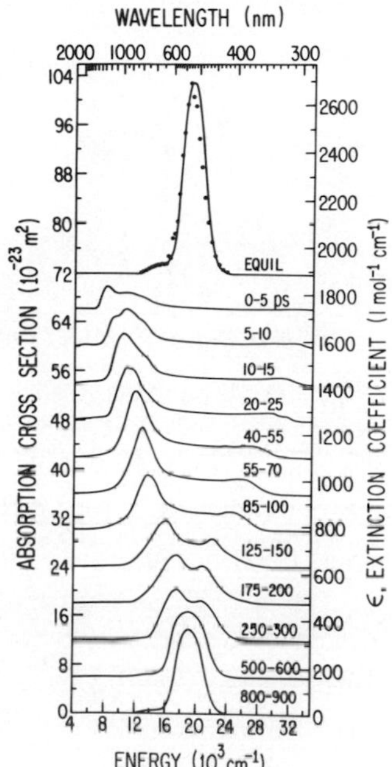

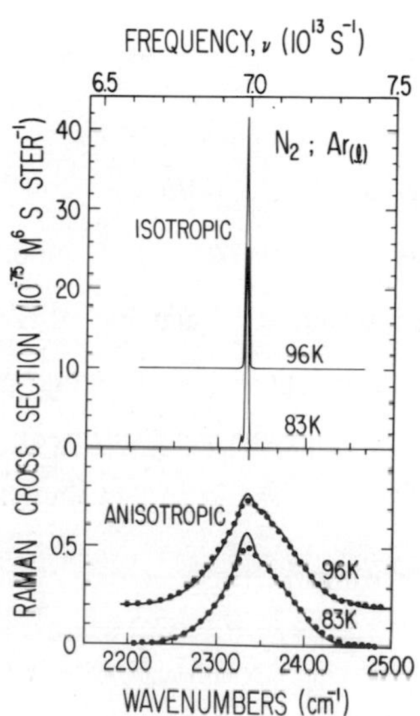

FIGURE 3. The left panel shows the electronic absorption spectrum for I_2 in liquid Xenon, made up of transitions from the ground state to the excited A, B, and B'' states(5). The computed equilibrium room temperature spectrum is shown at top, with the gas phase I_2 band contour as measured by Tellinghuisen(6) shown as dots. Below are the computed transient spectra from 0 to 900 picoseconds for the reaction sequence beginning with photodissociative excitation to the A state, then either I atom escape or solvent caging followed by recombination to form a highly vibrationally excited I_2 molecule which loses its energy to the solvent. The right panel shows the isotropic and anisotropic Raman spectra of N_2 in liquid Ar. The lines show the spectra computed from molecular dynamics(3) and the dots show the measurements of Hanson and McTague(7).

We have shown, for systems for which the potential energy and the appropriate connection to the radiation field (dipole moment or polarizability) are known sufficiently accurately, that infrared, electronic, and nonresonance Raman spectra can be computed from classical molecular dynamics followed by simple quantum corrections to the spectra. Note that no adjustable parameters are needed for any of the spectra presented here. These essentially classical spectra can be compared to experimentally measured spectra to check that the underlying computed dynamics are correct, and the agreement illustrated here indicates that a basically classical view of the atomic motions involved in these spectra is a useful one, in harmony with our well calibrated physical intuition.

References

1. Gordon, R. G., *Adv. Magn. Reson.* **3**, 1 (1968).
2. Berens, P. H. and Wilson, K. R., *J. Chem. Phys.* **74**, 4872 (1981).
3. Berens, P. H., White, S. R., and Wilson, K. R., *J. Chem. Phys.* **75**, 515 (1981).
4. Armstrong, R. L. and Welsh, H. L., *Canad. J. Phys.* **43**, 547 (1965).
5. Bado, P., Berens, P. H., Bergsma, J. P., Wilson, S. B., Wilson, K. R., and Heller, E. J., in *Picosecond Phenomena III*, edited by K. Eisenthal, R. Hochstrasser, W. Kaiser, and A. Laubereau (Springer-Verlag, Berlin, 1982) p. 260.
6. Tellinghuisen, J., *J. Chem. Phys.* **76**, 4736 (1982).
7. Hanson, F. E. and McTague, J. P., *J. Chem. Phys.* **72**, 1733 (1980).

STIMULATED RAMAN TECHNIQUES FOR TIME-RESOLVED SPECTROSCOPY OF GAS-PHASE MOLECULES

P. Esherick, A. J. Grimley and A. Owyoung

Sandia National Laboratories
Albuquerque, New Mexico

I. INTRODUCTION

The development of stimulated Raman spectroscopy (SRS) for gas-phase applications has enabled a variety of studies demonstrating unprecedented sensitivity and spectral resolution[1]. In earlier work we concentrated on purely spectroscopic applications demonstrated the ultra-high spectral resolution (0.002 cm^{-1}) capabilities of SRS. For example, this high instrumental resolution has allowed us to observe fully rotationally resolved Q-branch spectra in both light spherical top (e.g. CH_4, SiH_4) and heavy spherical top (e.g. CF_4, SF_6) molecules. More recent work has been directed at exploiting both the spatial and temporal resolving power of the technique.

Any Raman technique utilizing a pulsed pump laser has several distinct advantages over other approaches for time-resolved measurements. First, since Raman is an instantaneous light scattering process, the temporal behavior of the scattered signal directly follows that of the incident pump radiation. Since excited state lifetimes are not involved, there are no collisional quenching effects, and the scattered signal is unaffected by

*This work supported by the U. S. Department of Energy.

ISBN 0-12-066280-9

individual state lifetimes other than in observed linewidths. (In fact, the resolved Raman linewidth can, by itself, yield basic information on collision rates.) A second important advantage of Raman techniques is that they are, in principle, applicable to all molecules, including homonuclear diatomics. Visible sources can be used which readily cover the entire range of vibrational and rotational Stokes shifts. Since the scattered signal is also in the visible, numerous detectors with fast response times, high quantum-efficiency, and low noise are available.

The major disadvantage of Raman techniques is the very small scattering cross-sections which are characteristic of the Raman process. This has limited most applications of Raman scattering, particularly time-resolved applications, to studies in condensed phases. Recently, however, the development of high-resolution stimulated Raman spectroscopy has overcome this disadvantage and has demonstrated sufficiently high sensitivity to allow routine studies of molecular gases at low pressures (1-10 torr) both in static cells and in supersonic free-expansion jets. Since SRS uses a pulsed pump laser, it has been quite natural to extend the technique into time-resolved applications, such as rotationally specific studies of photoexcited species.

As a prototypical system we have studied(2) CO_2-laser-excited SF_6. In this work we clearly demonstrate the ability of SRS to monitor both the depletion of the ground state and population of vibrationally excited states of this system, with excellent temporal and spectral resolution.

The energy level diagrams shown in Fig. 1 illustrate the two basic configurations which have been used in these studies. In both cases the experiment involves two distinct steps. First, the system is excited in the ν_3 vibrational mode by an infrared laser Second, the effect of this excitation is probed via the SRS process. The difference between the two cases centers on the particular transition probed by SRS. In Fig. 1a, the effects of the excitation process on the ground-state populations are detected by monitoring the ν_1 fundamental. Alternatively, as shown in Fig. 1b, SRS can be used to probe the ν_1 transition originating from the v_3=1 level. By this means, the CO_2-laser-excited state population can be monitored in this level, or even higher levels.

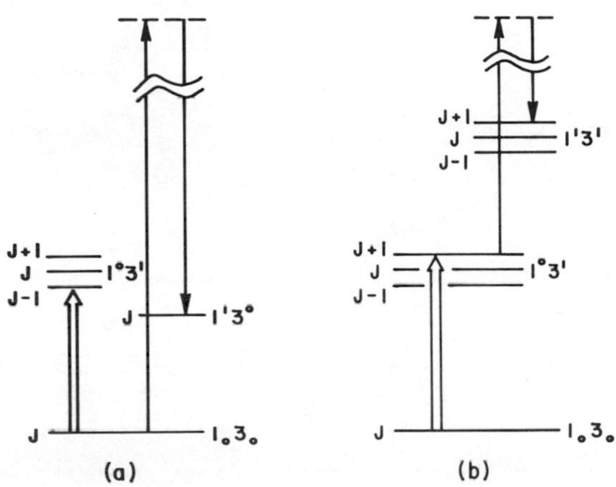

FIGURE 1. Energy level schematic of the CO_2 laser excited SF_6 system, with SRS monitoring of (a) ground state depletions and (b) excited state (v_3=1) populations.

II. EXPERIMENTAL TECHNIQUE

The experimental apparatus is shown schematically in Fig. 2. A CO_2 laser is used to excite the ν_3 vibration of SF_6. The effect of IR excitation on the Raman spectrum is then monitored via stimulated Raman gain spectroscopy. The SF_6 gas was cooled to ~90 K in a pulsed molecular jet to reduce the collision rate while maintaining a reasonably high ($1.3 \times 10^{17}/cm^3$) molecular density.[3]

The SRS technique uses a high-power pulsed laser (the "pump") to induce stimulated Raman gain at the Stokes shifted frequency. A gated cw laser (the "probe") is used to monitor this transient gain. Scanning the frequency of the pump laser relative to the fixed frequency probe results in a direct display of the Raman gain profile, i.e. the Raman spectrum.

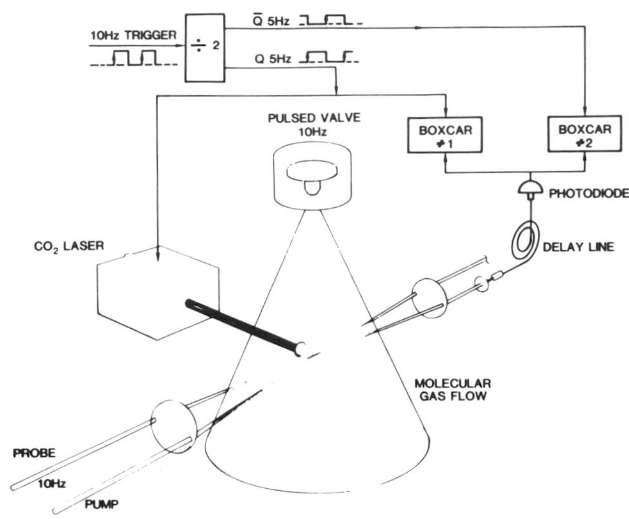

FIGURE 2. Schematic representation of the experiment.

In our experiments the pulsed pump laser is a single-mode electronically scannable cw dye laser which has been pulse amplified up to 1 MW by three Nd:YAG-pumped dye amplifiers. This hybrid dye laser system operates at a 10-Hz repetition rate and emits pulses of 12-nsec duration with $\sim$60-MHz spectral width (FWHM). The probe source is a cw krypton-ion laser operating in a single longitudinal mode with a nominal output of 0.6 watt at 647.1 nm and a spectral width of 2 MHz. The two laser beams are crossed at a common focus in the molecular jet, with the size of the sampled volume determined by the spatial overlap of the two optical beams. The output of the CO_2 TEA laser is directed into the sample chamber at right angles to the molecular jet and SRS beams as shown in Fig. 2. Relative timing between the CO_2 laser pulse and the Raman pump laser is set by an adjustable delay pulse generator.

After passing through the sample chamber, the probe beam is transmitted to a fast photodiode through a 150-m-long fiber-optic cable. The transient signal is delayed in the optical fiber, allowing its detection by two boxcar integrators (PARC 162/165) after rf noise from the CO_2 laser discharge has dissipated. An alternate pulse data collection scheme (Fig. 2) allows us to collect two sets of data simultaneously, one with the CO_2 laser on and one with the CO_2 laser off, thereby giving a direct measurement of the effect of the IR excitation on the SF_6 spectrum.

III. RESULTS AND DISCUSSION

The ν_1 Raman spectrum of SF_6 probed in these experiments is marked by its simplicity, particularly at reduced temperatures as in a molecular free-expansion jet. The origin of the fundamental band occurs at 774.544 cm^{-1}, with the Q-branch structure unfolding towards lower Stokes shift. The band exhibits a rigid-rotor type spectrum that can be fit readily with only a single rotational parameter, making the ν_1 band very attractive for monitoring population movement between individual J levels.

The most easily observed effect of the IR excitation of SF_6, illustrated in Fig. 3, is the depletion of rotational state populations in the ground vibrational level. The upper figure shows the pronounced resonance depletion that occurs during excitation of the SF_6 near J=28 in the ν_3 R-branch by the CO_2 P(14) line at 947.479 cm^{-1}. The appearance of a distinct "hole" in the rotational distribution is typical of spectra obtained at short delay times. At longer delay times (Fig. 3b) the effect of collisions becomes apparent in the re-equilibration of the rotational distribution and the filling-in of the initially observed hole. The overall band depletion persists on a μsec time scale,

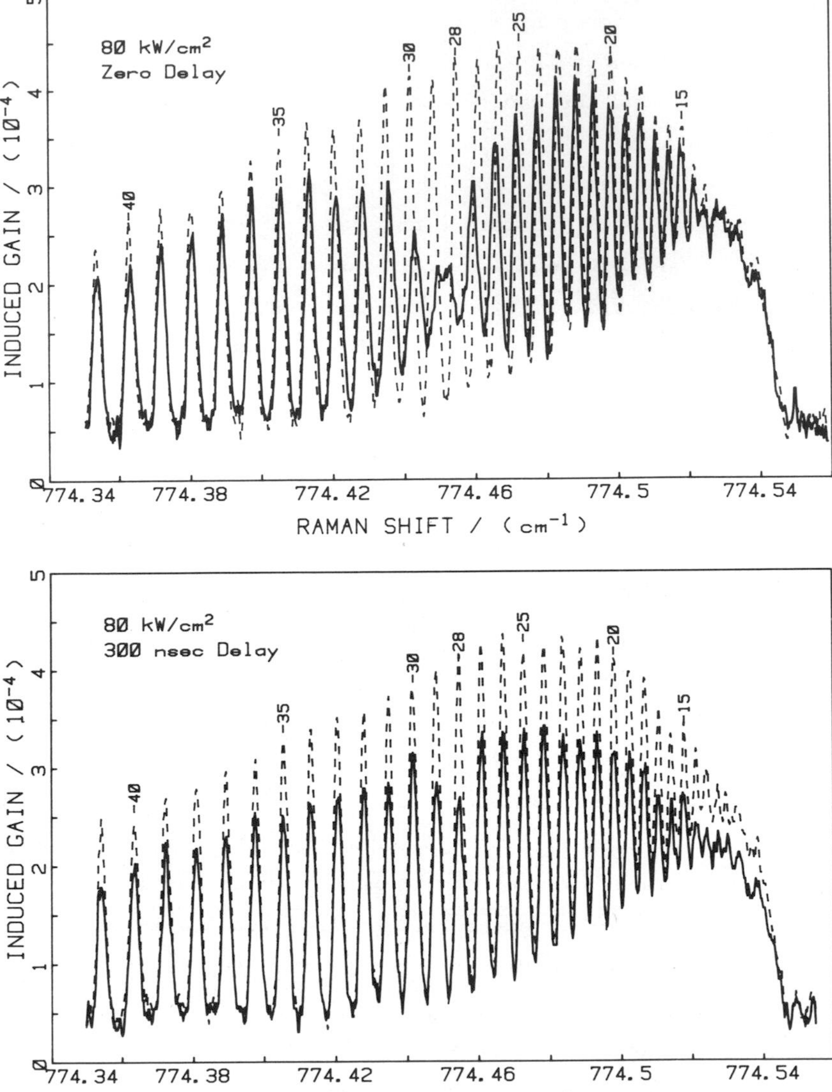

FIGURE 3. Raman gain spectra of the ν_1 fundamental of SF_6 showing the effect, at two different delay times, of excitation of ν_3 near P(33) by the P(18) line of the CO_2 laser. The CO_2 laser "perturbed" spectrum is shown as a solid line, while the "normal" spectrum is shown as a dashed line.

indicating that relaxation of vibrational energy back down to the ground state is relatively slow.

For the most part, excitation by other CO_2 laser lines produces the expected results. The P(16) line, for example, excites a wide range of rotational levels near J=55 in the ν_3 Q branch, producing an overall depletion of the ground state but no well defined rotational-state-specific depletions. The CO_2 P(18) line, which is resonant with P branch transitions at J=32 and 33, depletes both of these levels, as expected. Unanticipated, however, is the simultaneous appearance of a very strong selective depletion of the band in the vicinity of J=16. Indeed, this observation could not be accounted for on the basis of direct excitation of the ground state via any known one-, two-, or three-photon processes at the CO_2 laser frequency(4).

A key observation leading to the eventual explanation of this phenomenon was the complete absence of any depletion, or otherwise unusual effects near J=16, in the IRDR studies of Dubs et al.(5) We have subsequently explained our unusual observation on the basis of a resonant AC Stark effect involving the upper level of the Raman transition(3). The observed "depletion" is actually the result of shifts in the energies of the v_1=1 state rotational levels near J=16. These shifts are induced by the CO_2 laser because of a close resonance between the frequencies of the CO_2 P(18) line and the SF_6 R(16) line of the $\nu_3+\nu_1-\nu_1$ transition. This explanation is supported spectroscopically by our measurement, via the $\nu_1+\nu_3-\nu_3$ spectrum, of the anharmonic shift X_{13}=-2.907 cm^{-1}.(3)

In addition to detecting depletion of the ground state populations, SRS techniques can also yield both spectroscopic and dynamic information on the excited species. The scheme used to probe population in the v_3=1 level via the $\nu_1+\nu_3-\nu_3$ hotband transition is diagramed in Fig. 1b. Signal levels present in these spectra were considerably lower than in the ground state spectra, requiring the use of a modest multi-pass configuration for the SRS beams. A signal increase of 3.56 was realized in this manner by retroreflecting the SRS beams through the sample a total of 5 times. The free-expansion jet becomes an essential part of this experiment because of the necessity to eliminate thermally populated hotbands which would, at room temperature, obscure the presence of the CO_2-laser-pumped species.

In Fig. 4 we illustrate a typical $\nu_1+\nu_3-\nu_3$ spectrum, observed 200 nsec after excitation near R(28) of ν_3 by the CO_2 P(14) line. The spectrum shows an apparent Boltzman rotational distribution, similar to the ground-state, with a distinct added hump in the region initially pumped by the CO_2 laser. At longer delay times times this added hump disappears, leaving only the Boltzman distributed intensity. Increasing the probe delay time even further, the $\nu_1+\nu_3-\nu_3$ band intensity is found to decay exponentially with a time constant of 0.7 μsec. Very similar behavior is observed with excitation by the CO_2 P(18) line, with the exception that the intensity "hump" appears in the vicinity of J=32, as expected.

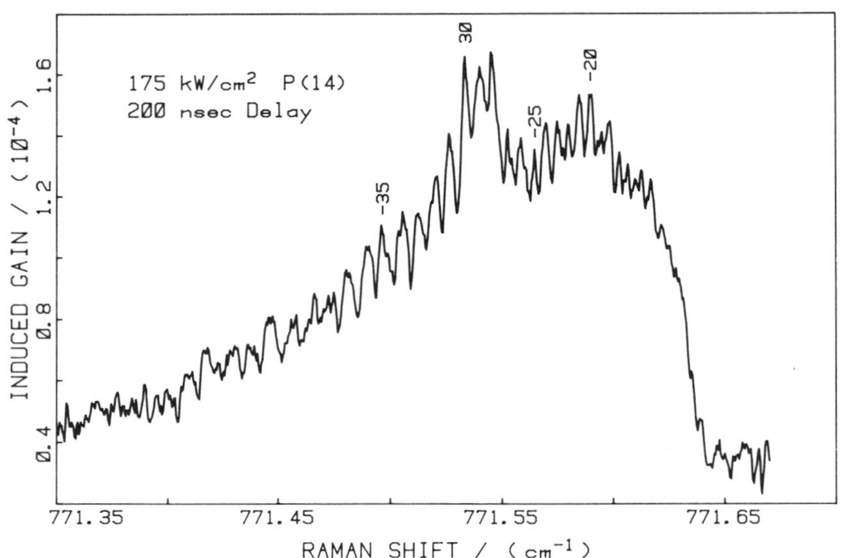

FIGURE 4. Raman spectrum of the $\nu_1+\nu_3-\nu_3$ transition in SF_6, observed during excitation of ν_3 by the CO_2 P(14) line.

III. CONCLUSIONS

These preliminary results clearly demonstrate the applicability of SRS techniques to dynamical studies of multi-photon excitation processes in gas-phase molecular systems. This capability for obtaining time-resolved spectral information with SRS should lead the way to eventual Raman probing of molecular photo-fragments or other transient species.

REFERENCES

1. P. Esherick and A. Owyoung, in "Advances in Infrared and Raman Spectroscopy" (R.J.H. Clark and R.E. Hestor, ed.), Vol. 9, p. 130. Heyden and Son Ltd., London, 1982.
2. P. Esherick, A.J. Grimley, and A. Owyoung in "Laser Spectroscopy V" (A.R.W. McKellar, T. Oka, and B.P. Stoicheff, ed.), p. 229. Springer-Verlag, New York, (1981).
3. P. Esherick, A.J. Grimley, and A. Owyoung, to be published in Chem. Phys.
4. C.W. Patterson, private communication.
5. M. Dubs, D. Harradine, E. Schweitzer, J.I. Steinfeld and C.W. Patterson, to be published in J. Chem. Phys.

TIME-RESOLVED STUDY OF SUBPICOSECOND ORIENTATIONAL RELAXATIONS IN MOLECULAR LIQUIDS*

C. L. Tang
J-M. Halbout+

Cornell University
Ithaca, New York

ABSTRACT

Subpicosecond orientational relaxations of molecules in the liquid state are observed using the recently developed femtosecond laser interferometric technique. A new rapid relaxation process is observed in CS_2 and nitrobenzene. This is interpreted as due to the relaxation of the molecules within the librational wells in the liquids. It should be present in other molecular liquids also.

I. INTRODUCTION

Most of the information on the orientational relaxation of molecular rotations in liquids has come from Rayleigh scattering studies in the past[1] and more recently from spin resonance studies[2]. The problem is complicated by the fact that the relaxation is not governed by a single exponential process. It is known from the extensive studies carried out by, for example, Fabelinskii[1], Starunov, and others that there can in fact be

* Work supported by the National Science Foundation through the Materials Science Center of Cornell University, Ithaca, NY.

+ Now at Bell Telephone Laboratories, Murray Hill, NJ.

ISBN 0-12-066280-9

three different regions in the spectral profile of the Rayliegh scattered light. In the case of CS_2, for example, there is a Lorentzian central component characterized by a Debye anisotropy relaxation time. This central component merges continuously into a wing region which in the time domain corresponds to the sub-picosecond regime. The signal in the Rayliegh wing region is often weak and the line shape is complicated and difficult to interpret; the corresponding subpicosecond dynamics of the molecules are, therefore, not well known. In this paper, we describe an interferometric technique[3] making use of the recently developed femtosecond laser [3,4] that allows the subpicosecond orientational relaxation of molecules in liquids to be measured directly in the time domain.

We summarize first in Table I the general characteristics of the mode-locked subpicosecond lasers achieved in our laboratory. The pulses obtained are among the shortest reported so far. In our initial experiments, nontunable passively locked lasers were used; Fig. 1 shows an auto-correlation trace of the laser pulses

Table I

	Non-Tunable	Tunable
Wavelength (nm)	620	580-620
Pulse Length (fsec)	60	400
Average Power (mW)	30	120
Peak Power (KW)	5	3
Repetition Rate (s^{-1})	10^8	10^8

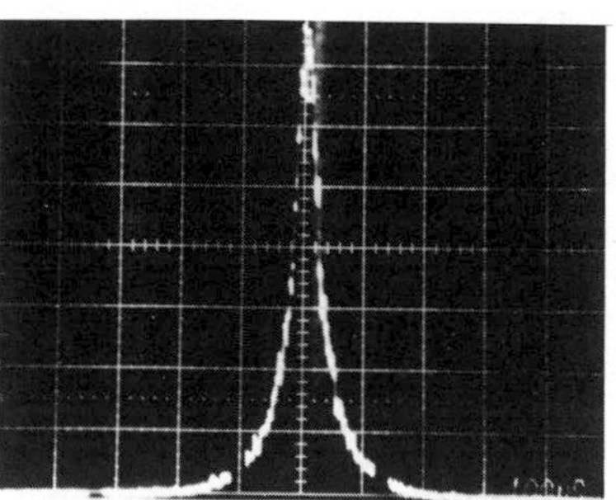

Fig.1- Auto-correlation trace; FWHM~100 fs. Pulse FWHM 60fs. Upper trace: 1 ps/div. Lower trace: 0.2 ps/div.

obtained with a FWHM of 100 femtosecond corresponding to a pulse width of approximately 60 fs.

The orientational relaxation of the molecules can be measured by aligning the molecules via the optical Kerr effect using an intense short pulse of pump light and observing the subsequent decay of the light-induced change in the indices-of-refraction of the medium with a probe pulse. Consider, for example, a linear molecule such as CS_2. The polarization $\vec{P}$ induced in the molecule by a linearly polarized E-field in the $\hat{z}$ direction is:

$$\vec{P} = E(\alpha_{11} \cos^2\Omega + \alpha_{\perp} \sin^2\Omega)\ \hat{z} + \frac{E}{2}(\alpha_{\perp} - \alpha_{11})\ \sin 2\Omega\ \hat{x} \tag{1}$$

where α_{11} and $\alpha_{\perp}$ are the parallel and perpendicular components of the linear electronic polarizability of the molecule and Ω is the angle between the molecular axis and the $\vec{E}$-field as shown schematically in Fig. 2. The potential energy due to the induced polarization $\vec{P}$ in the $\vec{E}$-field is:

$$V(\Omega) = -\frac{1}{2}E^2\,[(\alpha_{11} + \alpha_{\perp}) + (\alpha_{11} - \alpha_{\perp})\cos 2\Omega] \tag{2}$$

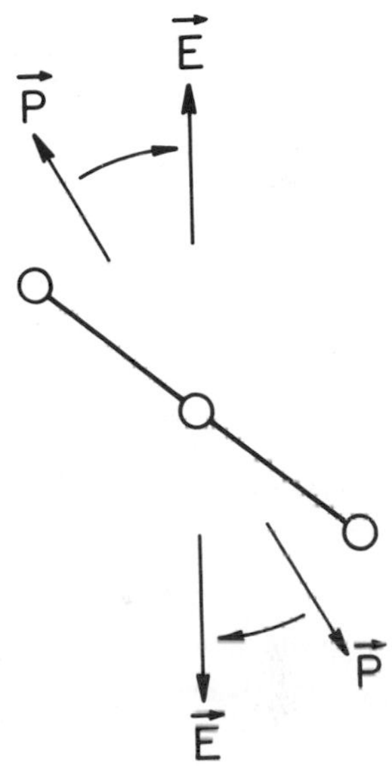

Fig.2-Optical Kerr effect. The linear molecule will tend to rotate to lower the potential energy due to the induced polarization P in the E-field.

and the molecule will rotate in order to minimize this energy. The resultant optical alignment of the molecule is, therefore, proportional to the incident intensity. Because the molecule is nonspherical, or $\alpha_{11} \neq \alpha_{\perp}$, this results in a light induced index-of-refraction change in the medium parallel and perpendicular to the $\vec{E}$-field:

$$\Delta n_{11}(t) = \frac{2\pi}{n}\int_{-\infty}^{t} x^{(3)}_{zzz}(t - t')E^2(t')dt' , \qquad (3a)$$

$$\Delta n_{\perp}(t) = \frac{2\pi}{n}\int_{-\infty}^{t} x^{(3)}_{xzz}(t - t')E^2(t')dt' , \qquad (3b)$$

respectively, where n is the index-of-refraction of the medium and $x^{(3)}$ is the third-order optical susceptibility of the medium. If the incident pulse width is much shorter than the orientational relaxation time, the light-induced index-change and birefringence will persist after the passage of the "pump" light until the molecular distribution in the orientational space is thermalized again; the decay of the light-induced index-change corresponds to the orientational relaxation of the molecules in the liquid.

In addition to the orientational contribution, there may be other contributions to the light-induced index-change.[5] The molecule may also have an electronic hyperpolarizability third-order in the electric field in addition to the linear polarizability $\tilde{\alpha}$. The electronic contribution to the light-induced index-change is, however, instantaneous and should disappear with the pump pulse. If the molecular shape can change, there will also be a similar contribution to the light-induced change in the index-of-refraction due to molecular distortions. For CS_2, the corresponding relaxation time or response time is known to be on the order of 20 ps.[6] With a pump pulse width of less than 0.1 ps, this contribution is negligible. Thus, in CS_2 any persistent index-change after the passage of a pump pulse less than 0.1 ps is primarily due to the rotational contribution of the molecules.

II. EXPERIMENTAL

Figure 3 shows a schematic of the interferometric setup we used to measure the light-induced index-change and its subsequent relaxation in liquids such as CS_2. The optical path length change due to the index change in the sample cell is measured using a white-light Mach-Zehnder interferometer. Because the total phase-shift through the sample is relatively small (~5×10^{-3} rad), the interferometer is feedback stabilized at the phase-quadrature point. Other details of the setup are given in Ref. 3.

With the sample illuminated by a modulated pump pulse train, the interferometer is interrogated with a delayed CW probe pulse train. The amplitude of the ac output of the interferometer is proportional to the phase-shift in the sample induced by the pump pulse integrated over the probe pulse:

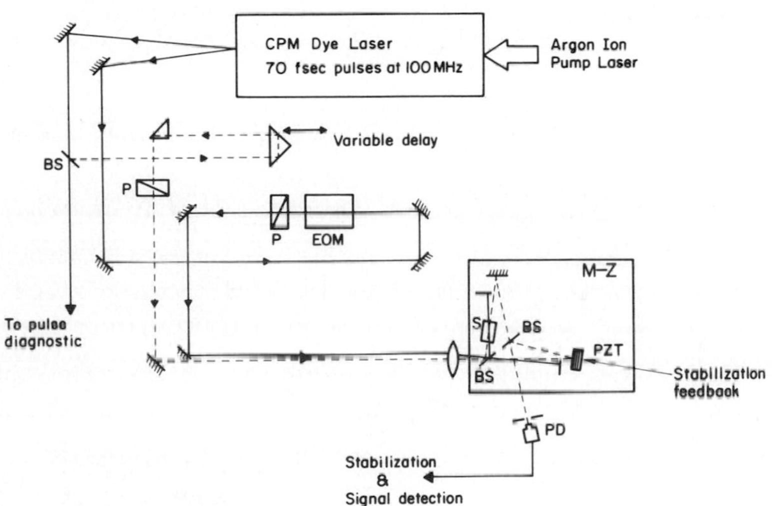

Fig. 3. Schematic of the experimental setup. BS:beam splitter, EOM:electro-optic modulator, M-Z:Mach-Zehnder interferometer, P:polarizer, PD:photo-detector, S: sample cell.

$$\Phi(\tau) = \iint_{\text{pulse}} I_{probe}(t - \tau)\chi^{(3)}_{zzz}(t - t')I_p(t')dtdt' \qquad (3)$$

where $I_{probe}(t)$ and $I_p(t)$ are the probe and pump pulse intensities, respectively, τ is the delay time between the pump and probe pulses, and $\chi^{(3)}_{zzz}$ is the third-order optical susceptibility describing the light-induced change in the direction of the applied field $\hat{z}$.

If the response time and the relaxation time of the medium are instantaneous or $\chi^{(3)}_{zzz}(t-t')\sim\delta(t-t')$, $\emptyset(\tau)$ is a measure of the cross-correlation of the pump and probe pulse intensities. If the relaxation time T_r of the medium is long compared with the pulse widths, $\emptyset(\tau)$ gives the relaxation time of the medium; in the case of an exponential relaxation process, for example,

$$\phi(\tau) \sim e^{-\tau/T_r} \int_{\text{pulse}} I_{probe}(t)dt \int_{\text{pulse}} I_p(t)dt \qquad (4)$$

in the region where τ and T_r are both large compared with the pulse widths.

The general characteristics of the laser used in the experiment are summarized in Table I. The pump and probe pulse trains were derived from the two outputs of a colliding-pulse ring-laser. The width of the pulses are 70 fs (Fig. 4) at the output of the laser. The measured cross-correlation at the sample had a FWHM of 200 fs due to a time jitter between the two pulse trains and certain broadening in the optical components traversed by the pulse trains. The average power of the pump beam was ~10mW and that of the probe beam was ~0.7 mW at the sample and the beam diameter was $\sim 10^{-4}$ cm^2.

III. RESULTS AND DISCUSSION

Typical measured phase-shifts for two liquids, CS_2 and nitrobenzene, are shown in Fig. 4. The peak values have been renormalized to facilitate comparison; the actual peak signal for $C_6H_5NO_2$ was three times smaller than that for CS_2 under the same experimental conditions.

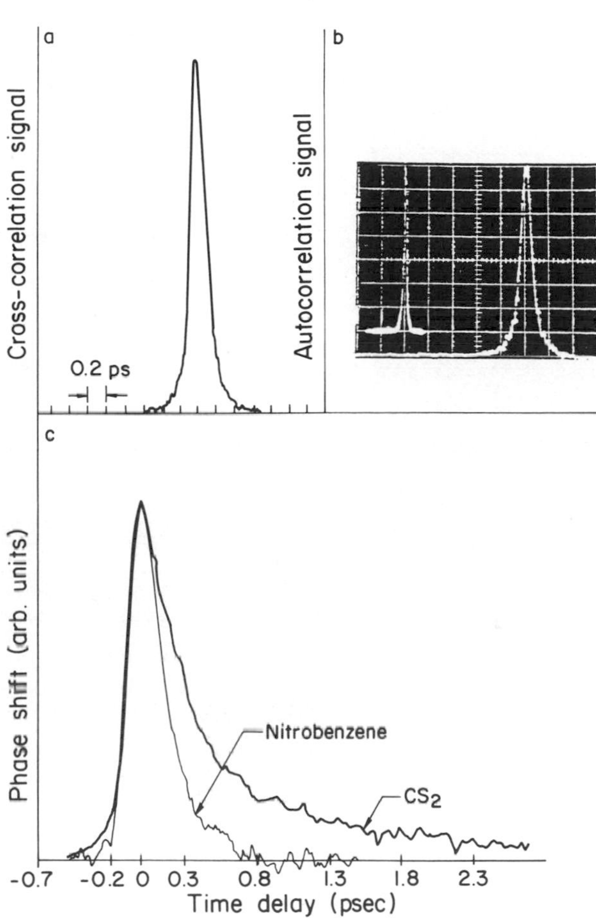

Fig. 4 - (a) Cross-correlation between pump and probe pulses; FWHM 200 fs. (b) Auto-correlation of laser pulse. Upper trace: 1ps/div. Lower trace: 0.2 ps/div. (c) Measured light-induced phase-shifts in CS_2 and nitrobenzene.

Consider first the result for CS_2. Comparison of the time dependence of the measured phase-shift with the cross-correlation curve shows that there is very little electronic contribution to the light-induced index-change, since the trailing edges of the two deviate from each other almost from the very peak. The measured decay characterizes, therefore, the orientational relaxation of the CS_2 molecules in the liquid state, there being very little molecular distortional contribution excited with a 70 fs pulse and a vibrational relaxation time of CS_2 of about 20 ps. There are clearly two rotational relaxation time constants: a longer time constant T_ℓ = 2 ps and a shorter time constant T_s = 330 fs after taking into account the finite width of the pump and probe pulses. The 2 ps relaxation time in CS_2 is clearly the Debye anisotropy relaxation time well-known from previous Rayleigh scattering studies.[1] The 330 fs relaxation time has never been observed before, to our knowledge. A possible explanation of this fast rotational relaxation time is the following.

In liquids, the mean-field potential in the orientational space imposed by the surrounding molecules on a molecule consists of a series of hills and wells of different depths and widths as shown schematically in Fig. 5(a), but the average potential is independent of the molecular orientation in an isotropic liquid. Within each well, there is a series of librational levels.[7] The

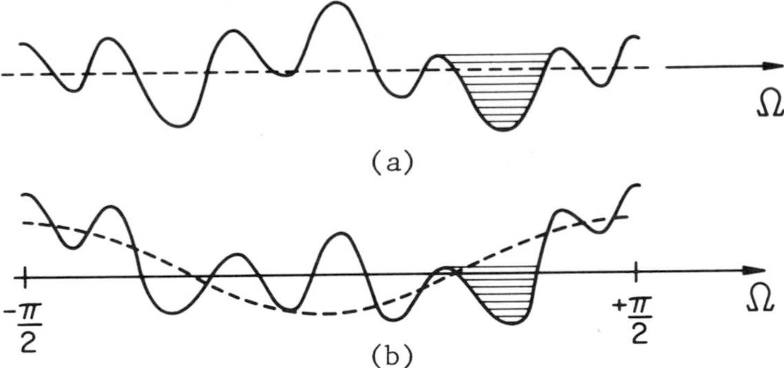

Fig.5- Schematic of the potential energy of molecules in liquids in the orientational space. (a) In the absence of any applied field and (b) with applied field.

molecules in the same environment are distributed over all the wells and librational levels according to the usual considerations in statistical mechanics and the linear optical properties of the medium is isotropic. In the presence of an optical field, the potential in the liquid is distorted as shown schematically in Fig. 5(b), through the addition of the potential term Eq.(2). The corresponding anisotropic equilibrium distribution of the molecules is of course different from that for the case without the optical field and leads in turn to an optically induced index-of-refraction change in the liquid. When the applied field is suddenly removed, the molecular distribution relaxes in two stages from that corresponding to the situation shown in Fig. 5(a) to that shown in Fig. 5(b). It first relaxes rapidly within the librational manifold of each well due to collisions with surrounding molecules and then relaxes from well-to-well through a rotational Brownian motion. The redistribution from well-to-well corresponds to the classical anisotropy diffusion process and the corresponding time constant is the Debye relaxation time or the long decay process we observed. The averaged decay within the librational wells is the new relaxation time T_s we observed for the first time in CS_2.

In terms of this interpretation of our experimental results, the rapid orientational relaxation observed in CS_2 should be a characteristic feature of other molecular liquids as well. The results on nitrobenzene shown in Fig. 4 substantiate this model. The viscosity of nitrobenzene is more than five times that of CS_2. Since the Debye relaxation time is proportional to the viscosity, T_ℓ for nitrobenzene ($\gtrsim$10 ps) is much longer than the excitation pulse width; therefore, there should be very little optically induced redistribution of population from well-to-well in the orientational space and the long decay component should be absent in our experiment. A careful analysis of the data on nitrobenzene shown in Fig. 4 shows that there is very little decay corresponding to the anisotropy diffusion process; the observed decay fitted very accurately to a single exponential with a short relaxation time T_s of ~150 fs, after taking into account the finite excitation pulse width. This is one of the shortest relaxation phenomena ever directly measured. It corresponds to the relaxation within the librational wells.

In conclusion, the experiment here has demonstrated that time-domain experiments using the recently developed femtosecond laser interferometric technique are uniquely suited for studying the subpicosecond molecular dynamics in the liquid state. A new rapid orientational relaxation process has been observed in CS_2 and nitrobenzene and should be present in other molecular liquids as well. It is interpreted as due to the relaxation within the librational wells in the liquids in a mean-field approximation picture. Similar studies on the effects of solvents, temperature, pressure, etc. should give new insights into the molecular interactions that influence the relaxation mechanisms in liquids.

REFERENCES

1. I. L. Fabelinskii, Molecular Scattering of Light, "Nauka", Moscow (1965); V. S. Starunov, "Optical Studies in Liquids and Solids", Proc. of the P. N. Lebedev Phys. Inst. (Consultants Bureau, New York, 1969).

2. See, for example, S. A. Zager and J. H. Freed, Parts I and II, J. Chem. Phys. Vol. 77, p. 3344 (Sept. 15, 1982).

3. J. M. Halbout and C. L. Tang, App. Phys. Lett. 40, 765 (1982).

4. R. L. Fork, B. I. Green, and C. V. Shank, App. Phys. Lett. 38, 671 (1981).

5. R. W. Hellwarth, Progress in Quantum Electronics, (Pergamon, New York, 1977), 5 and references therein.

6. W. Clements and B. P. Stoicheff, App. Phys. Lett. 26, 92 (1975).

7. See, for example, G. M. Korenowski and A. C. Albrecht, Chem. Phs. 38, 239 (1979).

DYNAMIC MEASUREMENTS OF GAS PROPERTIES BY VIBRATIONAL RAMAN SCATTERING: APPLICATIONS TO FLAMES

Michael C. Drake
Marshall Lapp
Robert W. Pitz
C. Murray Penney

General Electric Research & Development Center
Schenectady, New York

The driving forces of efficient utilization of fuels, pollutant emissions, and the need to use "alternate" fuels, have accelerated research in areas related to high temperature chemical kinetics and mechanisms, turbulent fluid mechanics, and multi-phase flow phenomena - particularly, in the *interaction* of these effects and the production of analytical flame models that can accommodate them. Said more directly by way of example, the determination of the structures present in a turbulent flame illustrates a problem of substantial complexity, potential technical importance, and opportunity for the application of advanced experimental methods.

In this vein, the use of light scattering spectroscopy for dynamic studies of the properties of gas-phase systems is receiving increasing attention because it can provide data essential for interpreting these systems that is unavailable by other means. Thus, the capability to probe reactive as well as non-reactive media and to obtain the thermodynamic and flowfield properties simultaneously, if required, has permitted us to attack a range of fundamental problems in fluid mechanic and combustion research.

Here, we discuss the study of flame systems in laboratory configurations designed to probe features essential to the creation and validation of combustion models. The experimental

ISBN 0-12-066280-9

elements of the work that bring new insight into the flame studies revolve around the *temporal* and *spatial precision* and the *simultaneity* of temperature, composition, density, and velocity data. Clearly, the inhomogeniety and dynamic character of turbulent mixing accompanied by chemical reactions mandate the requirement for time and space precision. Just as important, current analyses of flames, in order to have any reasonable power, need to have several variables measured at the same time, to evaluate critical features that often depend upon products of various fluctuation properties. For example, important transport property terms in conservation equations are determined by the product of fluctuation values of velocity times density, a major species concentration, or temperature.

For these reasons, we have developed a program of flame research based upon the use of pulsed vibrational Raman scattering (VRS) and laser Doppler velocimetry (LV), to which laser-induced fluorescence (LIF) is being added currently. The VRS is used to provide simultaneous values (taken once per second with microsecond resolution) of flame temperature and all major species concentrations (H_2, O_2, N_2, and H_2O; and in work underway, CO, CO_2, and CH_4 as well). The density is then calculated from the major species concentrations. The LV instrumentation can provide two orthogonal components of velocity, and the LIF system will provide capability to measure important flame radicals (initially, OH). All of these probes can be operated in a triggered fashion in order to determine all variables in a near-simultaneous fashion.

As an illustration of this area of work, the Raman measurement system developed in our laboratory is shown in Fig. 1, where the LV and LIF probes are omitted for the purpose of clarity.

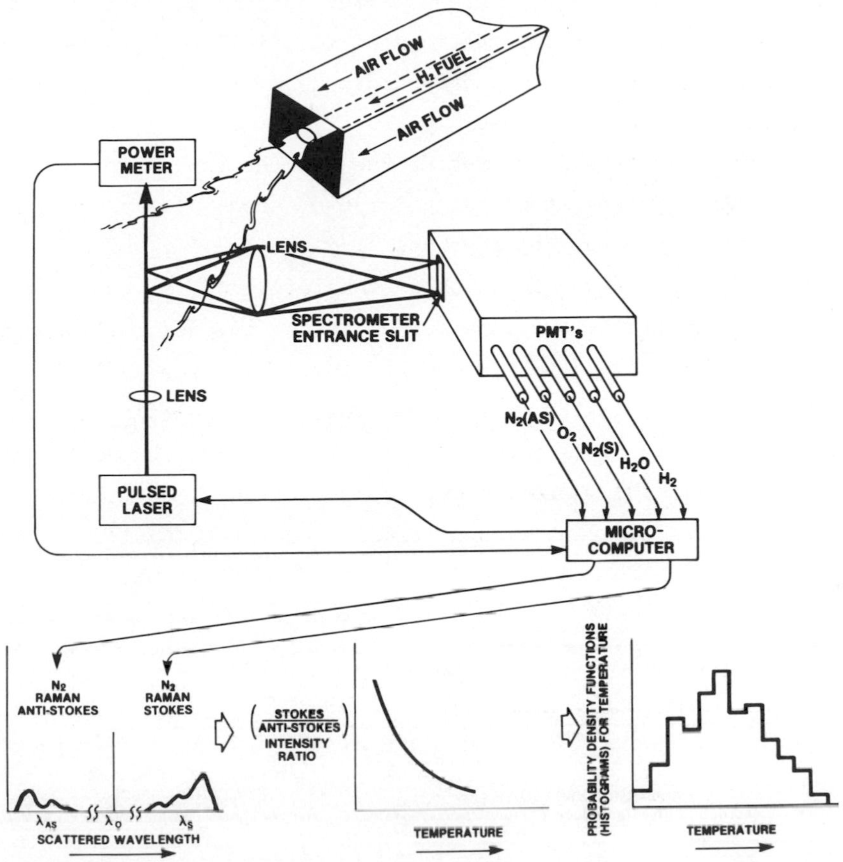

FIGURE 1. Schematic of turbulent combustor geometry (square 15 x 15-cm cross section with central 3.2-mm diameter fuel tube) and Raman data acquisition system.

The various photomultiplier tube channels produce the density, composition, and temperature information. The temperature channels are detailed further in order to indicate the measurement method used here, based upon the ratio of Stokes (S) to anti-Stokes (AS) signals. Also indicated schematically in this figure is the data presentation adopted for the use of modelers, viz., probability density functions (pdf's) of the variables, illustrated here for temperature determined from N_2 signals.

This presentation is chosen because many current combustion models describe the turbulent fluctuations of flowfield properties in terms of one or more pdf's, and assumptions of their shapes then can be validated by VRS measurements.

In Fig. 2 we show illustrative results for the use of VRS for determining temperature pdf's for two cases: (a) a laminar premixed flame produced on a porous plug burner, for which an expected sharply-peaked function is obtained, and (b) the edge of a turbulent diffusion flame burned in the combustor illustrated in Fig. 1, for which a very broad pdf is found. The variation of the temperature pdf with radial position y for the same axial station as in Fig. 2 (x/d=50, or 50 fuel-tip-diameters downstream) is shown in Fig. 3, where the development of the intermittent ambient temperature peak with distance y is seen clearly.

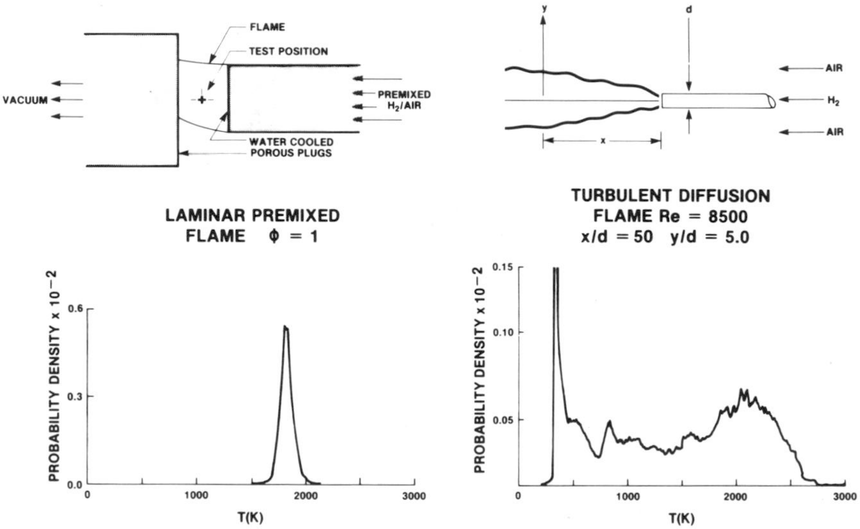

FIGURE 2. Left-hand side: Temperature pdf (500 measurement points) for isothermal zone of post-flame gases from laminar, premixed, stoichiometric H_2-air flame. Right-hand side: Temperature pdf (2000 measurement points) in mixing layer of H_2-air turbulent jet diffusion flame. For the 3.2-mm diameter fuel tube used, the axial measurement station x/d corresponds to 160 mm downstream, and the radial position y/d to 16 mm.

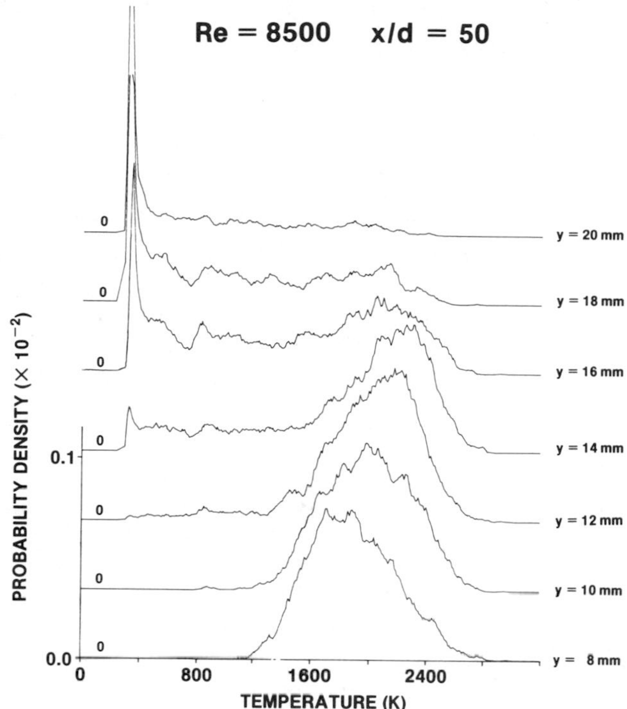

FIGURE 3. Temperature pdf's (2000 measurement points each) across the mixing layer of H_2-air turbulent jet diffusion flame. Note that the curve for y=16 mm corresponds to the data shown at the right-hand side of Fig. 2.

The sharp spikes at low temperatures seen in Figs. 2 and 3 correspond to the fact that the flame/external-air-flow boundary passes through this measurement position in an intermittent fashion, so that ambient gas temperature is found for a certain fraction of the experimental measurements. The ambient air (non-turbulent fluid) and the flame gas (turbulent fluid) contribute individually to combustion model analyses, and thus must be considered separately. With multispecies VRS, this can be accomplished by utilizing the fact that the simultaneously-obtained *composition* characterization of the flame permits one to apply criteria to *each temperature* data point that lead to ascertaining whether or not the volume sampled corresponded to flame-generated species or to ambient air. The criterion used here is based on the mixture fraction (the hydrogen element mass fraction normalized to be unity in the fuel stream and zero in the oxidant stream - determined here as the conserved scalar formed, to within reasonable accuracy, from the sum of the H_2 and H_2O species concentrations). If this fraction is less then an assumed small "critical" value, then the fluid element is considered to be non-turbulent ambient fluid, while values of the mixture fracture greater than the assumed critical value indicated flame gas. (The conditionally-averaged results are relatively insensitive to the actual choice of the small cut-off value.) This conditional sampling separation is shown in Fig. 4, and the conditional averages of some of the specific variables are shown in Table 1.[3] Here, species mass fraction is given by X; the density is given by ρ; the mixture fraction is denoted by ξ; the Favre-averaged (density-weighted) mixture fraction is denoted by $\tilde{\xi}$; and the overall probability of obtaining data corresponding to turbulent flame gas or non-turbulent ambient air is shown in the last row.

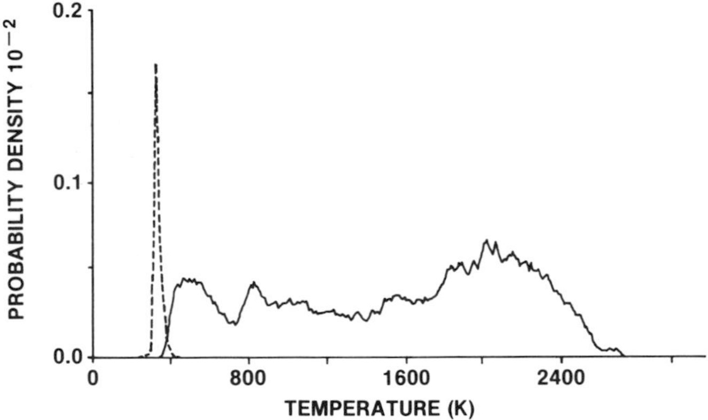

FIGURE 4. Separation of the temperature pdf shown in Fig. 2 (and, for y=16 mm, in Fig. 3) into ambient and flame gas portions, by conditional sampling.

TABLE I. Conditional Averages Corresponding to the Data Shown in Fig. 4

	Turbulent Fluid	Non-turbulent Fluid	All
T	1560	330	1350
$X(N_2)$	0.69	0.78	0.71
$X(H_2O)$	0.20	0.01	0.16
$X(H_2)$	0.02	--	0.02
$X(O_2)$	0.09	0.20	0.11
ρ	0.297	1.071	0.437
ξ	0.017	--	0.014
$\tilde{\xi}$	0.010	--	0.006
Probability	0.83	0.17	1.00

Thus, the time resolution obtained in this repetitively-pulsed experiment permits one to resolve different classes of sampled fluid, a fact significant for finding characterizations of turbulent flames meaningful for modeling. This capability should prove to be especially useful in the more complicated flow configurations often studied in the development of practical fluid mechanic devices.

We have considered here some details of only one class of time-resolved light-scattering flame diagnostics, based upon use of pulsed laser sources and vibrational Raman scattering. In this approach, data describing the time dependence of the flame parameters are not obtained. Other types of scattering diagnostics can provide that type of information, to a greater or lesser extent, but no one method has been developed to date which is capable of providing all the desired requirements, viz., pdf's for a variety of species and temperature simultaneously, frequency spectra,[4] and the additional highly desirable capability of spatial mapping.[5] Thus, complimentary time-and space-resolved measurement probes interconnected into a compatible measurement system is a prime focus for the advancement of combustion diagnostics - and for increased understanding in combustion science.

REFERENCES

1. Drake, M.C., Lapp, M. and Penney, C.M., Use of the Vibrational Raman Effect for Gas Temperature Measurements, *to be published in* "Temperature: Its Measurement and Control in Science and Industry," Vol. 5. Amer. Inst. Physics, New York, (1982); Drake, M.C., Lapp, M., Penney, C.M., Warshaw, S., and Gerhold, B.W., Measurements of Temperature and Concentration Fluctuations in Turbulent Diffusion Flames Using Pulsed Raman Spectroscopy, *Eighteenth Symposium (International) on Combustion,* p. 1521. The Combustion Institute, Pittsburgh, (1981); Drake, M.C., Lapp, M., Penney, C.M., Warshaw, S., and Gerhold, B.W., Probability Density Functions and Correlations of Temperature and Molecular Concentrations in Turbulent Diffusion Flames, AIAA Paper No. 81-0103(1981).

2. Lapp, M. and Penney, C.M., Instantaneous Measurements of Flame Temperature and Density by Laser Raman Scattering, *in* "Proceedings of the Dynamic Flow Conference 1978 on Dynamic Measurements in Unsteady Flows" p. 665. *published by* Proceedings of the Dynamic Flow Conference 1978, P.O. Box 121, DK-2740 Skovlunde, Denmark,(1979).

3. Drake, M.C., Pitz, R.W., Lapp, M., and Penney, C.M., Applications of Raman Scattering Probes to Turbulent Diffusion Flames," *presented at* CLEO '82, IEEE/OSA Conference on Lasers and Electro-Optics, Phoenix,(April 13-16, 1982).

4. *For Raman data, see* Pealat, M., Bailly, R. and Taran, J.P.E., Real Time Study of Turbulence in Flames by Raman Scattering, *Opt. Comm. 22,* 91(1977). Rayleigh scattering is an example of a scattering technique for density data well-suited to produce frequency spectra.

5. *For Raman data, see* Sochet, L.R., Lucquin, M., Bridoux, M., Crunelle-Cras, M., Grase, F., and Delhaye, M., Use of Multi-channel Pulsed Raman Spectroscopy as a Diagnostic Technique in Flames, *Comb. and Flame 36,* 109(1979). Laser-induced fluorescence is an example of an imaging technique based on light scattering well-suited to produce spatial maps.

TIME RESOLVED PHOTOACOUSTIC DETECTION OF COLLISIONAL RELAXATION OF VIBRATIONALLY EXCITED HD MOLECULES[1]

Jack Gelfand
Richard B. Miles

Department of Mechanical and Aerospace Engineering
Princeton University
Princeton, New Jersey

Eric Rohlfing[2]
Herschel Rabitz

Department of Chemistry
Princeton University
Princeton, New Jersey

In this paper we review the use of the time resolved photoacoustic technique for relaxation measurements. The first detailed measurements of the mechanism and rates of vibrational relaxation from the upper vibrational levels of deuterium hydride (HD) are discussed as an illustration (1-3). In the time resolved photoacoustic relaxation method a pulsed laser is used to populate a specific vibration-rotation state. In this experiment we used a pulsed visible wavelength dye laser to populate the $v = 4$, 5 and 6 states of HD via direct overtone absorption. The subsequent collisional relaxation produces changes in the translational

[1]*This research was supported by the National Science Foundation and an unrestricted Grant from Exxon Research and Engineering Corporation.*
[2]*Present address: Exxon Research and Engineering, Linden New Jersey.*

ISBN 0-12-066280-9

energy of the gas which is monitored in the time domain by acoustic detection of the accompanying pressure changes.

In order to obtain unambiguous relaxation data from photoa-acoustic measurements, certain design criteria must be met to avoid interferences from acoustic resonances, noise from absorption in the cell windows, and diffusion of excited molecules to the cell wall. One can demonstrate that, for time scales short compared to the characteristic time for thermal diffusion to the wall, the average pressure change that the microphone senses is directly proportional to the thermal changes in the bulk gas due to the relaxation process independent of the laser beam geometry used to excite the gas [3]. The effects of acoustic ringing can be minimized if the fundamental acoustic resonant frequencies of the cell are not on the time scale of the relaxation. One must consider the radial, azimuthal, and longitudinal resonant frequencies. Consideration of both the thermal diffusion and acoustic aspects of the system place an upper and lower bound on the cell diameter (3).

The beam from a Candela LFDL-1 dye laser is passed ten times through the cylindrical cell containing HD. A Knowles BT-1834 microphone placed at the cell wall serves as the pressure transducer. The laser frequency is alternately switched on and off the HD absorption and routed to separate memories in a signal averager. These are subtracted to produce the final relaxation signal. The photoacoustic cell consists of two pyrex end arms with a total length of 2.6 meters. The extreme length of the photoacoustic cell is chosen to minimize the effects of the noise associated with the laser beam striking the windows (2,3). The arrival of the window noise at the microphone is delayed by a time proportional to the speed of sound in the gas. Since the relaxation begins immediately after the laser pulse, this creates a region in time between the laser firing and the window noise arrival. Photoacoustic signals taken on and off of the HD 5-0

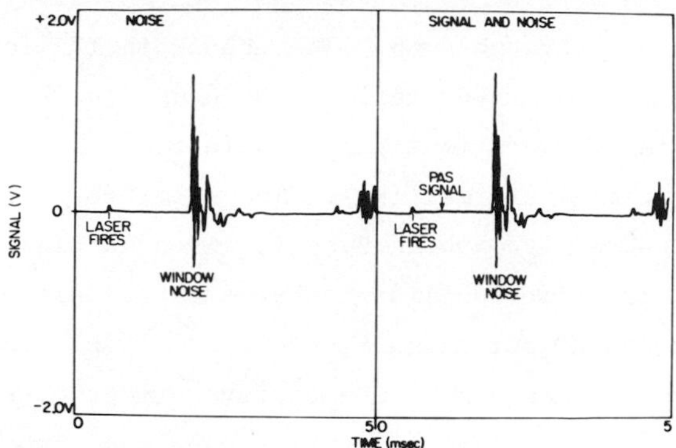

FIGURE 1. Background noise recorded off the HD v = 5 absorption line (left) and signal plus noise recorded on the line (right) at 600 torr and 298°K.

R(1) transition frequency are shown in Fig. 1 to illustrate this effect.

Time resolved photoacoustic data has been taken at 298°K for relaxation of highly vibrationally excited HD with several collision partners. Typical curves of translational energies versus time following initial pumping into the labelled vibrational state are shown in Fig. 2. These data were taken with approximately 250 torr of pure HD in the photoacoustic cell. Analysis

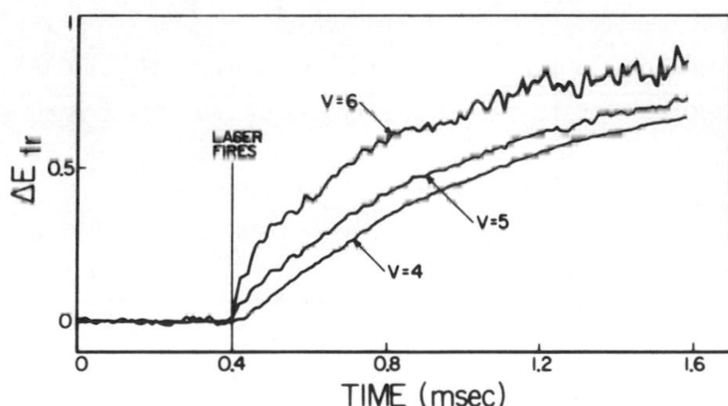

FIGURE 2. Relaxation data from HD in v = 4, 5 and 6 in pure HD at 298°K. The signals are normalized to the total amount of energy deposited in the gas.

of the time domain relaxation curves has yielded self relaxation rates of HD in v = 1 through v = 6. The change in the form of the initial portion of the curves reflects a change in the mechanism to a predominance of vibrational to translational relaxation from the higher vibrational states (1-4). The rates for self-relaxation for HD extracted from these data is given in Fig. 3. Both the V-V and V-T rates for collisions between the molecules in the labelled states and HD molecules in v = 0 are given. The analysis of this experiment makes use of the energy corrected sudden scaling relations for vibrational rates to reduce the number of fitted rate parameters to two fundamental rates (1,4,5). All of the other rates required to describe the relaxation from the pumped

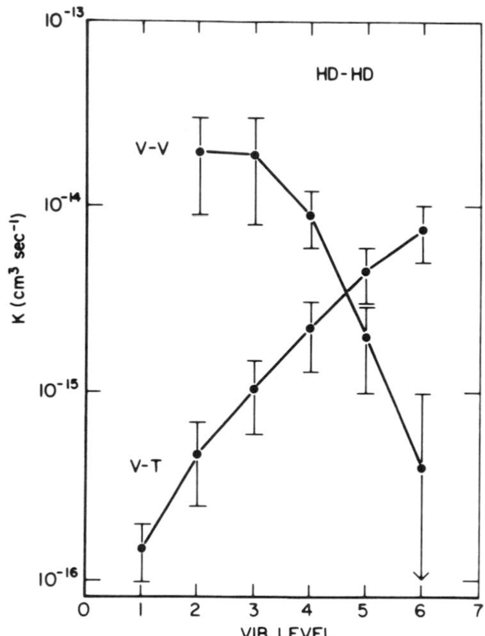

FIGURE 3. The HD-HD vibration to vibration (V-V) and vibration to translation (V-T) relaxation rates obtained from fitting the data in Fig. 2. These are for collisions in which one collision partner is in the labelled state and the other is in v = 0.

state are generated by the ECS scaling relations. Vibrational to translational relaxation rates for HD-He collisions have been extracted from similar measurements on HD-He mixtures (3,4).

REFERENCES

1. E.A. Rohlfing, J. Gelfand, R.B. Miles and H. Rabitz, *Chem. Phys.* *51*, 121 (1980).
2. E.A. Rohlfing, J. Gelfand, R.B. Miles and H. Rabitz, *J. Chem. Phys.* *75*, 4893 (1981).
3. E.A. Rohlfing, J. Gelfand and R.B. Miles, *J. Appl. Phys.* *53*, 5426 (1982).
4. E.A. Rohlfing, J. Gelfand, R.B. Miles and H. Rabitz, *J. Chem. Phys.* in preparation.
5. A.E. DePristo and H. Rabitz, *Chem. Phys.* *44*, 171 (1979).

STROBOSCOPIC INTERFEROMETRY: A MEANS TO FOLLOW BROADBAND INFRARED PHENOMENA USING AN FT-IR SPECTROMETER

James L. Chao

IBM Instruments, Inc.
P.O. Box 332
Danbury, Ct. 06810

T. Kumar

Kansas State University
Department of Chemistry
Manhattan, Kansas 66506

I. INTRODUCTION

The use of a Fourier transform infrared spectrometer to obtain time-resolved infrared spectra is reported here. While a variety of laser Raman techniques exist to investigate time-dependent vibrational phenomena, these techniques are normally limited to a very narrow spectral bandwidth or else require the use of costly array detectors. This form of spectral multiplexing ... i.e., the use of dispersive elements with optical multi-channel analyzers, photodiode arrays, or CCD-type devices, differs considerably from the type which is used in interferometry, commonly referred to as the Fellgett Advantage. There are many other differences as well. In general, with an FT-IR, the system measures directly the absorption of infrared wavelengths whereas in Raman systems, inelastic scattering of visible wavelengths is observed. This causes differences in the requirements for both the types of materials one chooses for optics as well as on the selection of detectors. Furthermore, the selection rules differ for the two processes.

When one next goes beyond spectral multiplexing, one may then move towards experiments which include simultaneous temporal multiplexing. This is much more straight forward in Raman spectroscopy. By varying laser pulse lengths and employing optical delay lines one can measure directly very short vibrational relaxation processes. However, it is much more involved when one works in the realm of interferometry. The system operates in the time-domain automatically when a Fourier-transform method is employed. It is because of a desire to measure vibrational phenomena that occur on a time scale short compared to the interferometer's measurement time that this technique was developed. Experiments which can be repeated reproducibly and can be triggered asynchronously at greater than 10 Hz become candidates for this method. The useful time range for the technique is from about ten microseconds to several hundred milliseconds.

ISBN 0-12-066280-9

II. THEORY

The implementation of this technique involves the use of a hardware interface which serves to synchronize the strobing of the experimental process with the interferometer's scanner. Figure 1 illustrates use of two parallel Michelson interferometers, one of which uses a HeNe laser to mark the spatial positioning of the moving mirror (MM). Points of equally spaced optical path difference are determined from the HeNe laser zero crossing points (XA) which are used to trigger A/D conversion of the detector signal. By moving the scanner at constant velocity, the XA pulses generated from the HeNe laser are equally spaced in time and can, therefore be used as a master clock for the system.

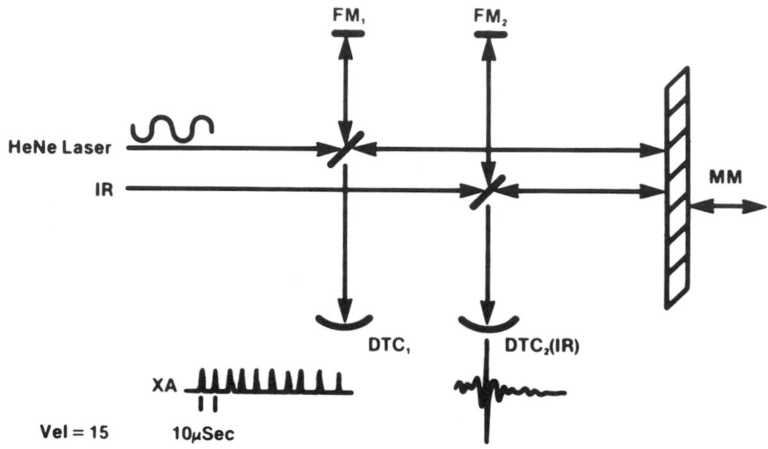

FIGURE 1. Two parallel path Michelson interferometers. Pulse train (XA) generated from laser interference fringes is used to mark position of moving mirror (MM). Phase-locking assures constant velocity allowing A/D conversion of IR signal to be equally spaced in position.

An example of the methodology of stroboscopic interferometry is shown in Figure 2. Suppose we wish to measure the interferogram intensity at ten equally spaced mirror positions. As shown in this figure, ten experimental trials are used to obtain the ten different mirror positions. The numbers in the array represent the time elapsed from the experimental triggering point or time=0. In trial #1 an experimental trigger begins when the constant velocity moving mirror arrives at mirror position 1. In each subsequent trial, the experiment is triggered one position earlier with respect to those mirror positions for which data collection actually occurs. As a result of ten such trials, we have generated every different time (0-9) for each mirror position. One then realizes that if one were interested in the spectrum related to time=7, one would Fourier transform an interferogram reconstructed from data points marked time=7. These points lie on the two diagonals shown in bold face. The sorting process required for this strobscopic system is written as a post-collection software subroutine. In typical experiments, the number of points collected in any given trial are, in fact, in the thousands (rather than ten) and the number of trials is between 20 and 100. Because of the large number of points, attention must be paid to the algorithms to speed up the sorting time.

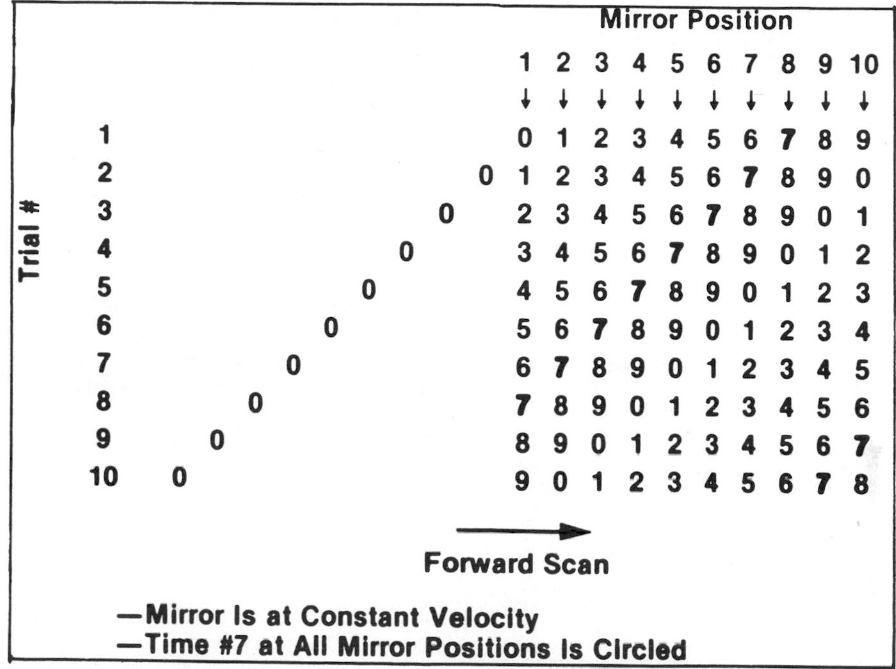

FIGURE 2. Diagram of stroboscopic interferometry experiment.

III. EXPERIMENTAL

Two areas of interest have been experimentally investigated. The first area is the study of the dynamics of polymer film stretching. The second area is the study of photochromophores in solution. Structural changes due to absorption of visible light generated from a Quanta-Ray pulsed Nd:YAG laser are measured with the use of an IBM Instruments, Inc. Fourier transform infrared spectrometer. Both systems are studied by synchronizing the beginning of the excitation cycle with the forward motion of the interferometer's scanner.

A. POLYMER FILMS

The study of the properties of polymer films has many very important industrial implications. In the past, the study of polymer morphology has been confined primarily to static systems or to very slowly varying responses. We have begun to examine vibrational properties during stretching in order to elucidate the dynamics of mechanical stress to materials. The time scales for these processes extend from the microsecond time scale on upwards. The hope of these studies is to improve predictions of the causes for damage or failure in polymers subjected to sudden stress. It should be noted that the fixtures we designed for polymer stretching are crude; however, we believe they are sufficient to allow us to arrive at some qualitative observations and to make some reasonable predictions.

The polymer stretcher used in these experiments was based on a solenoid powered by an audio power amplifier driven as a perfect square wave generated from the time=0 synch pulses. The film used here was a 5 to 10 cm^2 piece of polypropylene of approximately 1 mil thickness stretched at ~20 Hz repetition rate.

RESULTS

The preliminary results for the polymer stretching experiments have been surprising. For example, a literature search on the stretching of isotactic polypropylene has led to some insight into those bands which are stress sensitive. Two peaks at 974 and 988 cm^{-1} respectively are known to be stress sensitive. However, the earlier work differs from these experiments in that the greatest stretch possible at 20 Hz is less than 5 percent. In those experiments, stretch of polypropylene was as high as 25% where the polymer was firmly held.

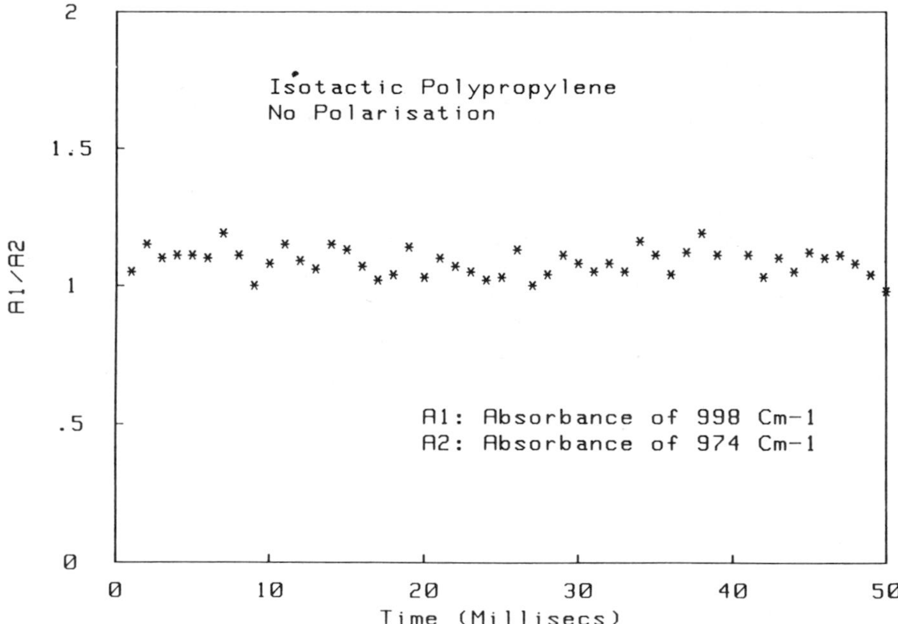

FIGURE 3. Band intensity ratio as a function of time from stretching of polypropylene film at 20 Hz. No stress sensitivity is seen here.

The criterion for a stress sensitive band is to compare the ratios of the absorbance intensities between the symmetric and asymmetric modes of identical vibrational transitions. Figure 3 shows this comparison for a 3% stretch along the crystalline axis at an approximately 20 Hz repetition rate.

In these real-time stretching experiments, these vibrational bands are not found to be stretch sensitive. These measurements are for much shorter impulse times (less than a few milliseconds) than in those experiments and, furthermore, for a much smaller percentage stretch.

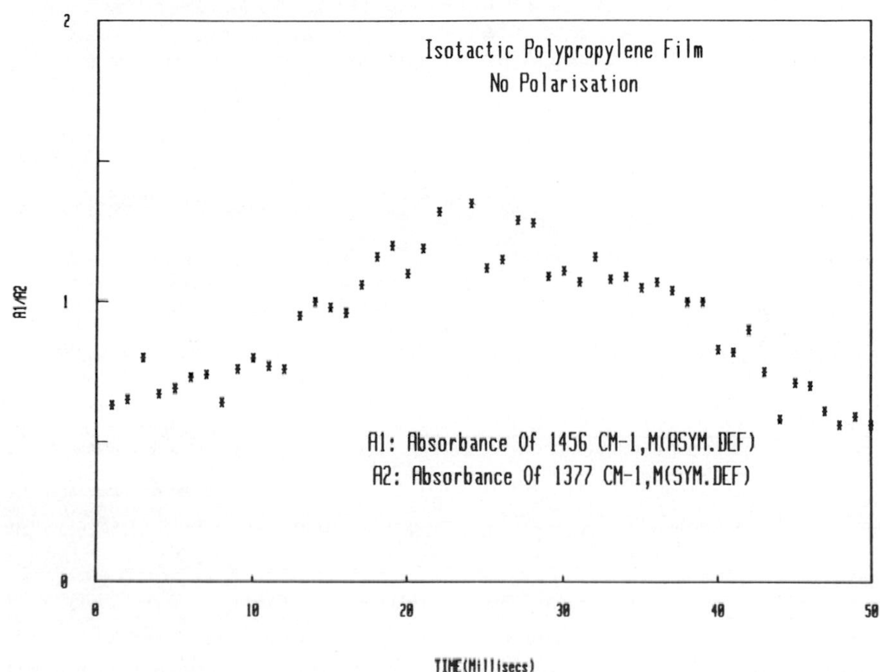

FIGURE 4. Band intensity ratios in isotatic polypropylene. Monotonic response function is seen with same period as stretching process.

In Figure 4, we find a pair of stretch sensitive bands at 1377 and 1456 cm^{-1} for the same experiment. In these band ratio plots against time, one sees the monotonic period that is indicative of a response to the stretching cycle.

B. PHOTOCHROMOPHORES

Experiments to look at the changes of laser excited photochromophores were performed. In these studies, metal dithizonates were dissolved in a variety of different solvents. These effects have been studied previously and have been related to structural changes relative to solvent interactions. These compounds exhibit extremely colorful dye changes upon irradiation with visible light. For example, for Hg dithizonate in CH_2Cl_2, the color changes from orange to bright violet upon exposure to light. While the colors exhibited upon irradiation can be changed under variety of conditions, pH and concentration were selected as the primary means by which the rates of relaxation to intense laser pulses were controlled.

Since the lifetime could be varied from several microseconds to minutes, the initial experiments were adjusted so that the concentration of Hg dithizonate in CH_2Cl_2 resulted in lifetimes that were extremely long (> 5 minutes). This allowed for the measurement of the infrared absorption of the excited state mixture.

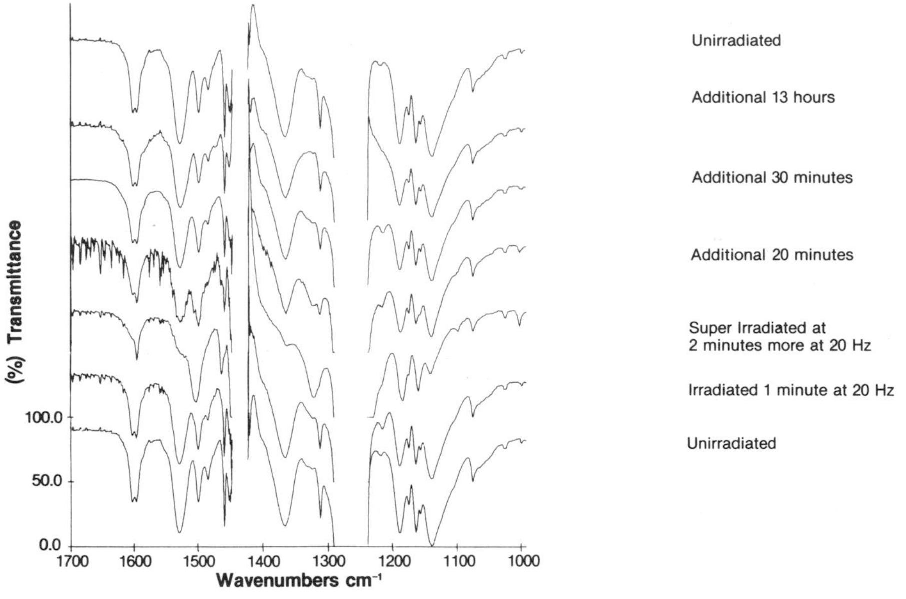

FIGURE 5. Steady-state results of Hg dithizonate in CH_2Cl_2 for various amounts of irradiation. Lifetime of compound is adjusted to determine which bands change and to demonstrate that changes are reversible.

The results of the steady state measurement of infrared spectra for Hg dithizonate from various amounts of irradiation and after allowing the compound to relax are summarized in Figure 5. After a long enough period, the compound relaxes to its original state. From these steady state experiments, we were able to define the solvent to be used, the metal dithizonate to be studied, its concentration, the amount of light needed to saturate the sample and the bands that would change.

A doubled Nd:YAG laser was externally triggered at about 12 Hz and approximately 25 millijoules of power per shot was focussed into a liquid IR cell. The concentration was then adjusted so that the lifetime would be in the millisecond time scale.

RESULTS

The results on the stroboscopic experiment are seen as successive time files in Figure 6. The data is ratioed against the time=0 file in order to accentuate the small changes. This can be attributed to limitations of the quantum efficiency as well as complete relaxation to the initial state. In time files 75-95, the predicted changes of the steady-state results are qualitatively seen. The ratio was used rather than spectral subtraction because it shows the changes more clearly.

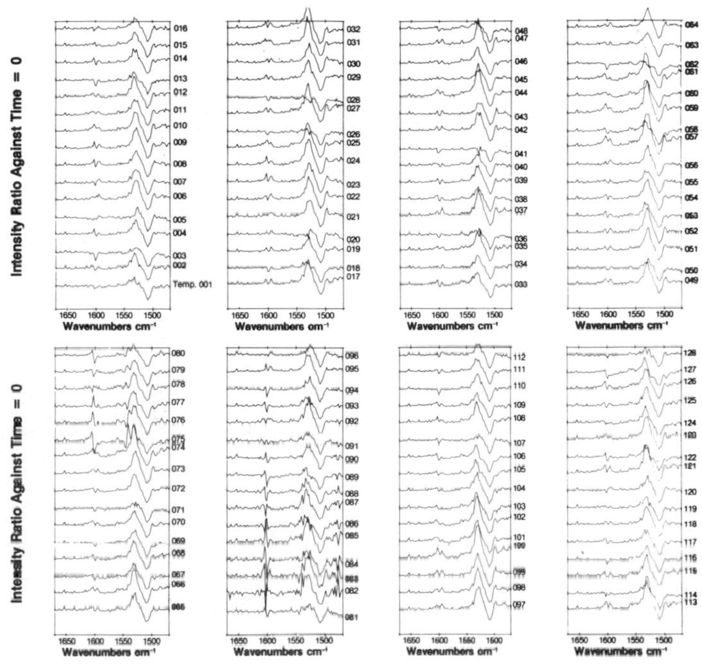

FIGURE 6. Stroboscopic interferometry results on Hg dithizonate every 500 microseconds. CHanges are seen by ratioing the spectra against one from an unirradiated sample.

ACKNOWLEDGEMENTS

The authors would like to thank Dave Hoenig for his early contribution to the design and testing of the sychronization apparatus and polymer stretcher, C.G. Venkatesh of Bruker Instruments, Inc. for the post-collection sorting routine, M. Paquette at Dow Chemical Research for his collaboration on the laser work, and, Prof. W. G. Fateley, at Kansas State University whose assistance made this work possible.

INFRARED MULTIPHOTON EXCITED VISIBLE LUMINESCENCE FROM HEXAFLUOROBENZENE

Michael T. Duignan

Naval Research Laboratory
Washington, D. C.

Shammai Speiser[1]

Department of Chemistry
Technion-Israel Institute of Technology
Haifa, Israel

I. INTRODUCTION

Infrared Multiphoton Excitation (MPE) of polyatomic molecules is frequently accompanied by visible luminescence (1-7). Such luminescence may arise from a variety of sources such as dielectric breakdown or chemiluminescent processes (e.g., radical recombination). A mechanism which proposes a unimolecular conversion of vibrational energy to electronic energy in the parent molecule is of current interest.

Nitzan and Jortner (8) have postulated two mechanisms leading to Infrared Multiphoton-Induced Luminescence (MPIL). The first one involves a non-radiative internal conversion from high vibrational states of the ground electronic state to low

[1]*To whom correspondence should be addressed.*

ISBN 0-12-066280-9

vibrational states of the fluorescing excited electronic state. This process designated as inverse electronic relaxation (IER) has rate which is much slower than the fluorescence rate of an excited electronic state produced by direct one-photon optical pumping. Another possibility is direct one-photon infrared excitation (DIRE) from the MPE produced high vibrational state of the ground electronic state into the emitting excited electronic state. In this case the fluorescence life time may match that of the conventionally produced electronically excited molecule.

However, a fundamental complication of the determination of origin of MPIL is that the energy of a visible photon is similar to a typical unimolecular reaction threshold--that is, within an ensemble of molecules undergoing MPE, if a reasonable fraction have enough internal energy to emit a visible photon, it is highly probable that some have sufficient energy to dissociate.

Several recent observations have indeed been interpreted as a collisionless production of electronically excited parent molecules in an MPE process, but it is difficult to provide definitive proof (6-9). In addition to the complications associated with decomposition, the normal fluorescence lifetime and spectrum are not necessarily characteristic of the parent substance if it is produced by MPE. The process must be collisionless, which is difficult to demonstrate even at extremely low pressures (6). Moreover, a prompt rise of luminescence during the laser pulse does not rule out collisional V $\rightarrow$ E excitation mechanisms (6).

We report here visible/UV luminescence accompanying MPE of hexafluorobenzene (C_6F_6). Our attempt to identify the source of this luminescence illustrates the complexity of the task. We intend to show that neither IER nor DIRE from parent hexafluorobenzene molecules is likely to be a major source of the observed luminescence.

II. EXPERIMENTAL

The IR laser pulse, at 1023 cm^{-1} consisted of a "spike" of 120 ns FWHM and a 1 μs tail. For some experiments a plasma shutter was used to provide truncated pulses at 40 ns FWHM with a steep (<1 ns) fall time. The beam was brought to a focus inside the luminescence cell and the resulting VIS/UV emission was monitored at right angles. The luminescence virtually disappeared when the laser was tuned off resonance from C_6F_6 absorption, showing that dielectric breakdown is not involved.

Kinetic measurements were performed using the truncated pulse, with system response time set at the minimum 12 ns level. In these measurements, luminescence was observed through 360 & 560 nm interference filter. Output from the photomultiplier was amplified and stored by means of an A/D Transient Waveform recorder; resulting data were averaged and printed by an interfaced calculator. Rate constants were obtained by least-squares analysis of semi-log plots.

For luminescence spectra, non-truncated IR pulses were used. A tunable monochromator with ~6 nm resolution and a system time response of 130 ns was chosen. The resulting luminescence signal was stored on the Biomation and integrated over the following successive time intervals: 0.3 μs (during which ~75% of the IR energy is delivered), 0.5 μs (after which the IR pulse is essentially over), 1, 2, 5 and 10 μs. The integrated signals were corrected for instrument response and for a weak emission component (probably from the KCl windows) found to be independent of IR wavelength.

III. RESULTS AND DISCUSSION

The temporal behavior of the MPIL of C_6F_6 shows at least three components. Fig. 1 shows luminescence decay monitored at

360 nm, as well as the profile of the exciting laser pulse. The 360 nm luminescence consists of a sharp spike henceforth referred to as the very short lifetime (VSL) component, followed by a component which shows a 600 ns exponential decay, pressure independent in the 0.145-1.46 torr range. Figure 1c shows luminescence decay monitored at 560 nm. The sharp spike is now smaller, the subsequent decay is longer than 600 ns and is pressure dependent, ranging from 1.56 μs at 0.43 torr to 2.8 μs at 2.22 torr.

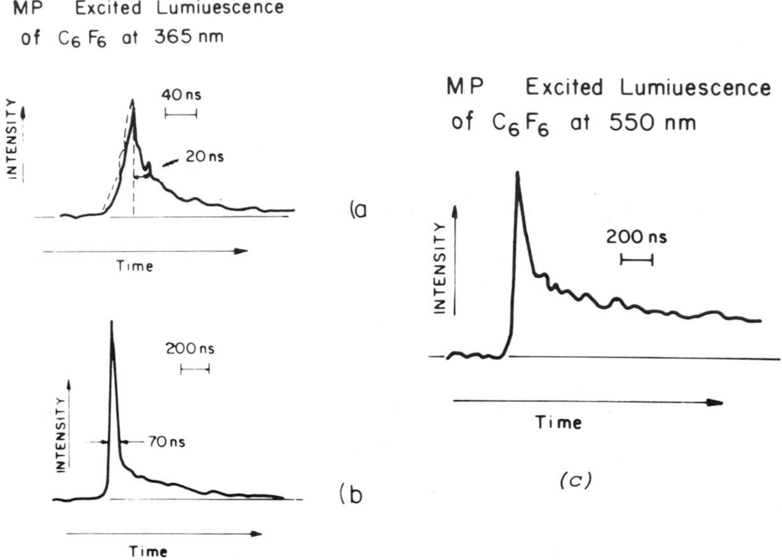

FIGURE 1. Oscilloscope traces of (a) profile of 1023 cm^{-1} IR laser pulse truncated with plasma shutter (broken curve) and synchronous (± 10 ns) initial luminescence emission at 360 nm from 1.2 torr C_6F_6 (solid curve). (b) Full temporal profile of luminescence emission at 360 nm. The slow decay time is 600 ns. (c) Temporal profile of luminscence emitted at 560 nm.

The difference in the temporal behavior allows us to separate the emission spectra by time-resolved spectroscopy. Figure 2 compares the regular fluorescence spectrum of C_6F_6 (11) with the luminescence spectrum of the short-lived species, and with that of the long-lived species. The normal fluorescence lifetime of C_6F_6 has been determined to be 3.6 ns (11). The lifetime of the VSL component was on the order of our instrument response time and was estimated to be 8 $\pm$ 5 ns. Momentarily assuming that the VSL emission component in the 325-425 nm region (Fig. 2) is from parent hexafluorobenzene, the fraction emitting was calculated to be $\sim 4 \times 10^{-4}$.

At fluences much smaller than those employed in these experiments C_6F_6 undergoes some decomposition to produce gaseous C_2F_4 and a smaller amount of $CF_3C_6F_5$ as well as an unidentified black solid residue (12). These same products were detected after our (high fluence) luminescence observation experiments by both IR and GC/MS. The character of these gaseous products are strong evidence for CF_2 formation. Thus we cannot rule out the proposition that the short-lived luminescence is due to electronically excited fragements rather than parent molecules.

We indeed observe that the long-lived emission (Fig. 2) includes the C_2 Swan band system, superimposed on an unassigned broad background which peaks towards the infrared. Our evidence for C_2 is similar to that found in other systems (4), and we also find a similar effect of oxygen. The pressure-dependent decay time and the quadratic pressure dependence of the fluorescence yield of the long-lived emission (Fig. 3) indicate a collisional production of electronically excited C_2 and other, unidentified fragments that might contribute to this emission.

Theoretical modeling for IR absorption behavior in hexafluorobenzene (13) completed subsequent to this study has allowed greater confidence in extrapolating absorption cross sections to fluences employed in these experiments. With a high degree of probability, virtually all C_6F_6 molecules within the focal

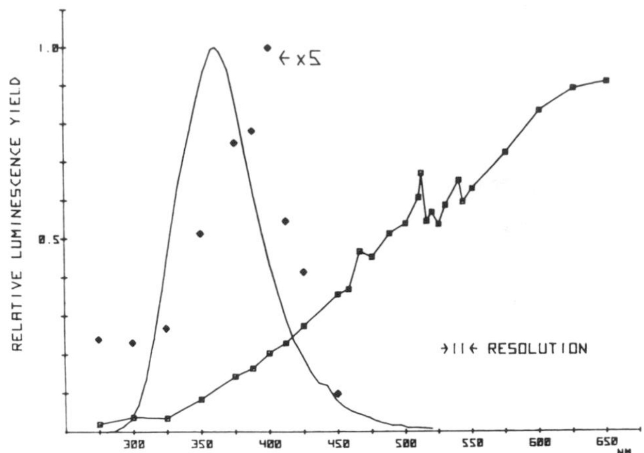

FIGURE 2. *Fluorescence spectrum of gaseous C_6F_6 at room temperature (solid curve); Corrected MPE-induced luminescence spectrum of the short-lived species, measured by integration during the laser pulse (solid diamonds); slow component (corrected) of the MPE-induced emission spectrum (open squares).*

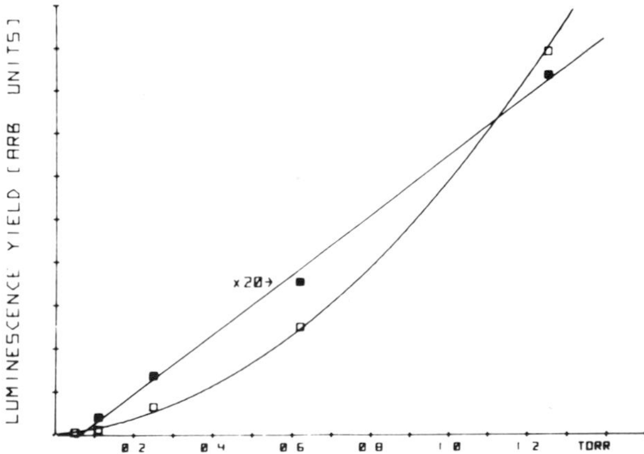

FIGURE 3. *MPE-induced luminescence yield as a function of C_6F_6 pressure (a) (linear fit-solid squares) due to short-lived species, monitored at 350 nm and integrated over the first 300 ns of the laser pulse; (b) (quadratic fit-open squares) due to long-lived species, monitored at 512 nm.*

volume are destroyed with each laser shot. Further absorption of IR energy by hot fragments seems likely. The pressure dependence of the luminescence yield and decay rate as well as the "black body" -- like broad emission background point to collisional mechanisms for the long-lived MPIL. Only C_2 could be positively identified as a source among the long-lived components. Evidence for unimolecular production of the VSL emission is based on both the linear pressure dependence of the luminescence yield (Fig. 3) and, to a lesser extent, the rapid quenching after the IR field is turned off. It seems, in fact, that the VSL luminescence is intimately linked to a high rate of vibrational up-pumping. Due to the very high vibrational state densities of S_0 near the origin of S_1 in C_6F_6, the measured decay rates are too fast and the luminescence yield too great for IER as described by Nitzan and Jortner (8) to offer a major contribution to any of the described emission components. Although the VSL component spectrum (Fig. 2) is curiously similar to the normal fluorescence spectrum of C_6F_6, one would expect DIRE to be a minor process at the high internal energies of parent molecules within the focal volume.

In this work we have demonstrated the advantages of using truncated IR laser pulses which facilitates time resolved spectroscopy measurement. We have shown that neither IER nor DIRE from parent hexafluorobenzene molecules are likely sources of major contributions to MPIL. More sophisticated experimental techniques, such as optical time of flight spectroscopy (6,14,15) will be needed to determine the identities of the emitting species or the mechanisms by which they luminesce.

ACKNOWLEDGMENTS

This paper is based in part on the Ph.D. Thesis of M.T.D. and was done at the Chemistry Department, Brandeis University. We are indebted to Prof. E. Grunwald for many stimulating discussions and for his kind hospitality during the sabbatical year of S.S.

Financial support was provided by the NSF and by the Edith C. Blum Foundation.

REFERENCES

1. R. V. Ambartzumyan, N. V. Chekalin, V. S. Doljikov, V. S. Letokhov and E. A. Ryabov, *Chem. Phys. Letters* 25,515 (1974).
2. R. V. Ambartzumyan, N. V. Chekalin, V. S. Doljikov, V. S. Letokhov and N. V. Kokhman, *J. Photochem.* 6, 55 (1978).
3. R. V. Ambartzumyan, Yu. A. Gorkhov, G. N. Makarov, A. A. Puretzki and N. P. Furzikov, *Chem. Phys. Letters* 45, 231 (1977).
4. M. L. Lesieski and W. A. Guillory, *J. Chem. Phys.* 66, 4317 (1977).
5. M. H. Yu, H. Reisler, M. Mangir and C. Wittig, *Chem. Phys. Letters* 62, 439 (1979).
6. H. Reisler and C. Wittig in: Photoselective Chemistry, eds. J. Jortner, R. D. Levine and S. A. Rice (Wiley, N. Y., 1981) and references therein.
7. Y. Haas and G. Yahav, *Chem. Phys. Letters* 48, 63 (1977); G. Yahav and Y. Haas, *Chem. Phys.* 35, 41 (1978); *Chem. Phys. Letters* 83, 493 (1981).
8. A. Nitzan and J. Jortner, *J. Chem. Phys.* 71, 3524 (1979).
9. Y. Haas in: Photoselective Chemistry, eds. J. Jortner, R. D. Levine and S. A. Rice (Wiley, N. Y., 1981).
10. G. A. Loper and E. K. C. Lee, *Chem. Phys. Lett.* 13, 140 (1972).
11. K. Al-Ani and D. Philips, *J. Phys. Chem.* 74, 4046 (1970); D. Philips, *J. Chem. Phys.* 46, 4679 (1967); our measurements reconfirm the authors' spectrum of C_6F_6.
12. M. T. Duignan, Ph.D. thesis, Brandeis University, 1981.
13. M. T. Duignan, G. Garcia and E. Grunwald, *J. Am. Chem. Soc.* 103, 7281 (1981).
14. J. D. Campbell, M. H. Yu, M. Mangir and C. Wittig, *J. Chem. Phys.* 69, 3854 (1978).
15. R. B. Ambartzumian, et al., *Apply. Phys.* 22, 77 (1980).

DIFFERENCE RAMAN SPECTROSCOPY WITH A STOPPED-FLOW DEVICE

Yoshifumi Nishimura
Tadayuki Uno
Masamichi Tsuboi

Faculty of Pharmaceutical Sciences
University of Tokyo
Hongo, Bunkyo-ku
Tokyo, Japan

Difference Raman spectroscopy can often provide a totally new type of information of biological systems (1). If it is combined with a stopped-flow device, this must be powerful for characterizing a reaction intermediate or for determining the rate constant of a reaction (2). A few pieces of work, recently made along this line on the redox reaction of cytochrome c, will be described below.

I. REDUCTION KINETICS OF FERRICYTOCHROME c WITH ASCORBATE

Ferricytochrome c in neutral aqueous solution was rapidly mixed with sodium ascorbate solution by the use of a Jasco SFC-5 stopped-flow device. The mixed solution was introduced into a rectangular tube with inside dimensions of 2 x 2 x 10 mm and with 4 mm thick glass walls. To excite the Raman scattering a continuous wave Ar^+ laser beam of 514.5 nm or of 488.0 nm was used. The beam was focused into the rectangular tube, aligned perpendicularly to the scattering plane. Scattered light was collected at 90°, and examined with a spectrograph (2) and a Tracor Northern IDARSS detector system, which consists of a TN-1223-4I intensified photodiode array head, a TN-1710-4K analyzer, and an XY-recorder.

With 514.5 nm excitation, Raman bands of ferrocytochrome c are much stronger than the corresponding bands of ferricytochrome

ISBN 0-12-066280-9

c, and as the reaction proceeds the intensities of them increase. A series of difference Raman spectra was obtained, by subtracting the Raman spectrum of completely reduced cytochrome c from each of the Raman spectra obtained at 1 sec, 21 sec, 51 sec, and 91 sec after the reducing reaction was initiated. By examining the peaks at 1314 and 1230 cm^{-1}, for example, in such a series of difference spectra, a kinetic plot of the Fe(III) → Fe(II) reaction of cytochrome c was obtained.

With 488.0 nm excitation, Raman bands of ferrocytochrome c are in general comparable to the corresponding bands of ferricytochrome c. In a difference spectrum of (intermediate state)-(completely reduced state), therefore, both of the Fe(III) and Fe(II) cytochrome c peaks are found. Thus, positive peaks at 1637 and 1373 cm^{-1} correspond to the oxidized form and negative peaks at 1622 and 1364 cm^{-1} to the reduced form.

II. ABSORPTION CORRECTION IN TIME-RESOLVED RESONANCE RAMAN SPECTROSCOPY

As Strekas et al. (3) pointed out, the wavelengths of the strong Raman scatterings of ferrocytochrome c at 1314 and 1230 cm^{-1} for the 514.5 nm excitation are in its strong 550 nm absorption band. Its molar extinction coefficient (ε) is as high as 28,000 (at 550 nm), and hence 1-mm path length of its 0.2 mM solution should transmit only 1/3 of the 550 nm light. In our reaction system, the concentration of ferrocytochrome c is a function of time and therefore the transmittance of the Raman scattering lights must be time dependent. Thus, the observed Raman intensity *versus* time curve would never reflect the actual product concentration *versus* time relation. Even if no apparent anomaly is seen in the observed kinetic curve, a proper correction may be needed to reach a correct rate constant.

Let us assume that, at time t, the mole fraction of the reactant (ferricytochrome c) is x and that of the product (ferrocytochrome c) is 1-x. If the total concentration of cytochrome c is c, effective path length of the scattered Raman beam is ℓ, molar extinction coefficient of the reactant is ε_r, and that of the product is ε_p, the apparent Raman intensity $I_{app}(t)$ at a proper wavenumber and at time t is related with x as

$$I_{app}(t)=[uxI_r + v(1-x)I_p]/u^x v^{1-x}, \tag{1}$$

where

$$\log_{10}u = \varepsilon_r c\ell, \quad (2)$$

$$\log_{10}v = \varepsilon_p c\ell, \quad (3)$$

and I_r and I_p are the apparent Raman intensities of reactant and product, respectively. If u/v is not greatly different from 1, a simple correction formula

$$\frac{1}{x} = \frac{u}{v}\left(\frac{1}{x_{app}} - 1\right) + 1 \quad (4)$$

can be used. Here,

$$x_{app} = (I_p - I_{app}(t))/(I_p - I_r) \quad (5)$$

is the apparent value of x (extent of reaction) obtained without taking self-absorption into account.

By the use of the simpler correction formula (Eq.(4)) the reaction rate constant of the reducing reaction of cytochrome c was determined to be 50 $sec^{-1}M^{-1}$ at pH 7 and at 25°C.

III. ALKALINE ISOMERIZATION OF FERRICYTOCHROME c

Ferricytochrome c is known to undergo a pH-induced subtle conformational isomerization near pH 9 (4), and Kihara et al. (5) showed, by the 400 - 700 nm absorption spectroscopy, that a transient intermediate appears during the isomerization. We have attempted to observe the Raman spectrum of this intermediate.

The isomerization reaction was initiated by a pH-jump method. A 159 μM ferricytochrome c solution in 10 mM phosphate buffer (pH 7.0) was rapidly mixed with a 0.2 M glycine-NaOH buffer, so that the final pH of the solution became 10.5. A transient Raman spectrum was observed by the use of a multi-detector system mentioned above. The Raman scattering from the sample solution emitted in the period of 30 msec ∿ 530 msec after the initiation of the reaction was memorized. This stopped-flow Raman spectrophotometry was repeated for 100 times, and the data was accumulated. The result showed that the transient Raman spectrum is appreciably deviated from any of the linear combinations of the stationary-state Raman spectra of ferricytochrome c at pH 7.0 and at pH 10.5. By a proper subtraction of such stationary-state Raman spectra, a Raman spectrum assignable to the intermediate species has been obtained. It showed Raman bands at 1635, 1579, and 1561 cm^{-1}.

ACKNOWLEDGEMENT

This work was partly supported by a grant-in-aid (no. 543032) from the Ministry of Education, Science, and Culture of Japan and a grant from Toray Science Foundation (1979).

REFERENCES

1. Shelnutt, J. A., Rousseau, D. L., Dethmers, J. K., and Margoliash, E., *Proc. Nat. Acad. Sci. U.S.A. 76*, 3865 (1979).
2. Nishimura, Y. and Tsuboi, M., *J. Raman spectroscopy 12*, 138 (1982).
3. Strekas, T. S., Adams, D. H., Packer, A., and Spiro, T. G., *Applied Spectroscy. 28*, 324 (1974).
4. Dickerson, R. E. and Timkovich, R., *in* "The Enzymes" (Boyer, P.D., ed) Vol. 11, pp.397-547, Academic Press, New York.
5. Kihara, H., Saigo, S., Nakatani, H., Hiromi, K., Ikeda-Saito, M., and Iizuka, T., *Biochim. Biophys. Acta 430*, 225 (1976).

HIGH OVERTONE VIBRATIONAL SPECTRUM OF SF_6: EVIDENCE FOR LOCALIZED VIBRATIONS

Harold B. Levene
David S. Perry
University of Rochester
Rochester, New York

I. INTRODUCTION

The local mode picture of molecular vibrations was developed to explain the high overtone spectra of hydrogen-containing molecules. Two local vibrations each involving only a single bond but sharing a common atom, are kinetically coupled by the off-diagonal elements of the G-matrix: $G_{ij}/G_{ii} = m \cos\theta_{ij}/(m+M)$. Here G_{ii} and G_{ij} are the diagonal and off-diagonal G-matrix elements respectively. The angle between the bonds is θ_{ij}. The masses are M for the central atom and m for the substituent atoms. It is clear when $m<<M$ then $G_{ij}/G_{ii}<<1$ making local behavior possible. Conversely when $m \sim M$ (e.g. SF_6), then local vibrations will be strongly coupled. Is any kind of local behavior possible in non-hydrogen-containing molecules?

We report here a measurement of the high overtone vibrational spectrum of SF_6 and interpret the results using a local cartesian oscillator model.

II. EXPERIMENTAL AND RESULTS

A pulsed Nd^{+++}YAG laser is used to pump an optical parametric oscillator (OPO). The lithium niobate crystal and grating of the

ISBN 0-12-066280-9

OPO are scanned by a microcomputer. The tunable infrared output of the OPO is focused into an optoacoustic cell at 192K. The microphone signal is then amplified, integrated, normalized for laser intensity, and digitized. Transitions from the vibrational ground state to the following levels are observed in the range 2800 to 4200 cm^{-1} (1): $3\nu_3$, $3\nu_2\nu_3$, $\nu_2 2\nu_3\nu_6$, $3\nu_1\nu_4$, $\nu_1 2\nu_2\nu_3$, $2\nu_1\nu_2\nu_3$, $\nu_2 2\nu_3\nu_4$, $3\nu_1\nu_3$, $\nu_1 2\nu_3\nu_4$, $\nu_2 3\nu_3$, $\nu_1 3\nu_3$, $2\nu_2 3\nu_3$. The four bands involving the low frequency bending modes, ν_5 and ν_6, which are predicted to occur in unobstructed regions of the spectrum are not observed. Four of the observed bands have a similar contour to $3\nu_3$ (e.g., Fig.1) and therefore are single vibrational components; the others do not appear to be multiple components.

III. A LOCAL CARTESIAN OSCILLATOR MODEL

The cartesian approach is inspired by the finding of Patterson et al that a cartesian basis is much more effective than the usual angular momentum basis in accounting for the positions and intensities of the two F_{1u} components of the $3\nu_3$ band (2). The physical basis of the local cartesian oscillator model can be seen by

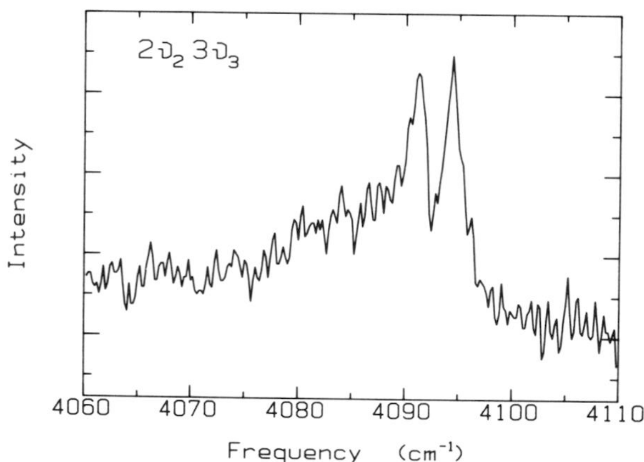

FIGURE 1. Spectrum of the highest transition observed.

examining the kinetic coupling (G_{ij}/G_{ii}). For opposing bonds, $\cos\theta_{ij} = \cos 180° = -1$, and the coupling of these vibrations is strong ($G_{ij}/G_{ii} = 0.37$). For vibrations of adjacent bonds, the kinetic coupling is zero ($\cos\theta_{ij}=0$) leaving only the much weaker potential energy interaction ($F_{ij}/F_{ii} = -0.06$).

As the main point of the model, we assume that the interaction of light with the bond dipoles produces excitations to levels which are predominantly stretching vibrations of 2 opposing bonds. This implies, in agreement with experiment, that transitions involving the excitation of the low frequency bending modes will not be observed. The selection of a single pair of opposing bonds means that all observed bands should be single components.

We do not present a local mode basis set for the purpose of diagonalizing the Hamiltonian, since the normal mode basis accurately accounts for the observed transition frequencies. The task is rather to identify which of the normal mode levels are approximately local cartesian oscillations. It should be pointed out that the vibrations here are less local than in the usual local mode description since two bonds receive substantial excitation. As well there may be small amplitude motion in other degrees of freedom.

In mapping the normal vibrations onto local cartesian vibrations, the identification of one cartesian axis as the excitation axis effectively reduces the symmetry from $O_h \longrightarrow D_{4h}$. The mapping of the six normal modes is as follows:

ν_1: $A_{1g} \longrightarrow A_{1g}$ $\quad$ ν_4: $F_{1u} \longrightarrow A_{2u} + E_u$

ν_2: $E_g \longrightarrow A_{1g} + B_{1g}$ $\quad$ ν_5: $F_{2g} \longrightarrow B_{2g} + E_g$

ν_3: $F_{1u} \longrightarrow A_{2u} + E_u$ $\quad$ ν_6: $F_{2u} \longrightarrow B_{2u} + E_u$

The symmetric and antisymmetric stretches of the 2 bonds on the excitation axis have symmetries A_{1g} and A_{2u} respectively. Assuming that the bond dipoles are approximately separable, then "allowed" transitions will involve combinations of A_{1g} and A_{2u} only. This implies that for combinations of ν_1, ν_2, ν_3 and/or ν_4

that there will be exactly one allowed component, which is in general lower in energy than the others. In all cases, the frequency of the single observed component agrees well with the calculated frequency of the lowest energy component. Similarly a transition involving ν_5 or ν_6 is "forbidden." Because ν_4 involves substantial bending as well as stretching, combinations involving ν_4 are observed to be less intense than the corresponding combinations involving ν_3.

IV. SUMMARY

The number and frequencies of the observed transitions in the SF_6 overtone spectrum are well described using the local cartesian oscillator model. The success of this approach is due to the nature of the excitation process as well as to the relatively weak coupling between orthogonal stretches.

ACKNOWLEDGMENTS

The authors gratefully acknowledge support from the National Science Foundation, Grant NSF CHE80-2317 and from the Donors of the Petroleum Research Fund of the American Chemical Society.

REFERENCES

1. Levene, Harold B. and Perry, David S., in preparation.
2. Patterson, C. W., Krohn, B. J., and Pine, A. S., *J. Mol. Spectrosc.*, 88, 133 (1981).

VIBRATIONAL ENERGY TRANSFER RATE CONSTANTS FROM THE THERMAL LENS

R.T. Bailey, F.R. Cruickshank, R. Guthrie,
D. Pugh, I.J.M. Weir

Department of Pure and Applied Chemistry
University of Strathclyde
Glasgow, Scotland

The details of the short time (∿ μsec) rise of the pulsed source thermal lens signal have been analysed experimentally and theoretically. Experimental results, in which the non-homogeneous density field is detected by the thermal lens effect on a second, continuous mode, probe laser are presented. The theoretical analysis, based on characteristic times for pressure wave propagation and vibrational-translational (V-T) energy transfer for the absorber molecules is also presented. These two analyses are then compared and an attempt is made to extract rate constants for the SF_6/cyclopropane vibrational energy transfer process.

INTRODUCTION

The time dependence of the pulsed source thermal lens effect in gases has already been examined and interpreted in terms of the V-T release and the thermal conduction (1,2). These two processes respectively control the rise and decay of the signals. A rigorous description of the theoretical analysis can be found in the literature (3) but here the following summary will suffice.

The absorbed laser pulse of gaussian radial intensity produces a similar distribution of excited absorber molecules. This decays with a first order, V-T rate constant, $^1/\tau_S$. The heat released by this process produces an immediate pressure change and on

ISBN 0-12-066280-9

expansion of the gas, which occurs in a characteristic time, τ_p, a density variation occurs. The theoretical treatment (3) yields equations to calculate the density change in time and the radial curvature of density on the axis, $(\delta^2\rho|\delta r^2)_{r=o}$. This quantity governs the strength, S, of the thermal lens signal where

$S = 1 - [1 + A\ (\delta^2\rho|\delta r^2)_{r=o}]^{-2}$ where A is a time dependent constant.

This theoretical approach produces thermal lens risetime curves as predicted. At low pressure, since τ_s is inversely proportional to pressure, $\tau_s >> \tau_p$. Hence the curve is V-T energy transfer controlled and the resultant thermal lens curve rose to a maximum with negligible pressure effects. At high pressures the pressure limiting behaviour with $\tau_s << \tau_p$ is approached. At these high pressures the risetime is entirely determined by the pressure wave propagation and the characteristic feature of an "overshoot" is clearly seen. The theoretical predictions of variation of shape and strength of the thermal lens signals as briefly described above are those which are later compared with experimental results.

EXPERIMENTAL

The schematic diagram of the apparatus can be seen in Figure 1. This arrangement produces a CO_2 laser pulse, radius 0.718mm and width of 1.3μs (F.W.H.M.). The probe beam is a He/Ne laser of wave-length 632.8nm. The thermal lens cell, surrounded by an air thermostated bath controlled to ±0.5^{0}C, was designed for the temperature range 20-150^{0}C. The lens signals were detected by an EMI 9558 photomultiplier tube and recorded using a Biomation, model 1010 waveform recorder. An Apple II microcomputer was programmed to average the signals and perform the calculations. The experimental arrangement has been described in more detail in reference 2.

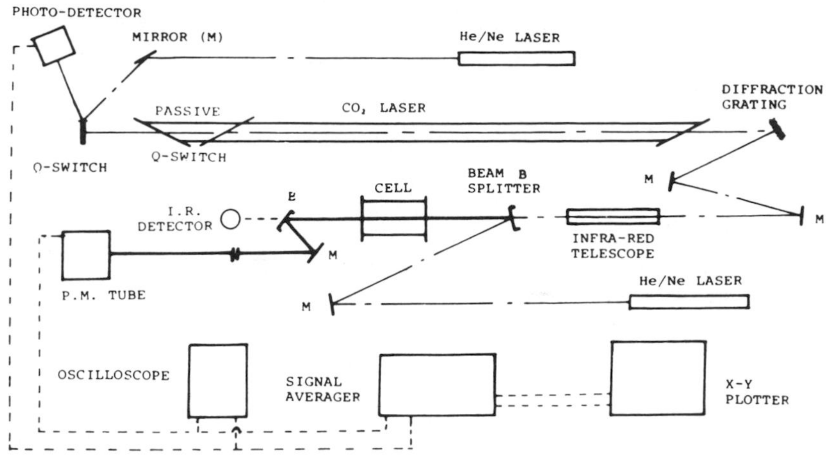

Figure 1. Schematic diagram of apparatus

RESULTS AND DISCUSSION

The apparatus as described was used to study the SF_6/cyclopropane system. Gas mixtures at 293, 317, 354 and 400K were studied. In each case several pressures in the range 25-600 torr were examined. At all times a constant SF_6 pressure of 1 torr was added as an absorber. At each temperature a series of curves was obtained. Those at 293K have already been presented (4). Risetimes at 354K are shown here. The time axis is calibrated in units of $S = {}^{k}/\tau_p$ where $\tau_p = {}^{Rg}/c$, with c = velocity of sound in gas mixture, Rg = radius of heating beam.

The behaviour is as predicted (see Figure 2). At low pressure the rise is V-T controlled and these curves are well reproduced by theory. This leads to accurate values of the V-T energy transfer rate constants. At higher pressures the characteristic limiting pressure wave form can be seen and good agreement is seen too beyond the overshoot. The further slow rise is probably due to a "tail" on the laser pulse not considered in the theoretical analyses.

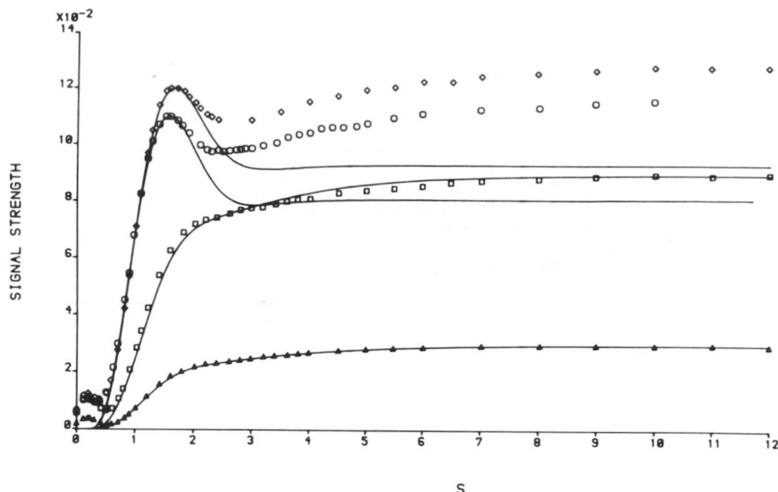

Figure 2. Results for SF_6/cyclopropane system at 354K. Symbols denote pressure in torr as follows:- △ = 26.15, □ = 42.1, ◇ = 206.4, O = 357.5. The points indicate experimental results and the solid lines attempts to fit the theory.

For a mixture of an absorber and buffer gas we expect the first order V-T rate constant to be expressable in the form

$$\frac{1}{\tau_s} = k_1 P(SF_6) + k_2 P(C_3H_6) \qquad (1)$$

where $P(SF_6)$ is a constant 1 torr. Hence a plot of values of $1/\tau_s$, obtained by fitting theoretical and experimental curves to each other, versus $P(C_3H_6)$ should be linear. Such a plot is shown in Figure 3. Since the processes involved depend only very weakly on T in this range, values at 317K and 354K are plotted together. The line drawn is a least squares fit to the pressures where no overshoot is seen.

This produced rate constants as follows:-

(1) intercept = k_1 = 86.1 $msec^{-1}$ $torr^{-1}$

(2) slope = k_2 = 3.85 $msec^{-1}$ $torr^{-1}$

The value of k_2 can be compared with the value of the V-T constant for pure cyclopropane of (8.0±1) $msec^{-1}$ $torr^{-1}$ (5).

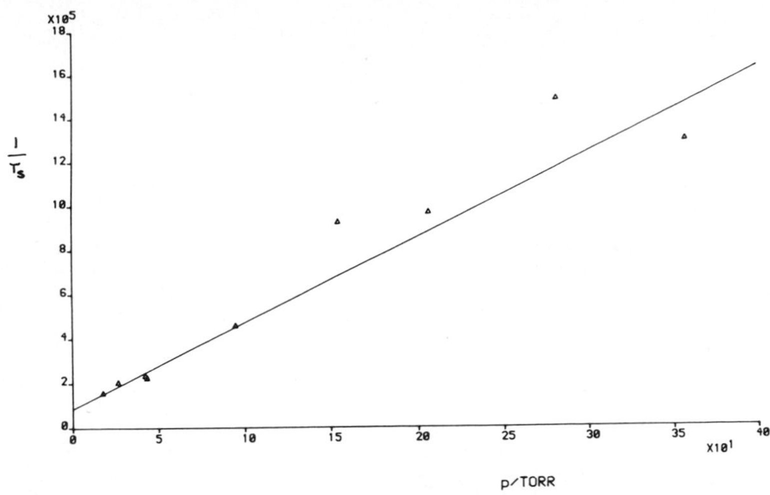

Figure 3. Results for the SF_6/cyclopropane system plotted according to equation (1).

REFERENCES

1. Bailey, R.T., Cruickshank, F.R., Pugh, D. and Johnstone, W., 1980, J. Chem. Soc. Faraday Trans. II, 76, 633.
2. Bailey, R.T., Cruickshank, F.R., Pugh, D. and Johnstone, W., 1981, J. Chem. Soc. Faraday Trans. II, 72, 1387.
3. Bailey, R.T., Cruickshank, F.R., Guthrie, R., Pugh, D. and Weir, I.J.M., Mol. Physics, in press.
4. Bailey, R.T., Cruickshank, F.R., Guthrie, R., Pugh, D. and Weir, I.J.M., 1982, J. Mol. Struct., 80, 433.
5. Fujimoto, G.T. and Weitz, E., 1979, J. Chem. Phys., 71, 5300.

SLOWING OF THE VIBRATIONAL DEPHASING TIME BY HYDROGEN BONDING IN GLYCEROL*

R. Dorsinville
W. M. Franklin
N. Ockman
R. R. Alfano

Ultrafast Spectroscopy and Laser Laboratory
Physics Department
The City College of New York
New York, N.Y.

The vibrational dephasing of the methylene CH_2 symmetric stretch mode in glycerol was directly measured over an extended temperature range using picosecond coherent Raman pump and probe spectroscopy. The dephasing time was found to increase dramatically as the temperature was lowered. This observation was attributed to the strong hydrogen bonding in glycerol which hinders the dephasing of the CH_2 vibration by the slowing down of molecular motions and fluctuations.

The knowledge of the vibrational and rotational motions of molecules in the condensed state is indispensable for a full understanding of the dynamics and interactions in the microscopic world of matter (1). A variety of experimental techniques have been employed to probe the dynamics of molecular motions. Recently, picosecond laser spectros-

*Research supported by NASA grant NAG 3-130

ISBN 0-12-066280-9

copy (2,3) has become available for directly measuring the kinetics of vibrational (3,4) and rotational (5) motion of molecules in the condensed state. Most of the picosecond work on vibrational dephasing in the liquid state has been pioneered by Laubereau and Kaiser (4). Direct measurement of the dephasing time of vibrations is achieved by first exciting the vibration of molecules by an intense picosecond laser pulse via stimulated Raman (6) or parametric beating (3,4). A weak probe pulse properly delayed with respect to the pump pulse can directly monitor the dephasing and population of the excited vibrations as a function of time (3,4). There have been several theoretical models proposed to account for dephasing (8-12). The mechanisms arise from a variety of interactions such as: Resonant energy transfer (7), collisions (8), hydrodynamic processes (9), coupling of high and low frequency modes (10), bath interactions, fast and slow repulsive and attractive interactions (11,12), and population changes, to name a few. The dephasing times predicted by these theories (8-12) were found in accord with experimental results of different simple non-associated, polyatomic liquids. Most of the direct measurements by picosecond laser spectroscopy have been concentrated on these types of liquids.

In this report, we present experimental data on the first direct measurement on the vibrational dephasing of a strongly associated liquid, glycerol, over an extended temperature range. Experimental evidence is presented which supports dephasing interactions in the strongly associated liquid are hindered by hydrogen bonding.

EXPERIMENTAL METHOD

Our experimental setup which is similar to the arrangement of Laubereau and Kaiser (4) is schematically shown in

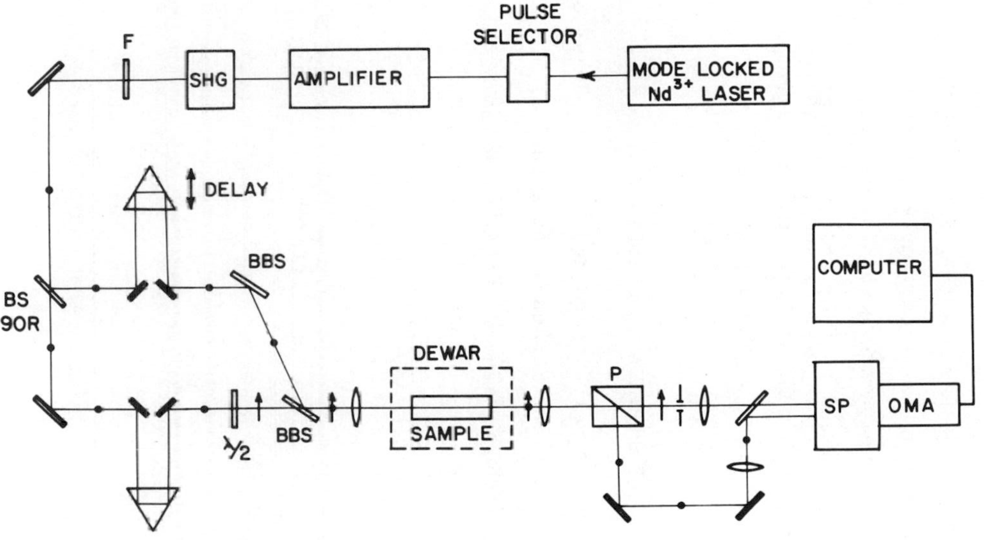

FIGURE 1. Picosecond coherent Raman pump and probe setup for dephasing time measurements.

fig. 1. A mode-locked Nd:glass oscillator, single pulse selector, three amplifiers, and a KDP SHG produces a 4 ps pulse at 530 nm. Various mirrors, prisims, beam splitters, and a $\lambda/_2$ plate divide up the 530 nm pulse into a pump pulse (90%) and a probe pulse (10%) polarized perpendicularly. Both the pump and probe pulses are collected and travel collinearly through a 10 cm sample, which was located in a dewar. When the temperature was lowered from 300 to 200 K the viscosity changed by four orders of magnitude. The symmetric CH_2 stretching mode at 2886 cm^{-1} underwent stimulated Raman scattering. The dynamic range was typically 100 to 1 ratio.

EXPERIMENTAL RESULTS

The dephasing times were determined from the decay of the profile of the coherent Stokes probe signal as a function of delay time between the pump and probe laser pulses. Table I lists the measured values of the dephasing times of the 2886 cm^{-1} of CH_2 vibration as a function of temperature from 300 to 200 K. As displayed in table I, the

TABLE I. Dephasing Times at different Temperatures for Glycerol

T_2 (psec)	T(K)
4 ± 1	290
9 ± 1	253
14 ± 2	238
16 ± 2	218
24 ± 4	203

vibrational dephasing time increases as the temperature decreases. In fig. 2, the dephasing time is plotted versus the inverse of the temperature on semi-log graph paper. The data is fitted to the equation:

$$T_2 = T_2(0)\ \exp\ (E_{AC}/kT)$$

where T_2 is the dephasing time, E_{AC} is the activation energy, T is the temperature, and k is the Boltzmann constant.

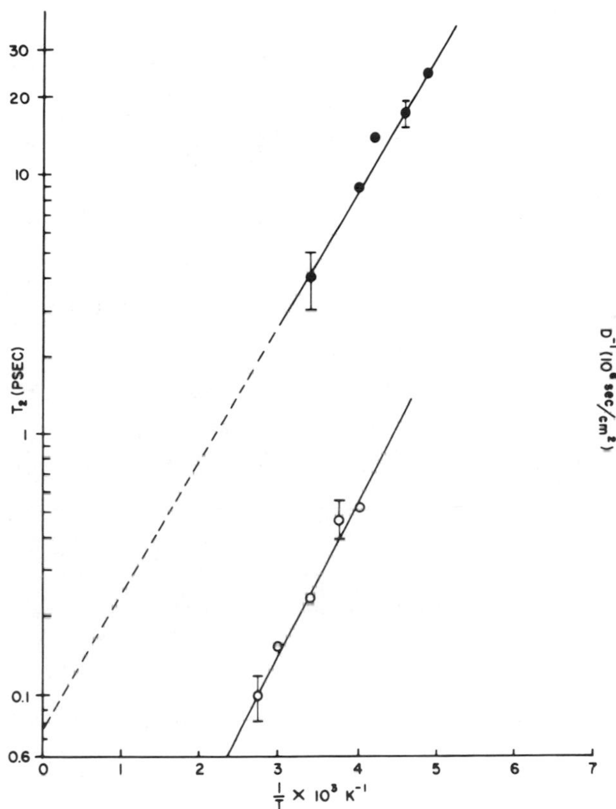

FIGURE 2. Dephasing time versus the inversus temperature (.); and inverse of the diffusion coefficient versus the inverse temperature (o) (13).

The solid line in fig. 2 represents the fit to eq. 1 yielding an energy $E_{AC} = 800\ cm^{-1} = 2.5 \pm 0.8$ Kcal/mol and dephasing time $T_2(0) = 80$ fsec at $T^{-1} = 0$.

DISCUSSION

The glycerol molecule has three hydroxyl groups per molecule which can hydrogen bond with other glycerol molecules to form a polymeric aggregate network which causes the high viscosity of this liquid. In fig. 2, the reciprocal of the rotational diffusion coefficient, D^{-1}, of glycerol is plotted from data of Larsson and Dahlborg (13). The salient feature of these curves is that the slope of T_2 and D^{-1} are parallel. The activation energy obtained from the slope of T_2 vs T^{-1} by fitting the data in fig. 2 is in excellent agreement with the hydrogen bond activation energy (13). The infinite temperature dephasing time $T_2(0) = 80$ fsec gives a measure of the correlation time of fluctuation without local restrictions.

Because of the high viscosity of glycerol at low temperatures, hard collisions can be neglected. The dephasing of CH_2 is most likely due to the participation of rotational rocking, torsions, and other low steric frequency modes (10). Our results strongly suggest that the dephasing interactions of the CH_2 stretch vibrational mode with low order molecular modes are significantly reduced by hydrogen bonding; that is, the hydrogen bonding inhibits and restricts the motion of the CH_2OH group - supporting the energy exchange mechanism (10). Our conclusions are further supported by the Raman measurements by Wang and Wright (14) which showed that the low frequency modes are inhibited by hydrogen bonding. Upon cooling the sample, the hydrogen bond breakage diminishes causing the formation of large aggregates which slow down local molecular motions and environmental fluctuations. This result accounts for the observed large increase in the vibrational dephasing time with decreasing temperature. This effect seems to counterbalance the expected shortening of the dephasing time due to the inhomogeneous broadening of the CH_2 linewidth.

ACKNOWLEDGMENT

We thank Dr. R. Lauver for continued interest in the research.

REFERENCES

1. S. Bratos and R. M. Pick, editors Vibrational Spectroscopy of Molecular Liquids and Solids, NATO Series B: Physics, Vol. 56, Plenum Press New York (1980).
2. A. J. DeMaria, D. A. Stetser, W. H. Glenn, Science 156 3782 (1972).
3. R. R. Alfano and S. L. Shapiro, Scientific American 228, 42 (1973); Phys. Rev. Lett. 26, 1247 (1971) and R. R. Alfano, Bull. Am. Phy. Soc. 15, 1324 (1970).
4. A. Laubereau and W. Kaiser, Rev. Mod. Phys. 50, 3 (1978) and D. Von der Linde, Ultrashort Light Pulses, editor S. L. Shapiro, Springer-Verlag, New York (1977)
5. P. P. Ho and R. R. Alfano, J. Chem. Phys. 67, 1004 (1977); 68, 4551 (1978); 74, 1805 (1981).
6. N. Bloembergen, American J. Phys. 35, 11 (1967); R. L. Carmen, F. Shimizu, C. S. Wang, N. Bloembergen, Phys. Rev. A 2, 60 (1970); and J. A. Giordmaine, W. Kaiser, Phys. Rev. 144, 676 (1966).
7. R. Abbott and D. E. Oxtoby, J. Chem. Phys. Lett. 70, 470 (1979).
8. S. F. Fisher and A. Laubereau, Chem. Phys. Lett. 35, 1 (1975).
9. D. W. Oxtoby, J. Chem. Phys. 70, 6 (1979).

10. S. Marks, P. A. Cornelius, C. Harris, J. Chem. Phys. 65, 8 (1976); S. M. George, H. Auweter, C. B. Harris, J. Chem. Phys. 73, 5573 (1980).
11. W. Rothschild, J. Chem. Phys. 65, 1 (1976); J. Chem. Phys. 65, 8 (1976).
12. K. S. Schweizer and D. Chandler, J. Chem. Phys. 76, 5 (1982).
13. K. E. Larsson, O. Dahlborg, Physica 30, 1561 (1964).
14. C. H. Wang and R. B. Wright, J. Chem. Phys. 55, 3300 (1971).

TIME RESOLVED V-T ENERGY TRANSFER MEASUREMENTS In IODINE MONOFLUORIDE†

P. J. Wolf*, R. F. Shea, and S. J. Davis

Air Force Weapons Laboratory
Kirtland AFB, NM

As part of a program to examine potential visible laser candidates we have been developing a technique for measuring energy transfer cross sections in the $B^3\Pi(0^+)$ state of interhalogen molecules[1]. Knowledge of V-T, V-V, and R-T transfer cross sections is essential for understanding the processes that are important in interhalogen $B^3\Pi(0^+) \rightarrow X^1\Sigma^+$ lasers. The technique consists of state selective time resolved laser induced fluorecence measurements. In this paper we describe the technique and give preliminary results for IF.

EXPERIMENT

A block diagram of the apparatus is shown in Figure 1. The IF molecules were produced in the reaction chamber by reacting I_2 with F_2. Laser excitation was provided by a Quanta Ray DCR1/PDL1 dye laser. Using Exciton Coumarin 480 dye, this laser

†Work supported by the Air Force Office of Scientific Research
*On assignment from the Air Force Institute of Technology, Wright Patterson AFB, Ohio

ISBN 0-12-066280-9

provided 1 mj pulses at 10 Hz with a pulse length of 8 nsec. The laser induced fluorescence was dispersed with a McPherson 0.3 meter monochromator whose output was coupled to a RCA model 31034 photomultiplier. The PMT output was fed into a Biomation 6500 transient digitizer which in turn was input into a Nicholet Lab-80 microcomputer.

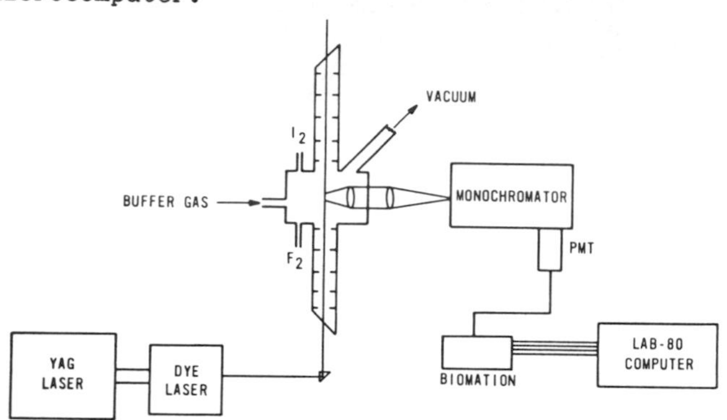

FIGURE 1. Block diagram of apparatus.

The experiment consisted of exciting a single (v', J') level in IF(B) with the dye laser. The monochromator was set to isolate a single v' → v" transition. This allowed the time evolution of the population of any chosen v' to be monitored. For the results reported here V-T transfer was studied for He as a bath gas.

RESULTS

A typical data trace when v' = 3 is excited by the dye laser is shown in Figure 2. The various curves depict the time evolutions of v' = 4, 2, 1, and 0 at a helium pressure of 3 torr. There are some qualitative features that can be observed from this figure. We note, for example, that the rate of production of v' = 1 is greatest when the population of v' = 2 is at its maximum. This implies that the population drains from v' = 3 to v' = 0 by a cascade through v' = 2 and v' = 1. There is also some upward transfer to v' = 4. Since the various curves represent raw data and have not been corrected for Franck-Condon factors, the relative magnitudes are misleading. The v' = 4 curve actually represents only a small (10%) amount of upward transfer. It is this process, however, which greatly complicates the analysis.

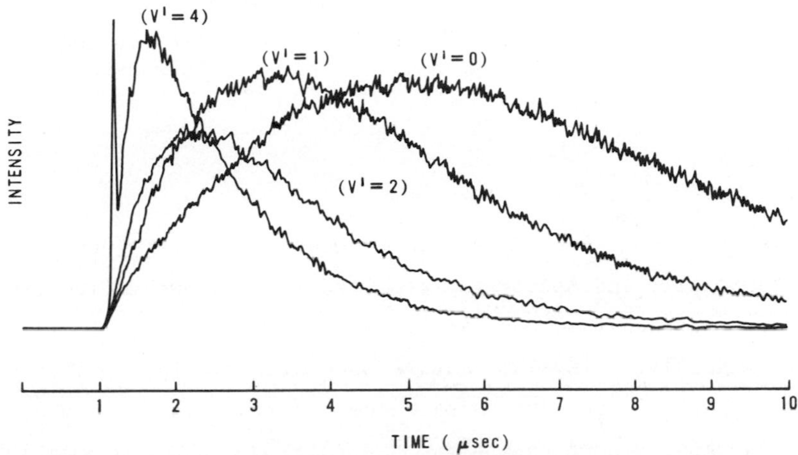

FIGURE 2. Fluorescence from various vibrational levels. Helium pressure is 3.0 torr.

In our analysis we have utilized the treatment of Montroll and Schuler which treats the V-T relaxation of a collection of harmonic oscillators. The model gives a closed form solution for the population of any v' level as a function of time. The only adjustable parameter is the relaxation rate which is determined by a least squares fit of the closed form solution to the data. By repeating the experiment at several He pressures one can obtain the transfer rate from a Stern-Volmer type plot. The results of this analysis slope give a rate constant of

$k_{4\to3}$ = (6.8 ± 0.5) x 10^{-12} cc/molecule-sec and $k_{3\ 2}$ = (5.0 ± 0.5) x 10^{-12} cc/molecule-sec. The ratio of $k_{4\to3}/k_{3\to2}$ = 1.36 is consistent with the relationship

$$k_{n\to n-1} = n\ k_{1\to0} \tag{1}$$

assumed by the Montroll and Schuler model. This treatment is more complex than that presented by Jurisich and Crim[3] for V-T transfer in HF. The difference in IF is that with its closer vibrational spacing there is considerable probability for $\Delta v = +1$ energy transfer while in HF this probability in negligible.

CONCLUSIONS

A technique has been developed to measure V-T rate constants using temporally and spectrally resolved laser induced fluorescence. It appears that one can follow the energy transfer in time through all v' levels. Future work will examine other collision partners and processes where the vibrational quantum number changes by more than ±1.

REFERENCES

1. S. J. Davis, L. Hanko, and R. F. Shea, (To be published in J. Chem. Phys).
2. E. W. Montroll and K. E. Schuler, J. Chem. Phys. 26, 454 (1957).
3. G. M. Jurisich and F. F. Crim, J. Chem. Phys. 74, 4455 (1981).

RAMAN SCATTERING FROM EXCITED STATES WITH LIFETIMES INTO THE FEMTOSECOND RANGE

M. Asano, L.V. Haley and J.A. Koningstein

Department of Chemistry
Carleton University
Ottawa, Canada

INTRODUCTION

In 1970 Shah, Leito and Scott[1)] reported the observation of a nonequilibrium phonon population from Stokes and anti-Stokes intensities of phonon lines if GaAs is exposed to cw radiation with a wavelength shorter than the band gap. The nonequilibrium was explained in terms of the emission of LO phonons when the electrons relax and these experimental data exhibit features of light scattering from non ground state systems. Early in 1977, Haley, Halperin and Koningstein[2)] described the observation of spontaneous Raman scattering from the excited E and $\bar{A}_1 + \bar{A}_2$ levels of ruby. Pumping and Raman probing was achieved with the radiation of one cw laser - monobeam pumping/probing technique. Later that year Wilbrandt et. al.[3)] recorded Raman scattering from the lowest lying triplet state of p-terphenyl in benzene. The triplets were prepared by electron radiation of a pulse of high energy electrons and a pulsed laser served as the Raman probe.

Experiments in our laboratory are aimed at recording light scattering spectra from short lived electronic excited states. The lifetime of the ^2E states of Cr^{3+} in αAl_2O_3 and Cr^{3+} in other hosts is for instance in the ms range and cw laser may be used to populate excited states with life times >1μs. Effective saturation is obtained if the number of photons delivered per life time τ is at least equal to the number of molecules in the irradiated volume. If conditions of effective population saturation exist, other laser photons are absorbed from the excited state. The consecutive absorption process and intersystem crossing) can be followed from a measurement of the laser intensity dependent value of the transmission of the sample.[4)] The

ISBN 0-12-066280-9

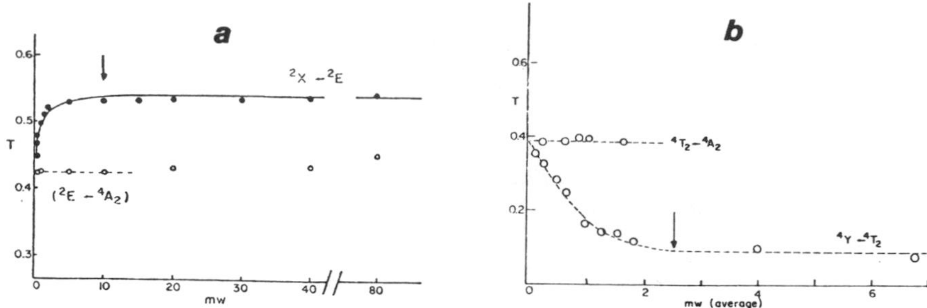

Fig. 1. Transmission studies at 457.9 nm for a) cw and b) pulsed radiation.

value of T increases or decreases if I_ℓ increases depending on decreases or increases in the absorption cross section (at λ_ℓ) of the transition from the excited state as compared to that of the transition from the ground state. Such monochromatic transmission studies enables one to obtain population numbers and lifetimes of the excited state. Also, the conditions for recording the Raman spectrum of the excited state follow from such a study - in particular if the energy level diagram is known. These conditions pertain to a) concentration b) absorption strength at λ_ℓ and c) laser power and if Raman probing and pumping is done with two different lasers, d) time dealy between pump and probe beams.

Results

In Fig. 1 we show the laser intensity dependent value of the transmission of an aqueous solution of Cr $(bpy)^{3+}$ (to which perchlorac acid was added to facilitate the solution of the salt) for cw and pulsed radiation at 457.9nm. One notes that a flux of $\sim 2.5 \times 10^{16}$ ph sec^{-1} causes effective saturation of a state with cw radiation. The transmission decreases if a 3.2×10^{-9} pulsed laser is used and another state is proposed and effective saturation occurs at 2.4mw which corresponds to 7.5×10^{22} ph sec^{-1}. This means that the lifetime of the state populated with the pulsed laser is reduced by a factor of 3×10^6. It is known that the λ_ℓ = 457.9nm of both lasers in resonant with $^4T_2$ of $Cr(bpy)^{3+}$ and that another lower lying ^2E state is at $\sim$14000cm^{-1} above the $^4A_2$ ground state. The $^4T_2 - {}^2E$ intersystem crossing is extremely effective >90%. The lifetime of ^2E is $\tau_{^2E} \sim 30\mu s$ for the solution in our experiment. This state can be populated with the photon flux of a cw laser. If this happens and $^4T_2$

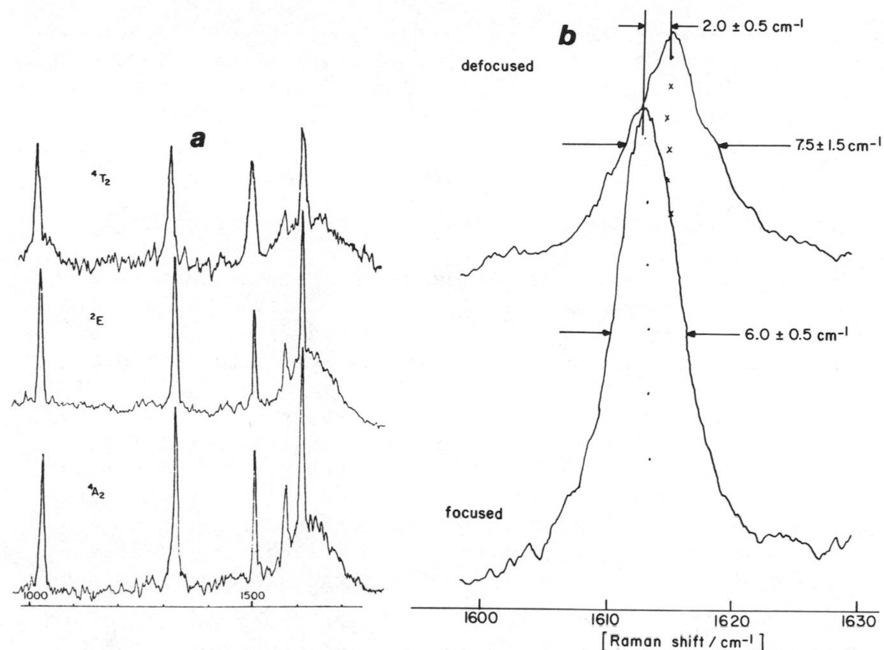

Fig. 2. Vibrational Raman scattering a) of 6ps $^4T_2$; 30 µs ^2E and $^4A_2$ ground state of $Cr(bpy)_3^{3+}$ and b) H comparison of positions of corresponding modes of $^4A_2$ (top) and ^2E (bottom) of $Cr(bpy)_3^{3+}$.

becomes populated with the pulsed laser then one calculates $\tau_{^4T_2} \sim 6$ps, which value is in good agreement with earlier predictions. On the other hand, $^2E\ Cr(bpy)_3^{3+}$ is rather reactive and photoaquation to $Cr(bpy)_3H_2O^{3+}$ can take place while even deligation can take place (above pH = 6). The latter is not the case in our experiments because transmission measurements at small light signals following measurements with a focused beam - even after prolonged periods - reveals no change in T. If we assume that ^2E:$Cr(bpy)_2H_2O^{3+}$ becomes populated with the cw laser and $^2E\ Cr(bpy)^{3+}$ with the pulsed laser then one obtains a lifetime of 90 sec for the former. This value is not acceptable because τ_E of the complex without water is 30µs. Also, it should be remarked here that the transmission measurements are most sensitive to those species which are in the majority and we conclude that the recording of a Raman spectrum of the solution with power levels of 10-20 mW cw and 2.5mW pulsed radiation at 457.9nm results respectively in the presence of excited state scattering of primarily ^2E and $^4T_2$: $Cr(pby)_3^{3+}$. The recorded Raman spectra-with cw and pulsed lasers at laser power levels which favour the maximum numbers of created ^2E and $^4T_2$ states - are shown in Fig. 2 and frequencies etc. are listed in Table 1. Also shown in Fig. 1 is the value of the cross section of the Raman shift at 1040 cm^{-1} as a function of I_ℓ (cw case). The increase in ^2E excited state Raman scattering is clearly present and the cross sections become smaller. Because T increases with I_ℓ, we find that the cross section for the absorption at λ_ℓ from ^2E is smaller and so is the resonance enhancement factor of the intensity if compared to that of the ground state. The above experiments clearly demonstrates the relevance of the transmission studies in the assignment of excited level Raman transitions.

As an example of vibrational Raman scattering from a femto lived excited state we mention here our results on all trans β-carotene in CS_2. The solution is characterized by a strong absorption at 490nm with side bands at 460 and 520nm while the fluorescence spectrum peaks at 580nm. The o-o transition between the singlet ground state S_o and a singlet excited state S_1 occurs at 19488cm^{-1} and pumping into the vibronic side band (∿490nm) results in population of S_1. The cross section of the transition at 490nm from S_1 is much smaller than that from S_o at 490nm and one observes a dramatic increase - by more than three orders of magnitude - is the value of the transmission of a 1 x 10^{-5} molar solution of all trans β-carotene in CS_2. From the laser intensity dependent value of T at 490.0nm one obtains the approximate value of τ_{S_1} <1ps. Pulses from the 3.2ns tunable dye laser can cause population of S_1 because of the extreme large value of the molecular extinction coefficient ($\varepsilon_{490.0} \sim 2.5 \times 10^5$). Time resolved Raman spectra were recorded in the following way. As a result of rise and decay times of the PMT and gated amplifier, the 3.2ns pulse of the laser is broadened to ∿9ns. By adjusting the 2ns aperture of the boxcar points in time before and after the pulse we establish experimental conditions which favour the recording of ground state Raman spectra. This because optical pumping of the S_1 state is least effective there. On the other hand if the aperture is adjusted in time where the laser pulse has reached maximum value $t = t_p$, optical pumping is most effective. From the behaviour of T verses I_ℓ we conclude that steady state conditions exists and theory suggest that $\gtrsim$65% of all molecules in the irradiated volume are in S_1. In Fig. 4, part of the time resolved vibrational Raman spectrum is shown. One observes a broadening of the band for $t = t_p$ while there does not appear to occur a shift. Frequencies and line widths of other modes are shown in Table I and one finds that the widths of bands recorded at $t = t_p$ is ∿11 + 2cm more than those recorded under conditions which favour ground state vibrational Raman scattering. This means that the lifetime of S_1 is >265 femtosecond. The excited Raman spectrum of the ps S_1 state of methyl

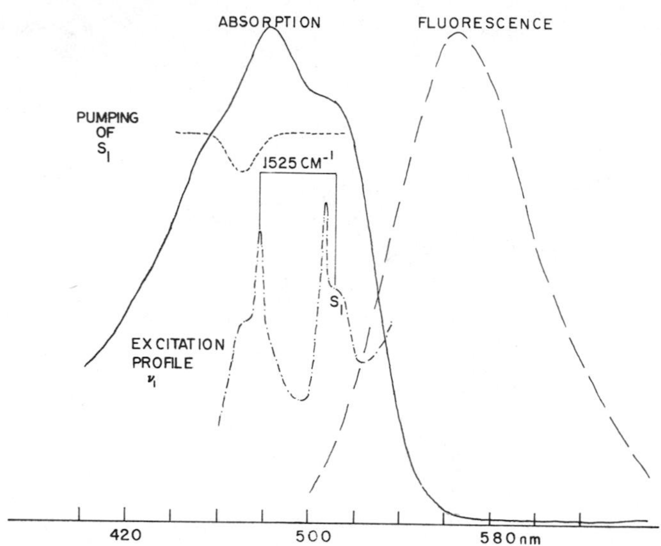

Fig. 3. Absorption, fluorescence and (time-resolved) excitation spectrum of the 1525 cm^{-1} mode of all trans β-carotene.

orange was recorded following the same procedure as outlined above - monochromatic transmission sutdies followed by monobeam pumping and Raman probing. Here we found changes by 5 cm^{-1} and the bands belonging to scattering of modes in S_1 did not show directly the effect of lifetime broadening.

A search for scattering from triplet states of methyl orange and all trans β-carotene was negative. The states are expected to have a long lifetime (>1μs), consequently the time evolution of the population of such states occurs at a point in time at which the intensity of the 3.2ns during laser pulse has vanishing intensity. Raman spectra of triplet all trans β-carotene were recorded after exposing the solution with an electron beam. Of interest here is the experimental result that weak Raman bands are observed in the cw excited resonance Raman spectrum of all trans β-carotene. The frequency of some of these bands are the same as those found of modes in the triplet spectrum. Such triplets could be populated if the samples are exposed to cw radiation and the vibrational Raman spectrum may become observable - see the results on $Cr(bpy)_3^{3+}$. The assignments of the triplet Raman spectrum can only be considered unambigious after the results of transmission experiments have become available.

Table I

Excited State Raman Transition

Compound	Lifetimes Excited State	Raman shifts cm^{-1}
$\alpha A\ell_2O_3$		29
$YA\ell G$		19
Cr^{3+} $ZnA\ell_2$ $_4$ $ms(^2E)$		6 electronic Raman transition $m^2E(\overline{E}2\overline{A})$
Beryl		64
		Vibrational Raman
$Cr(bpy)_3^{3+}$	30 μs (^2E)	1039.8; 1324.8; 1504.2; 1571.5; 1600.0 cm^{-1}
	6 ps $(^4T_2)$	1038.5; 1325.4; 1502.2; 1571.0; 1610.0 cm^{-1}
	Ground State S_0	1041.0; 1326.6; 1505.0; 1573.0; 1610.0 cm^{-1}
Methyl Orange	~10 ps S_1	1180; 1194; 1274; 1295; 1375; 1408; 1498; 1600; 1618 cm^{-1}
	Ground State S_0	1185; 1194; 1274; 1295; 1375; 1408; 1508; 1607; 1629 cm^{-1}
All trans	260fs S_1	1000; 1130; 1159; 1193; 1214; 1524 cm^{-1}
β carotene	Ground state S_0	1003; 1130; 1157; 1193; 1214; 1524 cm^{-1}

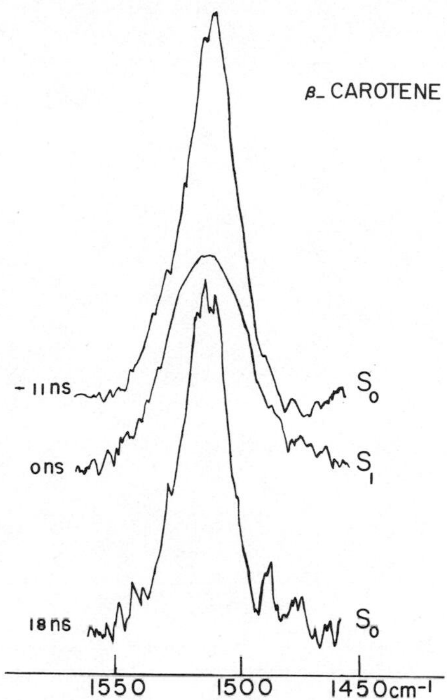

Fig. 4. Time-resolved vibrational Raman scattering from the 265 fs S_1 and S_0 ground state of all trans β-carotene.

References

1) J. Shah, R. C. C. Leite and J. F. Scott, Solid State Commun. 8, 1089 (1970).

2) L. V. Haley, B. Halperin and J. A. Koningstein, Chem. Phys. Letters, 54, 389 (1978).

3) R. Wilbrandt, N. H. Jensen, P. Pagsberg, A. H. Sillesen, K. B. Hansen, Nature, 276, 167 (1978).

4) B. Halperin and J. A. Koningstein, Can. J. Chem., 59, 2792 (1981).

RESONANCE RAMAN SPECTROSCOPY OF THE ELECTRONICALLY EXCITED STATES OF TRANSITION METAL BIPYRIDINE COMPLEXES

W. H. Woodruff[1]
R.F. Dallinger
M.Z. Hoffman
P.G. Bradley
D. Presser
V. Malueg
R.J. Kessler
K.A. Norton

Department of Chemistry
The University of Texas at Austin
Austin, Texas

Time-resolved resonance Raman (TR^3) studies of the electronically excited states of transition metal complexes having the general formula $M(bpy)_3^{n+}$ (bpy = 2,2'bipyridine) are reported. Depending upon the identity of M^{n+}, the nature of the lowest energy excited state may be ligand field ($M^{n+} = Cr^{3+}$), charge transfer ($M^{n+} = Ru^{2+}$, Os^{2+}) or $\pi \rightarrow \pi^*$ ($M^{n+} = Rh^{3+}$, Ir^{3+}). Thus these systems offer a rich opportunity for systematic TR^3 study of most of the types of excited states which are important in condensed-phase photochemistry.

I. $Ru(bpy)_3^{2+}$

The first systems in this class to be studied in our laboratories (1), indeed the first transition metal excited states to be studied by TR^3 in solution, were the metal-to-ligand charge-transfer (MLCT) electronic states of $Ru(bpy)_3^{2+}$, $Os(bpy)_3^{2+}$, and related complexes (1,4). The photochemical and photophysical properties of $Ru(bpy)_3^{2+}$ have been the subject of intense recent

[1]Author to whom correspondence should be addressed. Due to the obligatorially abbreviated nature of this paper much detail is omitted and this will be supplied upon request.

ISBN 0-12-066280-9

interest (7-13) due to the potential of this complex in photochemical energy conversion, its interesting photophysics, and its attractive chemical properties. The structures of the excited states of this complex are clearly of great interest.

At least two possibilities for the structure of the MLCT state exist. It may be formulated as $Ru(III)(bpy)_2(bpy^{\bar{\cdot}})^{2+}$, which has maximum symmetry of C_2, or the heretofore commonly presumed $Ru(III)(bpy^{-1/3})_3^{2+}$, which may have D_3 symmetry. We shall refer to the former structure as the "localized" model of the excited state, and the latter as the "delocalized" model. The experimental details of this study are presented elsewhere (4).

The electronic absorption and emission spectra of the ground state and MLCT excited state of $Ru(bpy)_3^{2+}$ have been determined (see ref. 4 for details). The ground state absorption in the near ultraviolet is considerable. Therefore, laser pulses in this wavelength range are suitable for generating good yields of the excited state. Also, the resonance enhancement of Raman scattering in the MLCT excited state should be very favorable because these wavelengths are near the excited state absorption maxima at 360 nm and 430 nm (see Fig. 3 or 4 for the $^*Ru(bpy)_3^{2+}$ absorption spectrum).

The excited state TR^3 results are obtained by illuminating the sample with a 5-7 ns duration laser pulse which contains greater than tenfold excess of photons over solute molecules in the illuminated volume. The excited state lifetime, ca. 600 ns, is much longer than the laser pulse width. Therefore, the absorbed photons from the laser pulse produce essentially complete saturation of the excited state. Other photons from the <u>same laser pulse</u> are Raman scattered by the sample, predominately by the excited state because of the extreme saturation condition. Complementary ground state spectra were obtained using c.w. ultraviolet illumination (350.7 or 356.4 nm) from a krypton ion laser.

The c.w. and TR^3 spectra of $Ru(bpy)_3^{2+}$ are compared in Fig. 1. The peak at 984 cm^{-1} in Fig. 1 is the internal intensity reference, the symmetric stretching vibration of $SO_4^{=}$ (0.50 <u>M</u> Na_2SO_4 added). The spectra were obtained using a Shriver rotating cell (14) and 135° backscattering illumination geometry. All of the Raman peaks are polarized in both the c.w. and TR^3 spectra.

The ground state spectrum exhibits typical features of the Raman spectrum of a bipyridine complex (1). Seven relatively intense peaks dominate the spectrum, representing predominately the seven symmetric C-C and C-N stretches of bipyridine wherein the two pyridine rings are related by symmetry.

The TR^3 spectrum of the MLCT state is shown in Fig. 1 (lower trace). This MLCT species has been variously denoted "triplet charge-transfer" (^3CT) (9), "$d\pi^*$" (7) or, as we shall use, simply $^*Ru(bpy)_3^{2+}$ (8). The lifetime of this state is ca. 600 ns (room temperature, aqueous solution). The TR^3 spectrum exhibits increased complexity compared to the ground state RR spectrum (Fig. 1, upper trace). Luminescence intensity measurements as a func-

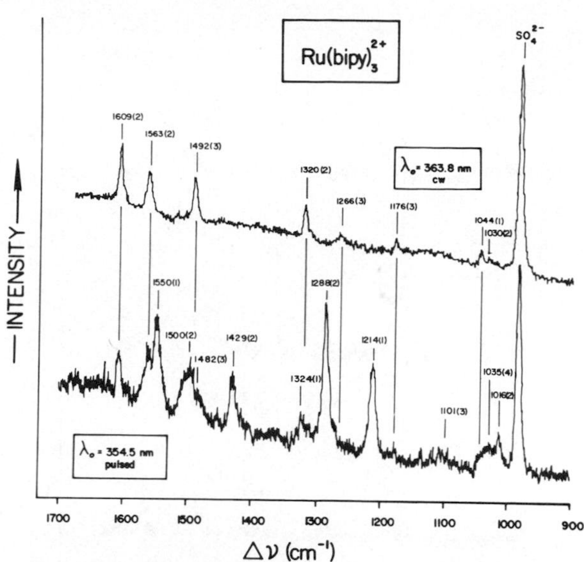

FIGURE 1. Comparison of the c.w. and TR^3 spectra of Ru(bpy)$_3^{2+}$ (top) and *Ru(bpy)$_3^{2+}$ (bottom).

tion of laser pulse energy confirm that the *Ru(bpy)$_3^{2+}$ state is greater than 90% saturated under the TR^3 illumination conditions in Fig. 1. Therefore, no TR^3 peaks observed are due to the ground state. It can be shown (4) that two sets of bpy frequencies are being observed in *Ru(bpy)$_3^{2+}$, one set at approximately the ground state frequencies and another set at considerably lower frequencies. This observation alone allows rejection of the possibility of D_3 symmetry for *Ru(bpy)$_3^{2+}$.

If the "localized" formulation of the structure *Ru(bpy)$_3^{2+}$ as Ru(III)(bpy)$_2$(bpy$^{\bar{\cdot}}$)$^{2+}$ is realistic, the resonance Raman spectrum of *Ru(bpy)$_3^{2+}$ can be predicted. A set of seven prominent symmetric modes should be observed at approximately the frequencies seen in Ru(III)(bpy)$_3^{3+}$. A second set of seven prominent Raman modes at frequencies approximating those of bpy$^{\bar{\cdot}}$ should also be observed. This prediction has been shown to be correct (4). The seven *Ru(bpy)$_3^{2+}$ peaks which show substantial (average 60 cm^{-1}) shifts from the ground state frequencies may be correlated one-for-one with peaks of Li$^+$(bpy$^{\bar{\cdot}}$) with an average deviation of 10 cm^{-1}. It appears clear that the proper formulation of *Ru(bpy)$_3^{2+}$ is Ru(III)(bpy)$_2$(bpy$^{\bar{\cdot}}$)$^{2+}$.

These TR^3 results demonstrate that the localized model of *Ru(bpy)$_3^{2+}$ is valid on the timescales of electronic motions and molecular vibrations. It is virtually certain that delocalization occurs on some longer timescale. The present experiments are mute as to the timescale on which delocalization may occur. It is possible that either time-resolved EPR or temperature de-

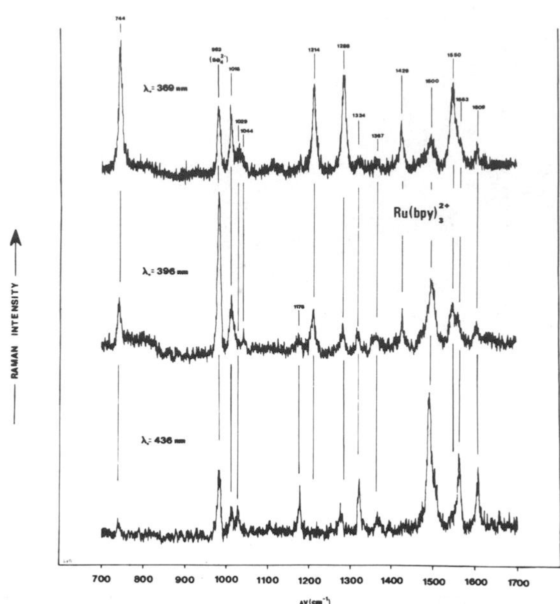

FIGURE 2. Comparison of TR^3 spectra of *Ru(bpy)$_3^{2+}$ acquired using laser excitation at 436 nm, 396 nm and 369 nm.

pendent fluorescence depolarization experiments may establish the timescale of delocalization in *Ru(bpy)$_3^{2+}$.

In this system, we have been able to observe the first resonance Raman excitation profile of an electronically excited state. The availability of a number of stimulated Raman shifted (SRS) frequencies between 340 nm and 460 nm, and also tunable sum frequencies adding the Nd:YAG fundamental to the Rhodamine 6G dye laser fundamental, allows the acquisition of the excitation profile of *Ru(bpy)$_3^{2+}$. The SRS frequencies were generated by focussing the laser beam inside a cell containing ca. 100 psi of H_2 or D_2 gas. Representative TR^3 spectra at three different wavelengths are shown in Fig. 2. It is seen that all of the "bpy$^{\bar{\cdot}}$" Raman peaks show greatly increased intensity at the short excitation wavelength, while the "neutral bpy" peaks increase in relative intensity at the long excitation wavelength. Three of the "bpy$^{\bar{\cdot}}$" modes in the excited state also show long-wavelength enhancement, and these are analogous to three modes of chemically reduced bipyridine radical anion which are also enhanced with long-wavelength excitation (17).

The resonance Raman excitation profiles are shown in Figs. 3 and 4. The radical modes clearly peak with the electronic transition of *Ru(bpy)$_3^{2+}$ at ca. 360 nm (Fig. 3), demonstrating a bpy$^{\bar{\cdot}}$ $\pi^* \rightarrow \pi^*$ assignment for this absorption. The neutral bpy modes peak with the weaker transition at ca. 430 nm (Fig. 4), suggesting

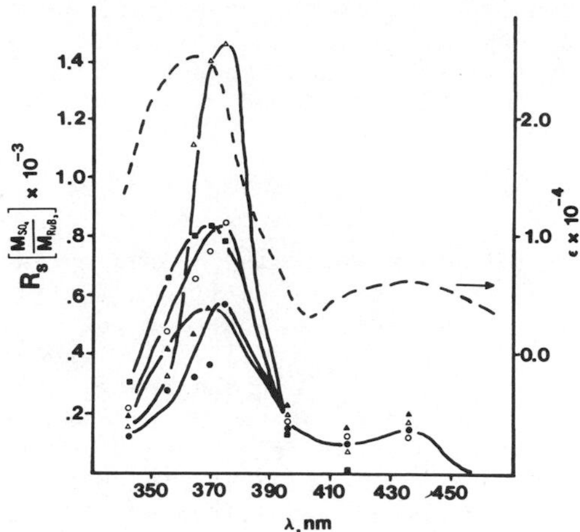

FIGURE 3. RR excitation profiles of the "$bpy^{\overline{\cdot}}$" modes of $*Ru(bpy)_3^{2+}$. Legend: △ = 744 cm^{-1}; o = 1214 cm^{-1}; ■ = 1288 cm^{-1}; ● = 1429 cm^{-1}; ▲ = 1550 cm^{-1}. The dashed line is the electronic absorption spectrum of $*Ru(bpy)_3^{2+}$.

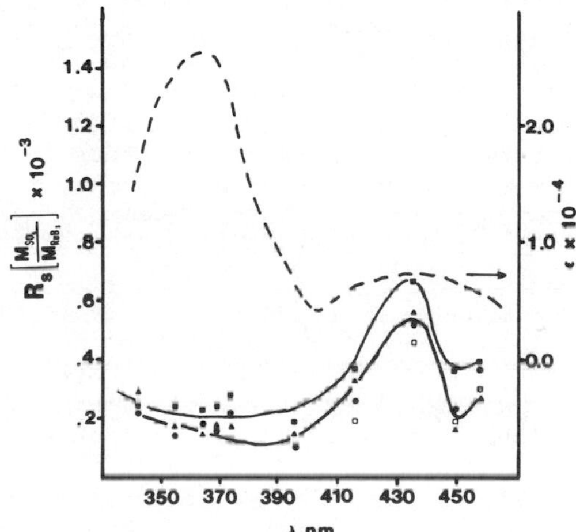

FIGURE 4. RR excitation profiles of the "neutral bpy" modes of $*Ru(bpy)_3^{2+}$. Legend: □ = 1176 cm^{-1}; ● = 1320 cm^{-1}; ■ = 1566 cm^{-1}; ▲ = 1609 cm^{-1}. The dashed line is the electronic absorption spectrum of $*Ru(bpy)_3^{2+}$.

that this absorption is actually the superposition of two $*Ru(bpy)_3^{2+}$ transitions, one being the Ru(III) LMCT mentioned above and the other a $bpy^{\overline{\cdot}}$ transition analogous to the long-wavelength absorptions of the chemically produced radical (17).

II. $Cr(bpy)_3^{3+}$

Unlike the MLCT situation which is the appropriate description of the lowest excited states of the Ru(II) and Os(II) polypyridine complexes, the lowest excited states of the analogous Cr(III) complexes are ligand-field in nature (18,19). The ground electronic state of $Cr(bpy)_3^{3+}$ is $^4A_2$, and the absorption of a photon produces the lowest spin-allowed state, $^4T_2$. The $^4T_2$ state probably has a lifetime of 10 ps or less (20), and undergoes intersystem crossing with near unit efficiency to the doublet manifold, promptly producing the thermally equilibrated $^2T_1$ and ^2E states which are approximately 13,000 cm^{-1} above the ground state. The ^2E state is the lower energy of this pair, therefore "^2E" is used hereinafter to denote the thermally equilibrated lowest doublet. The lifetime of this state is 63 μs (room temperature, deaerated aqueous solution) (18).

The photophysics of $Cr(bpy)_3^{3+}$ in aqueous solution is complicated by photochemistry which is dependent upon pH (18). Above pH 2, flash photolysis studies show the appearance of the ^2E state with its characteristic pH-independent lifetime, followed by a longer-lived "secondary transient" whose yield and lifetime both increase with pH. Below pH 6, the secondary transient decays with millisecond lifetimes to reform ground state $Cr(bpy)_3^{3+}$. Above pH 6, an increasing fraction of the secondary transient undergoes deligation to form $Cr(bpy)_2(OH)_2^+$. The limit overall quantum yield for this photodeligation reaction is 0.15 and is reached at approximately pH 9. The identity of the secondary transient has been suggested to be seven-coordinate $Cr(bpy)_3(H_2O)^{3+}$ (18). It is important for the forthcoming discussion to note that the extinction coefficient of the ^2E state at 457.9 nm is <u>greater</u> than that of the ground state (19) while $\varepsilon_{457.9}$ of the secondary transient is <u>smaller</u> (18) than that of the ground state.

Recent reports (21,22) have claimed observation of the TR^3 spectra of the excited states of $Cr(bpy)_3^{3+}$. The experiments were performed by simply dissolving ca. 10^{-3} M $Cr(bpy)_3(ClO_4)_3$ in water (22), therefore the pH, while neither specified nor controlled, could hardly have been lower than 5. Under these conditions, in addition to the photophysical transients, the photochemical "secondary transient" will form and have a lifetime of many milliseconds (18). The existence of the secondary transient was overlooked in these reports (21,22). It is clear, however, that the facile formation and relatively slow decay of the secondary transient precludes the possibility of saturating the ^2E

state, as claimed (21,22), under photostationary conditions by c.w. illumination. Indeed, in the saturation absorbance measurements (22) which purported to demonstrate c.w. saturation of ^2E, the observed negative absorbance change suggests instead that the secondary transient is the major steady-state species being produced (18,19). Additionally, the calculations (22) of the photon fluxes necessary to saturate the ^2E and $^4T_2$ states require a laser beamwaist diameter which is at least ten times smaller than the diffraction limit. For the foregoing and other reasons it is appropriate to reexamine the TR^3 spectroscopy of $Cr(bpy)_3^{3+}$ under conditions (pH = 0) wherein formation of the secondary transient is negligible.

We employed a number of laser illumination conditions in this study in order to span situations from unambiguous observation of the ground state to unambiguous observation of the saturated ^2E state. The laser focussing lens used had a focal length of 10 cm, the same as reported in ref. 22. The diffraction limited focal beamwaist diameter for our c.w. laser and this lens is 30μ (λ = 457.9 nm). We estimate our actual focussed spot diameter to be conservatively larger (100 μ) due to thermal lensing of the absorbing solution and other nonideal optical conditions. Average photon fluxes for our c.w. experiments at 457.9 nm and 356.4 nm ranged from 8 x 10^{-4} to 0.4 photons absorbed per molecule per ^2E lifetime (63 μ).

The pulse-excited spectra using the Nd:YAG third harmonic (354.7 nm) result in unambiguous saturation of the ^2E state because, on the average, every molecule in the illuminated volume encounters 15 photons during the 7 ns laser pulse and the ^2E lifetime is 63 μs. At the same time, this photon flux has no chance of producing any detectable concentration of the $^4T_2$ state, which has a lifetime of 10 ps or less. Indeed, even with the much higher photon flux available from our pulsed laser compared to that reported in ref. 22, no illumination conditions can be achieved which will result in appreciable population of the $^4T_2$ state with nanosecond-timescale excitation. This is because the $^4T_2$ state quantitatively and in picoseconds produces the ^2E state, which is a kinetic dead end on the nanosecond timescale with regard to further generation of quartet states. Only if the ^2E state returned to the ground state on the timescale of the laser pulse could an appreciable population of $^4T_2$ states be produced.

Five of the seven in-plane C-C and C-N stretching modes of bipyridine noted in the previous $Ru(bpy)_3^{2+}$ section retain appreciable intensity in $Cr(bpy)_3^{3+}$. The frequencies of these modes are listed in Table I for ground state (c.w.-illuminated) $Cr(bpy)_3^{3+}$ and for ^2E (pulse-excited) $Cr(bpy)_3^{3+}$. Within experimental error, there are no shifts in the frequencies of these modes between the ground and ^2E states. Formation of the ^2E state in the pulse-excited case is experimentally apparent only in changes in the relative intensities and linewidths of the Raman peaks between 356.4 nm c.w. illumination (ground state) and 354.7 nm pulsed

TABLE I. Resonance Raman Frequencies (cm^{-1}) for Ground State (C.W.-Excited) and ^2E (Pulsed-Excited) $Cr(bpy)_3^{3+}$ [a]

Ground state $Cr(bpy)_3^{3+}$ (average all c.w. excitation)	^2E $Cr(bpy)_3^{3+}$ (354.7 nm, 15 hν/molecule/pulse)
1040.0 ± 1.4	1040.0 ± 1.1
1324.7 ± 0.7	1325.1 ± 0.9
1503.2 ± 0.6	1502.2 ± 1.1
1569.8 ± 0.8	1568.5 ± 1.2
1607.6 ± 0.9	1608.3 ± 0.6

[a] $[Cr(bpy)_3^{3+}]$ = 2-12 m<u>M</u>, 1.0 <u>M</u> HCl.

illumination (^2E). The absence of frequency shifts in ligand modes as a result of production of the ^2E state of this complex is not surprising considering that both the ground state ($^4A_2$) and the ^2E state represent the same electronic configuration.

We suggest that the "^2E" TR^3 spectrum observed by c.w. illumination in the previous studies (21,22) is actually the "secondary transient" of Balzani, Hoffman, et al. (18), produced in photostationary yield in preference to the ^2E state because of the relative lifetimes involved. The bleaching observed as c.w. laser power is increased (22) is consistent with this suggestion, and specifically inconsistent with production of the ^2E state. We suggest that the "$^4T_2$" TR^3 result (22) is actually the ^2E state, with the small frequency shifts observed (22) either within experimental error or due to partial yield of "secondary transient" or photodeligated products. In this case, the increased absorbance as a function of increasing per-pulse laser energy (22) is consistent with the production of ^2E $Cr(bpy)_3^{3+}$.

III. CONCLUDING REMARKS

This study and others in this series (1-6) as well as results from other laboratories clearly establish TR^3 as the most versatile and generally applicable probe presently available which provides direct information on the structures of electronically excited states in fluid media, the relevant conditions for essentially all photobiology as well as much photochemistry and photophysics. Such structural information is crucial (a) to test (as in the present case) conclusions on excited state parameters drawn from less structure-specific experimental probes or from theoretical approaches, (b) to establish excited state potential surfaces experimentally under chemically relevant conditions, and (c) in general to understand the mechanisms whereby light is converted into chemical energy. Our efforts to extend both the temporal range and the chemical range of applicability of TR^3 and related laser spectroscopies are continuing.

ACKNOWLEDGEMENTS

The authors are grateful for support of this work by the National Science Foundation (grant CHE-8109541) and the Robert A. Welch Foundation (grant F-733).

REFERENCES

1. Dallinger, R.F. and Woodruff, W.H., J. Am. Chem. Soc. 101, 4391 (1979).
2. Woodruff, W.H. and Farquharson, S., Science 201, 831 (1978).
3. Dallinger, R.F., Farquharson, S., Woodruff, W.H. and Rodgers, M.A.J., J. Am. Chem. Soc. 103, 7433 (1981).
4. Bradley, P.G., Kress, N., Hornberger, B.A., Dallinger, R.F. and Woodruff, W.H., J. Am. Chem. Soc. 103, 7441 (1981).
5. Dallinger, R.F., Gaunci, J.J., Woodruff, W.H. and Rodgers, M.A.J., J. Am. Chem. Soc. 101, 1355 (1981).
6. Dallinger, R.F., Miskowski, V.M., Gray, H.B. and Woodruff, W.H., J. Am. Chem. Soc. 103, 1595 (1981).
7. (a) Hager, G.D. and Crosby, G.A., J. Am. Chem. Soc. 97, 7031 (1975); (b) Hager, C.G., Watts, R.J. and Crosby, G.A., ibid 7037; (c) Hipps, K.W. and Crosby, G.A., ibid 7042; (d) Crosby, G.A. and Elfring, W.H., J. Phys. Chem. 80, 2206 (1976).
8. Meyer, T.J., Acc. Chem. Res. 11, 94 (1978).
9. Balzani, F., Bolletta, F., Gandolfi, M.T. and Maestri, M., Topics Curr. Chem. 75, 1 (1978).
10. DeArmond, M.K., Acc. Chem. Res. 7, 309 (1978).
11. Sutin, N. and Creutz, C., "Advances in Chemistry Series, No. 138," Am. Chem. Soc., Washington, D.C., 1978, pp. 1-27.
12. Hipps, K.W., Inorg. Chem. 19, 1390 (1980).
13. Felix, F., Ferguson, J., Gudel, H.U. and Ludi, A., J. Am. Chem. Soc. 102, 4096 (1980).
14. Shriver, D.F. and Dunn, J.B.R., Appl. Spect. 28, 319 (1979).
15. Paskuch, B.J., Lacky, D.E. and Crosby, G.A., J. Phys. Chem. 84, 2061 (1980).
16. Motten, A.G., DeArmond, M.K. and Hauck, K.W., Chem. Phys. Lett. 79, 541 (1981).
17. Hornberger, B.A., Master of Science Thesis, The University of Texas at Austin, 1980; Woodruff, W.H. and Hornberger, B.A., manuscript in preparation.
18. Maestri, M., Bolletta, F., Moggi, L., Balzani, V., Henry, M. S. and Hoffman, M.Z., J. Am. Chem. Soc. 100, 2694 (1978).
19. Serpone, N., Jamieson, M.A., Henry, M.S., Hoffman, M.Z. and Maestri, M., J. Am. Chem. Soc. 101, 2907 (1979).
20. Kirk, A.D., Hoggard, P.E., Porter, G.B., Rockley, M.G. and Windsor, M.W., Chem. Phys. Lett. 37, 199 (1976).
21. Asano, M., Mongeau, D., Nicollin, D., Susseville, R. and Koningstein, J.A., Chem. Phys. Lett. 65, 293 (1979).
22. Asano, M., Koningstein, J.A. and Nicollin, D., J. Chem. Phys. 73, 688 (1980).

SPATIAL AND TIME-RESOLVED RESONANCE RAMAN SPECTRA OF EXCITED ELECTRONIC STATES OF THE URANYL ION

M. Asano and J.A. Koningstein

Department of Chemistry, Carleton University
Ottawa, Ontario, Canada

I. INTRODUCTION

The spectroscopic studies of uranyl ion in solutions or solid media have received considerable attention in the past. Particularly, the absorption, luminescence and Raman spectra were well studied (1), and the electronic and vibrational structures of the ion were discussed and reported in many places (2).

In this study, the ground level Raman (G.L.R.) and excited level Raman (E.L.R.) spectra of the uranyl ion are reported. The first excited state of the uranyl ion was populated significantly at the focused point of the pulsed laser beam which delivers 10^{22} $\sim 10^{23}$ photons per second. The scattered light from the ions at the focused point was carefully recorded and the Raman bands of the ion in the excited state and the ground state were identified. These G.L.R. and E.L.R. spectra were time-resolved, because the time-aperture of the boxcar was synchronized to a certain point in time within the duration of the excitation laser pulse. With this time-resolved Raman technique, the optical-pumping scheme of the uranyl ion energy system under high incident photon density will be discussed.

II. THE GROUND LEVEL AND EXCITED LEVEL RAMAN SPECTRA

Hydrolysis of the uranyl ion in aqueous solution has been studied in the past (3). According to these studies, three principal hydrolyzed species must exist in the solution; i.e., mononuclear (UO_2^{2+}), dinuclear($(UO_2)_2(OH)_2^{2+}$) and trinuclear($(UO_2)_3(OH)_5^{+}$) species. The concentrations of each species can be calculated as a function of pH of the solution from the equilibrium quotient

ISBN 0-12-066280-9

reported elsewhere (3).

A part of the pulsed laser Raman spectrum of the 0.020M uranyl nitrate aqueous solution is shown in Fig.1. The intensity level and the wavelength (447.4nm) of the pulsed laser were chosen so that an excited state could be populated (4). The laser beam was focused into the sample solution by a microscope objective lens. The tunable pulsed laser and the detection system were described elsewhere (4). The pH of the solution was adjusted by adding NaOH. The studies of the hydrolysis revealed that almost 100% of the mononuclear species was present when the pH was 0.9. The studies also indicated that 70% of the UO_2 units must be in the trinuclear species when the pH was 4.5 (3,4). Taking into account the vibronic band spacings observed in absorption and luminescence spectra of the solution at each pH (1,4), the 873cm^{-1} band of the pH 0.9 solution, the 856cm^{-1} band of the pH 3.6 solution and the 842cm^{-1} band of the pH 4.5 solution (Fig.1) were assigned to the $\nu_1$mode Raman bands of the UO_2 unit in the mono-, di- and trinuclear species of the uranyl ion, respectively. These Raman bands have been studied and reported by Toth and Begun (5). Although the spectra in the present study showed good agreement with their spectra in the 800-1000cm^{-1} region, the spectra presented here have different features in the 700cm^{-1} region. The Raman bands with the shifts of 704cm^{-1} and 716cm^{-1} shown in Fig.1 were not observed when the laser beam was not focused into the sample solution. The Raman spectra of the pH 4.5 solution recorded under three different optical set-ups are shown in Fig.2. The lower spectrum was record-

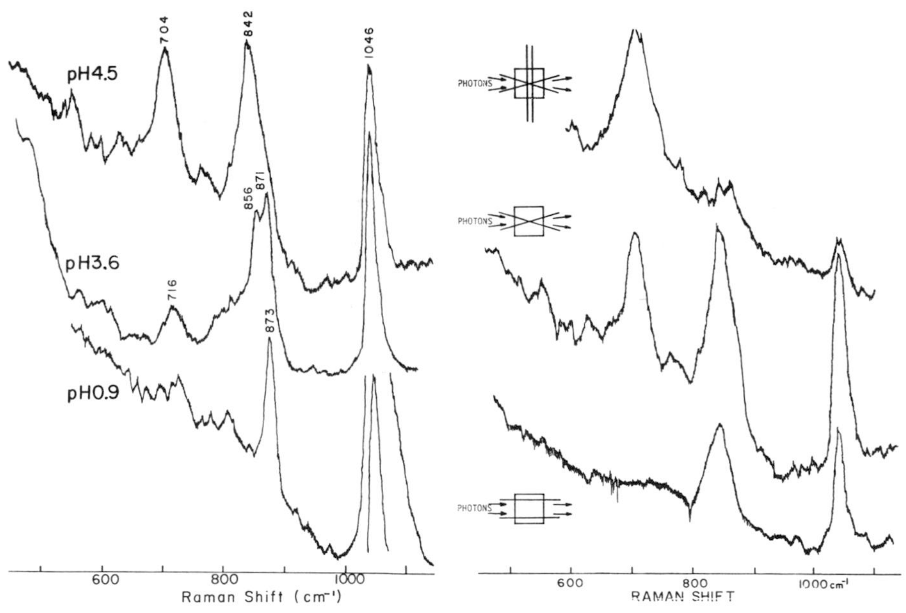

FIGURE 1. *FIGURE 2.*
Raman spectra of the uranyl aqueous solution at different pH's and different optical set-ups.

ed with the laser beam defocused and the middle spectrum with the beam focused into the solution (the same spectrum as the upper one in Fig. 1). A f=3.0cm microscopic objective lens was employed and the focal point was adjusted to be at the middle of the 1.0cm long quartz cell. The upper spectrum in Fig. 2 was recorded from scattered light which emerged from the focal point only. A spatial filter (a 0.03cm width slit) was placed in front of the sample cell, so that the scattered light from the defocused part of the beam was prevented from entering the detection system. By comparing these three spectra in Fig.2, it is evident that the intensity of the band at 842cm^{-1} (G.L.R.) was significantly reduced when only the light from the focal point was detected. On the other hand, the 704cm^{-1} band was not observed without detecting the light from the focal point. Relatively high population density of the excited state of the trinuclear species is expected in the focal point due to its high photon density. Thus, the 704cm^{-1} band was assigned to the vibrational E.L.R. band of the trinuclear species' ν_1 mode. In order to describe the E.L.R. in the context of the optical pumping scheme of the trinuclear species, a part of its energy level diagram is shown in Fig. 3. The assignments of the electronic states are based on those of UO_2^{2+} which have been reported as a part of the electronic structure of the ion that has been well studied (2).

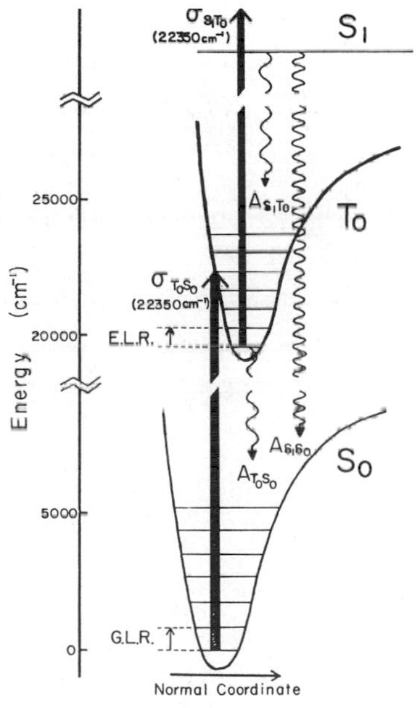

FIGURE 3. The energy level diagram of the trinuclear species of the uranyl ion.

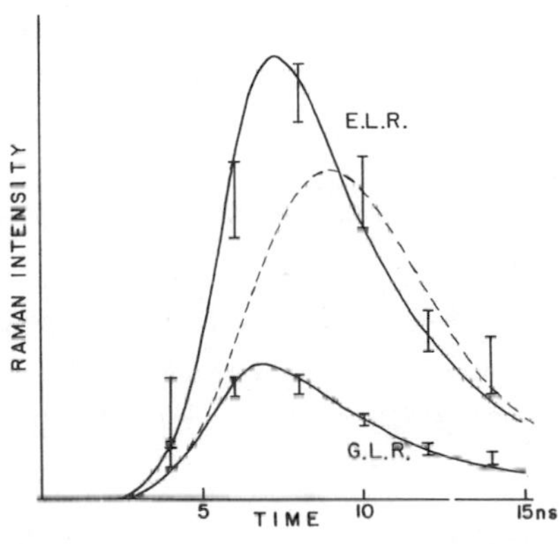

FIGURE 4. Time-resolved Raman band intensities of the ν_1 mode of the trinuclear species of the uranyl ion.

III. TIME-RESOLVED RAMAN SPECTRA

For the Raman spectra in Fig. 1 and 2, the gate (time-aperture) of the boxcar was simply fixed in time at the maxmum of the Raman scattering pulse. However, if the gate is time-scaned, the output signal of the boxcar is expected to reveal the shape of the Raman scattering pulse in time. The optical set-up with the microscope lens and the spatial filter (the upper spectrum of Fig. 2) was employed to to detect the Raman scattering pulse merely from the focused point of the beam. The observed G.L.R.($842cm^{-1}$) and E.L.R.($704cm^{-1}$) intensities from the trinuclear species were recorded at various points in time and shown by bars in Fig.4 with experimental uncertainties. A broken line in Fig.4 represents the excitation pulse of the laser. If one compares the pulse shape with the Raman pulse shapes, it can be found that both of the Raman pulses reach their maxima before the excitation pulse does. The Raman band intensity change in time can be linked to time-evolution of the ground and excited state population densities of the trinuclear species under the intense optical pumping.

The time-evolution of the population densities was calculated from a theoretical model assuming a simple three level system having two kinds of optical pumping and three kinds of relaxation processes as shown in Fig. 3. The G.L.B. and E.L.R. intensities must be proportional to these population densities of each state. This calculation also included deconvolution of a signal distortion due to the detection system time-delay. The absorption cross section of the secondary optical pumping ($\sigma_{S_1T_0}$, in Fig. 3) and the lifetime of the higher lying excited state ($\tau_{S_1}=1/(A_{S_1T_0} + A_{S_1S_0})$, in Fig. 3) were determined as curve-fitting parameters. The best fitted curves shown in Fig. 4 with solid curves were obtained with $\sigma_{S_1T_0} = 1.1 \times 10^{-18}$ cm^2 and $\tau_{S_1} = 1.5 \times 10^{-8}$ sec. The details of these calculations were described elsewhere (4).

References

1. E. Rabinowitch and R.L. Belford, "Spectroscopy and Photochemistry of Uranyl Compounds", Macmillan, N.Y. (1964).
2. J.T. Bell and R.E. Biggers, J. Mol. Spect. 18, 247 (1965); 22, 262(1967); 25, 312 (1968).
3. C.F. Baes and R.E. Mesmer, "The Hydrolysis of Cations", Wiley, N.Y., (1976) pp. 177-182.
4. M. Asano, Ph.D. Thesis, Chemistry Dept., Carleton University, Ottawa (1982).
5. I.M. Toth and G.M. Begun, J. Phys. Chem. 85, 547 (1981).

Time-Resolved Resonance Raman Spectra of Excited $^3B_2$ Phenanthrene

D. A. Gilmore and G. H. Atkinson

Department of Chemistry
Syracuse, University
Syracuse, N. Y.

I. INTRODUCTION

The value of time resolved resonance Raman (TR^3) spectroscopy for examining the vibrational degrees of freedom of transient photophysical intermediates has now been demonstrated for a variety of molecular systems (1-13). It has also been shown that TR^3 spectra can be utilized to obtain quantitative kinetic data on the formation and decay of photophysically-generated transients (12-13). TR^3 spectroscopy is especially useful for studying excited electronic states of polyatomic molecules at room temperature, many of which have not been directly observed previously. In this paper, we describe a study of the excited triplet $^3B_2$ electronic state of phenanthrene-h_{10} and -d_{10}. Specifically, the TR^3 spectra and a discussion of the influence of dissolved O_2 are presented.

ISBN 0-12-066280-9

II. EXPERIMENTAL

The experimental apparatus is in a pump-probe configuration using two pulsed dye lasers driven by a harmonic frequency of separate Nd:YAG lasers. The only significant difference from the apparatus described previously (12) is the use of a second dye laser for the initial pump excitation rather than the harmonic wavelength of the Nd:YAG output at 1.06 μ. The pump wavelength is 330 nm (7ns, FWHM pulse width), and probe wavelength is 485 nm (12ns, FWHM pulse width). The timing jitter is ±5ns. The phenanthrene-h_{10} was obtained from Aldrich and the phenanthrene-d_{10} was purchased from Stohler Isotope Chemicals. Both compounds were used without further purification. Sample solutions of $5x10^{-3}$M in phenanthrene were prepared in both spectrograde benzene and methanol.

III. RESULTS

The TR^3 spectra between 100 and 1700 cm^{-1} of electronically excited $^3B_2$ phenanthrene-h_{10} and -d_{10} in a benzene solvent are presented in Figures 1 and 2. These spectra were recorded 55 ns after the 330 nm excitation of phenanthrene into the S_1 state. The TR^3 bands assigned to the $^3B_2$ state of phenanthrene are labeled by their wavenumber positions while the unlabeled peaks derive from the normal Raman scattering of the solvent (i.e., benzene). The frequency positions of the bands in Figures 1 and 2 are given in Table I together with their relative intensities, polarizations, and depolarization ratios. Some of the TR^3 bands listed in Table I do not appear in Figures 1 and 2 because they are very weak and/or they are overlapped by the normal Raman bands of the benzene solvent. Analogous TR^3 spectra of $^3B_2$ phenanthrene using methanol as a solvent clearly revealed these obscured TR^3 peaks.

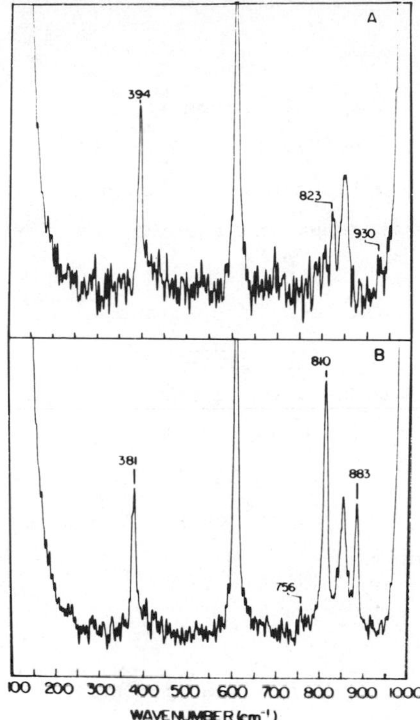

FIGURE 1. TR^3 spectra of electronically excited $^3B_2$ phenanthrene-h_{10} (A) and -d_{10} (B) in the 100-1000 cm^{-1} region. The wavenumber positions of Raman bands from $^3B_2$ phenanthrene are labeled. Unlabeled bands are due to the benzene solvent.

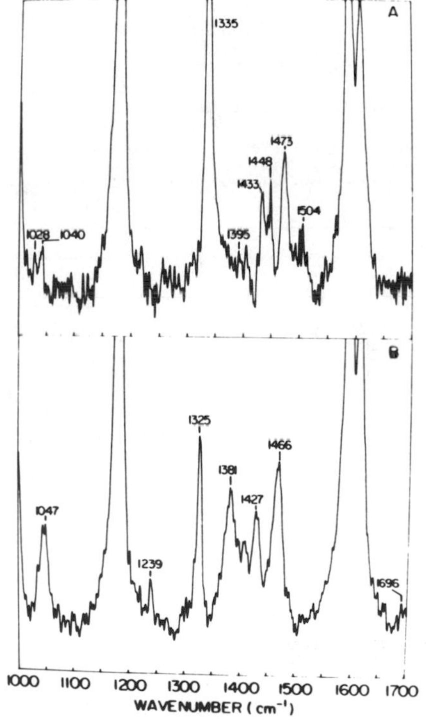

FIGURE 2. TR^3 spectra of electronically excited $^3B_2$ phenanthrene-h_{10} (A) and -d_{10} (B) in the 1000-1700 cm^{-1} region. The wavenumber positions of Raman bands from $^3B_2$ phenanthrene are labeled. Unlabeled bands are due to the benzene solvent.

TR^3 bands were identified specifically with the $^3B_2$ state of phenanthrene through their sensitivity to O_2 quenching, the frequency shifts observed upon deuteration, the absence of Raman bands from ground-state phenanthrene, and the time-resolved excitation profiles (14). The presence of O_2 has two effects since it is known to facilitate both $S_1 \rightarrow T_1$ and $T_1 \rightarrow S_0$ intersystem crossings (15). These phenomena are observed in the TR^3 spectra of O_2-saturated solutions as (1) an increase in the rate of appearance of $^3B_2$ bands, (2) a decrease in the fluorescence background, and (3) a decrease in the observed lifetime of T_1. The presence of O_2 did not affect the frequency positions of the TR^3 bands observed.

IV. DISCUSSION

The vibrational analysis of $^3B_2$ phenanthrene provided by the TR^3 spectra leads to several general characterizations with respect to ground-state phenanthrene. One finds some general correlations as well as distinct differences in the two types of spectra.

The major similarities are found in the frequency region normally assigned to the ring breathing and bending modes for aromatics (16, 17). Intense $^3B_2$ bands are observed for $-h_{10}$ and $-d_{10}$ at 1335 cm^{-1} and 1325 cm^{-1}, respectively while the analogous intense ground-state $-h_{10}$ and $-d_{10}$ bands appear at 1347 cm^{-1} and 1396 cm^{-1}, respectively. In the low frequency region, intense $^3B_2$ bands are seen at 394 cm^{-1} ($-h_{10}$) and 381 cm^{-1} ($-d_{10}$) while the analogous ground-state bands at 406 cm^{-1} ($-h_{10}$) and 392 cm^{-1} ($-d_{10}$) have similar relative intensities. The remaining intense bands found in the TR^3 $^3B_2$ spectrum occur at 981 cm^{-1} ($-h_{10}$) and 975 cm^{-1} ($-d_{10}$) for which there are no corresponding intense bands in the respective ground state spectra. Based on the normal coordinate assignments for these spectral regions, these results suggest that in the $^3B_2$ state there is increased motion in the ring vibrations comprised of C=C stretches and CCC bends over that found in the ground-state.

TABLE I Frequency positions (cm^{-1}) for $^3B_2$ phenanthrene-h_{10} and -d_{10} observed by time-resolved resonance Raman (TR^3) spectroscopy[a]

Phenanthrene-h_{10}				Phenanthrene-d_{10}			
Freq. (cm^{-1})	Rel. I	P[b]	$\rho_\perp$[c]	Freq. (cm^{-1})	Rel. I	P[b]	$\rho_\perp$[c]
394	(s)	P	0.30	381	(s)	P	0.30
				650[d]	(w)	-	-[f]
				756[d]	(vw)	-	-[f]
				810	(vs)	P	0.40
823	(m)	P	0.22	836	(vw)	-	-[f]
930	(w)	-	-[f]	883	(s)	P	0.24
981	(vs)	P	0.33	975	(vs)	P	0.41
1028[d]	(w)	P	-[e]				
1040	(w)	P	-[e]	1047	(m)	P	0.37
1166	(w)	P	-[e]	1183[d]	(vw)	-	-[f]
				1239	(m)	P	0.44
1335	(vs)	P	0.33	1325	(s)	P	0.38
1395	(w)	P	-[e]	1381	(m)	P	0.13
1433	(m)	P	0.27	1427	(m)	P	0.37
1448	(m)	P	0.25	1466	(s)	P	0.36
1473	(s)	P	0.31				
1504	(w)	-	-[f]				
				1565[d]	(vw)	-	-[f]
				1696[d]	(vw)	-	-[f]

a frequency positions accurate to ± 1 cm^{-1} except where noted

b P = polarization

c $\rho_\perp \equiv$ depolarization ratio ($=I_\perp/I_\parallel$); ±10%

d frequency positions accurate to ±2cm^{-1}

e band intensities permit measurement of polarization properties but not $\rho_\perp$

f band intensities too weak to permit measurement of polarization properties

Acknowledgement This research was supported by grants from the National Science Foundation.

References

1. Atkinson, G. H., C. R.-Conf. Int. Spectosc. Raman 7th 1980 182 (1980).

2. Atkinson, G. H., "Time-Resolved Raman Spectroscopy" in Advances in Infrared and Raman Spectroscopy, Vol. *9*, (R. J. N. Clark and R. E. Hester, eds.) p. 1. Heyden and Sons, London (1982).

3. Atkinson, G. H., "Time-Resolved Raman Spectroscopy" in Advances in Laser Spectroscopy Vol. *1*, (B. A. Garetz and J. R. Lombardi, eds.), p. 155. Heyden and Sons, London (1982).

4. Wilbrandt, R.; Jensen, N. H.; Pagsberg, P.; Sillesen, A. H.; Hansen, E. B. *Nature (London)*, *276*, 167 (1978).

5. Dallinger, R. F.; Woodruff, W. H., *J. Am. Chem. Soc.*,*101*, 4391 (1979).

6. Dallinger, R. F.; Guanci, J. J.; Woodruff, W. H.; Rogers, M. A. J. *J. Am. Chem. Soc.*, *101*, 1355, (1979).

7. Atkinson, G. H.; Dosser, L. R., *J. Chem. Phys.* *72*, 2195, (1980).

8. Asano, M.; Koningstein, J. A.; Nicollin, D., *J. Chem. Phys.* *73*, 688, (1980).

9. Jensen, N. H.; Wilbrandt, R.; Pagsberg, P. B.; Sillesen, A. H.; Hansen, K. B., *J. Am. Chem. Soc.* *102*, 7441,(1980).

10. Wilbrandt, R.; Jensen, N. H., *J. Am. Chem. Soc.*, *103*, 1036, (1981).

11. Atkinson, G. H.; Pallix, J. B.; Freedman, T. B.; Gilmore, D. A. and Wilbrandt, R., *J. Am. Chem. Soc.*, *103*, 5069,(1981).

12. Atkinson, G. H.; Gilmore, D. A.; Dosser, L. R., and Pallix, J. B., *J. Phys. Chem.*, *86*, 2305 (1982).

13. Atkinson, G. H. and Pallix, J. B., *J. Am. Chem. Soc.* (submitted for publication)

14. Gilmore, D. A. and Atkinson, G. H. (unpublished results)

15. Parmeter, C. S. and Rau, J. D., *J. Chme. Phys.* *51* 2242 (1969).

16. Schettino, V., Neto, N., Califano, S., *J. Chem. Phys.* *44*, 2724 (1966).

17. Schettino, V., *J. Chem. Phys.* *46*, 302 (1967).

TIME-RESOLVED RESONANCE RAMAN SPECTRA OF THE LOWEST TRIPLET STATE OF N,N,N',N'-TETRAMETHYL-P-PHENYLENEDIAMINE IN SOLUTION

Kenji Yokoyama[1]

The Institute of Physical and Chemical Research
Wako, Saitama, JAPAN

I. INTRODUCTION

N,N,N',N'-Tetramethyl-p-phenylenediamine (TMPD) is well known for the essential role played by the lowest triplet state (^{3}TMPD) in a biphotonic ionization to the cation radical ($TMPD^+$). The author studied the time-resolved resonance Raman spectra of this molecule and found a Raman band of ^{3}TMPD in solution at room temperature.

[1]Present Address: Resources and Environment Research Laboratories, NEC corporation, Miyazaki, Miyamaye-ku, Kawasaki, JAPAN

ISBN 0-12-066280-9

II. EXPERIMENTAL

An N_2 laser was used as an excitation source for preparing ^{3}TMPD and the frequency-doubled output of a Q-switched Nd:YAG laser was used as a Raman probe beam. The details of the system were reported previously (1). The output of a CW dye laser pumped with an Ar^+ laser and a Raman spectrometer were used in measurement of conventional Raman spectra. Deoxygenation of the sample solution was carried out through bubbling with dry N_2 gas 30 minutes just before the measurement.

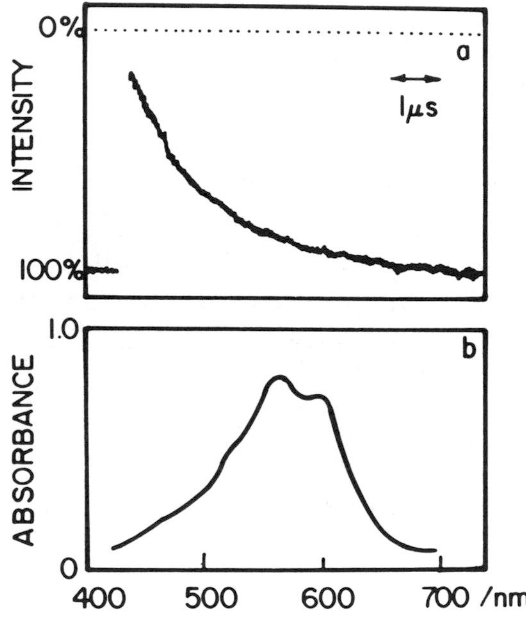

Figure 1. Decay curve and spectrum of the transient absorption of TMPD.

III. RESULTS AND DISCUSSION

Figures 1a) and 1b) show the time dependence of the transient absorption of TMPD in n-heptane solution at 532 nm with the N_2 laser excitation and its spectrum observed 150 ns after the excitation. The time dependence indicated a single exponential decay and the observed lifetime is 1.4 μs. When O_2 gas was bubbled into the sample solution, the transient absorption dissapeared. The spectrum in Fig. 1b) shows a good agreement with the results of the previous studies (2,3,4). Therefore, the transient absorption should be attributed to the $T_n \leftarrow T_1$ absorption of TMPD.

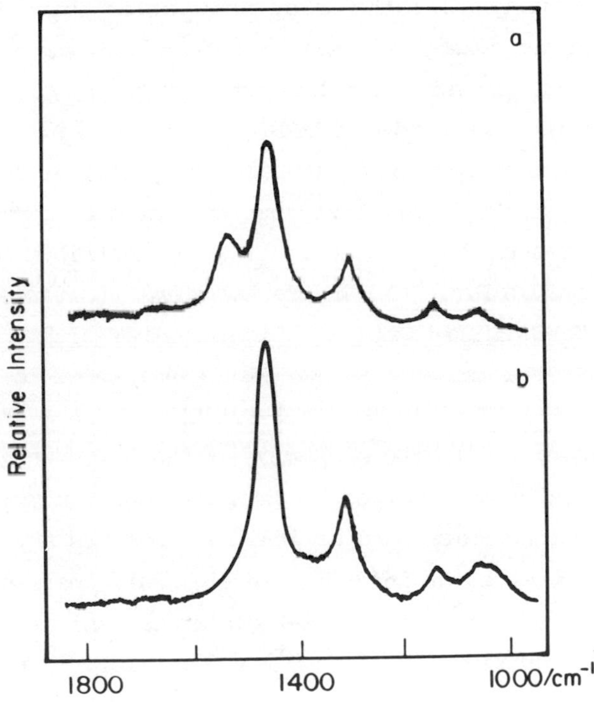

Figure 2. Resonance Raman spectra of the transient species of TMPD.

The frequency-doubled output of the YAG laser, 532 nm, satisfies the rigourous resonance condition to give a resonance enhancement of the Raman bands of ^{3}TMPD. Figures 2a) and 2b) show the Raman spectra of TMPD in n-heptane solution observed 150 ns after the excitation and without the excitation, in the region of 900-1800 cm^{-1}. With the N_2 laser excitation, a new Raman band appeared at 1530 cm^{-1}. In the other wavenumber region, no new Raman band was observed. The position of the band is reasonable for the skeletal vibrations of this molecule. When O_2 gas was bubbled into the sample solution, this band could not be observed. Figure 3 shows the resonance Raman spectra of TMPD at various delay times of the YAG laser pulse with respect to the N_2 laser pulse. The relative intensity of the band at 1530 cm^{-1} decreases gradually, when the delay time varies from 150 ns to 4 µs. This result agrees well with the fact that the lifetime of ^{3}TMPD is 1.4 µs. Therefore the band at 1530 cm^{-1} is assigned to the vibrational Raman band of ^{3}TMPD.

The absorption spectrum of $TMPD^+$ is very similar to that of ^{3}TMPD as shown in Fig. 4a). Therefore the resonance Raman spectrum of $TMPD^+$ was measured for investigating the contribution of $TMPD^+$ to the spectrum in Fig. 2a). Figure 4b) shows the resonance Raman spectrum of $TMPD^+$ in methanol solution together with that of methanol using the present system. Figure 4c) shows the resonance Raman spectrum of $TMPD^+$ in methanol solution using a conventional Raman spectrometer with the excitaion at 586 nm and the result agrees well with that reported by Jeanmaire and Van Duyne (5). $TMPD^+$ has a strong Raman band at 1640 cm^{-1} whereas any Raman band could not be observed at 1640 cm^{-1} in Fig. 2a). Therefore the Raman band at 1530 cm^{-1} in Fig. 2a) cannot be assigned to that of $TMPD^+$ and the contribution of $TMPD^+$ is negligible in the present study.

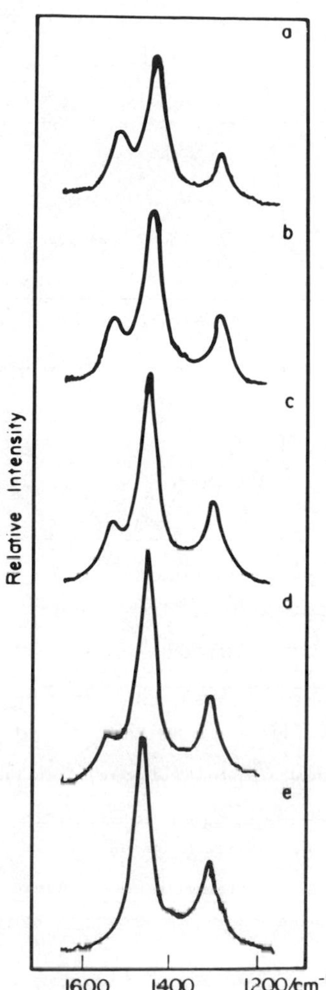

Figure 3. Resonance Raman spectra of TMPD, a) 150 ns, b) 500 ns, c) 1 μs, d) 2 μs, and e) 4 μs after the excitation.

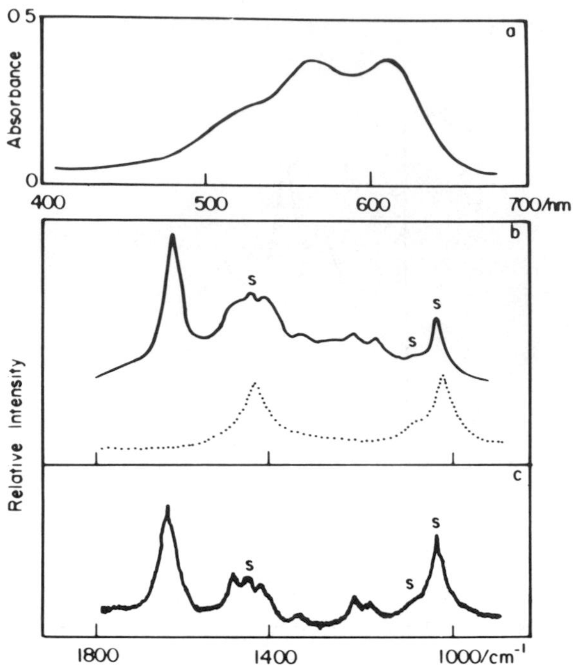

Figure 4. Absorption spectrum and resonance Raman spectra of $TMPD^+$ in methanol solution. a) Absorption spectrum, b) resonance Raman spectra with the excitation at 532 nm (solid line; $TMPD^+$ in methanol and broken line; methanol), and c) with the excitation at 586 nm. s; Solvent bands.

REFERENCES

1. Yokoyama, K.,and Kobayashi, T., Chem. Phys. Letters. 85, 175 (1982).
2. Cadogan, K. D., and Albrecht, A. C., J. Phys. Chem. 73, 1868 (1969).
3. Yamamoto, N., Nakato, Y., and Tsubomura, H., Bull. Chem. Soc. Japan 39, 2603 (1966).
4. Bagdasarjan, K. S., Kirjukhin, Y. I., and Sinitsina, Z. A., J. Photochem. 1, 225 (1973).
5. Jeanmaire, D. L., and Van Duyne, R. P., J. Electroanal. Chem. 66, 235 (1975).

OXIDATION REACTION OF EXCITED TRIPLET p-BENZOQUINONE

S. M. Beck
L. E. Brus

Bell Laboratories
Murray Hill, New Jersey

I. INTRODUCTION

Nature utilizes substituted quinones as electron carriers in bacterial reaction centers, photosystem II of green plants, and mitochrondria. We have thus attempted to understand the aqueous redox chemistry of the simplest parent, p-benzoquinone, via the technique of transient Raman scattering. Previous transient absorption studies (1,2) have shown that the intermediate species involved have severely overlapping optical spectra near 430 nm. An advantage of the Raman approach to kinetics is an ability to more clearly separate and identify multiple transient species. Our apparatus is described elsewhere in this volume, and in the literature (3).

II. OBSERVATIONS

Our technique in its general form is a pulse-probe experiment, with one pulse starting a chemical process, and a second pulse generating a resonance Raman spectrum at a fixed time delay. In the aqueous p-benzoquinone system, both functions can be carried out by one 416 nm pulse (4). A low flux 416 nm Raman spectrum of a 3 x 10^{-2} M solution shows two S_o ground state lines,

ISBN 0-12-066280-9

a C-H wag at 1163 cm^{-1} and a C=O stretch at 1676 cm^{-1} (4). However, at high flux this spectrum has a new dominant line at 1555 cm^{-1}, as shown in Figure 1a.

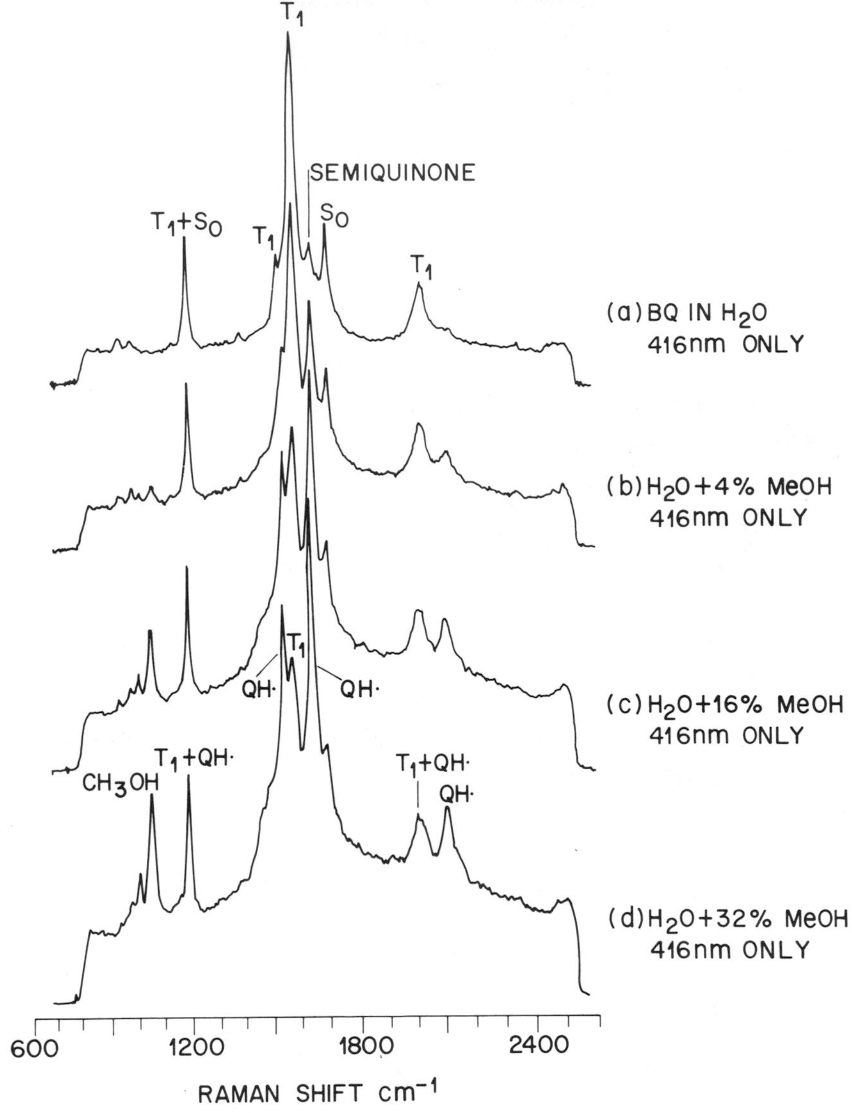

Figure 1. High flux 416 nm pulsed Raman spectra of p-benzoquinone solutions, as described in the text.

The lines labelled T_1 and semiquinone grow as the square of the 416 nm flux and represent photoproduct(s) created within the $\simeq$5 nsec pulsewidth. One photon is required to generate the photoproduct, and a second photon is required for Raman scattering. Absorption studies (1,2) have shown that the excited triplet T_1 is produced by photoexcitation, and has a moderate absorption ($\varepsilon \sim 5000$ ℓ/M^{-1}) at 416 nm. The 1550 cm^{-1} line in subsequent two pulse experiments was found to have a lifetime close to that previously observed for T_1. We assign lines labelled T_1 to triplet benzoquinone.

A weak photoproduct line near 1620 cm^{-1} is in the correct position to represent semiquinone radicals formed by excited-state oxidation of the water solvent. Bensasson et al. have shown that semiquinone absorption is present after decay of the triplet absorption (1). Alcohols oxidize more readily than water, and we investigated the effect of added methanol on these spectra (5). Figure 1b-1d show a strong growth of this line, as well as new photoproduce lines, as a function of added methanol. Independent experiments establish that these increasing lines unambiguously belong to the protonated radical (5). We are experimentally able to distinguish protonated semiquinone from the semiquinone radical anion. Semiquinone acid-base equilibrium does not occur within the laser pulse, and these spectra thus indicate that the net reaction at $\Delta t = 5$ nsec is abstraction of an H atom, and not an electron, from methanol. These mixed solvent spectra support the possibility that the $\simeq$1620 cm^{-1} peak in pure H_2O represents radical.

In time-delayed two pulse experiments, the 1620 cm^{-1} line is stronger and remains as T_1 decays. It appears that this line does represent semiquinone radical, formed not from vibrationally relaxed T_1 but as vibrationally "hot" excited quinone relaxes into T_1. There is also a two photon process for water oxidation observed at high ultraviolet flux (4).

These oxidation reactions nicely illustrate the potential of the Raman technique in chemical kinetics. In this example we observe the molecular precursor (S_o), the reactive intermediate (T_1), and the initial reaction products (semiquinones).

REFERENCES

1. J. C. Ronford-Haret, R. V. Bensasson, and E. Amouyal, J. C. S. Faraday I 76, 2432 (1980).
2. D. R. Kemp and G. Porter, Proc. Roy. Soc. A 326, 117 (1971).
3. S. M. Beck and L. E. Brus, J. Chem. Phys. 75, 4934 (1981).
4. S. M. Beck and L. E. Brus, J. Am. Chem. Soc. 104, 1103 (1982). This work was performed without knowledge of reference 1.
5. S. M. Beck and L. E. Brus, J. Am. Chem. Soc. 104, 4789 (1982).

RESONANCE RAMAN SPECTRA AND VIBRATIONAL ANALYSES OF REACTION INTERMEDIATES

Ronald E. Hester

Department of Chemistry
University of York
York, England

There are basically two reasons why Raman spectroscopy may be used to monitor the progress of a chemical reaction in solution. One of these is that the vibrational spectra of molecular species at ambient temperatures invariably are better resolved and more highly structured than electronic spectra (absorption or fluorescence) and thus provide less ambiguous "fingerprints" by means of which the kinetics of concentration changes may be measured. The other is that vibrational spectra may be analysed rather directly in terms of structures, symmetries and chemical bond characters of the molecular species involved in a reaction. This latter reason is of particular significance for reactions which proceed via short-lived intermediate species whose identities may hold the key to an understanding of the reaction mechanism. The relationship of structure and function may be explored in some considerable detail through vibrational analysis of transient intermediates in very many different reaction types. It is this structure and bonding aspect, rather than the kinetic one, which is examined in this paper.

During the past few years we have used Raman spectroscopy to study a wide variety of reaction intermediates, many of these being reactive free radical species (1,2). Our methods have been as varied as the chemical systems themselves and have included rapid mixing of solutions of reactions in both continuous- (3) and stopped-flow (4) devices, pulsed radiolysis (5) and ultraviolet photolysis both at ambient (6) and liquid nitrogen (77K) (7) temperatures. The bulk of this work has been done using c.w. laser excitation and conventional single-channel scanning spectrometers, although we also have used chopped (4) or pulsed (5) laser excitation and intensified vidicon multichannel detection (5,8) in some cases. In most cases these studies have employed resonance effects to increase the sensitivity of the Raman probe so that it can cope with reaction intermediates at sub-millimolar

ISBN 0-12-066280-9

concentrations (typically 10^{-4}-10^{-5} mol dm^{-3}). Since resonance-Raman enhancement of band intensities is specific to the species having an electronic absorption band coincident with the excitation laser wavelength, this effect also minimizes spectral interferences from solvents and other molecular species (e.g. excess reactants or products) which may be present at much higher concentrations than the intermediate species under investigation. This presentation will focus on recent studies and on the difficulties and limitations of the data analysis procedures.

An example of a system which has practical applications (in solar energy conversion schemes), and also serves to demonstrate how vibrational Raman spectra of an electronically excited state and of a free radical reaction intermediate may be obtained, is the photoreaction of the ruthenium(II)tris-2,2'-bipyridyl dication, $[Ru^{II}(bpy)_3]^{2+}$, with methylviologen (1,1'-dimethyl-4,4'-bipyridyl dication, MV^{2+}) in aqueous solution. In the absence of catalysts, this reaction proceeds by the following route (9).

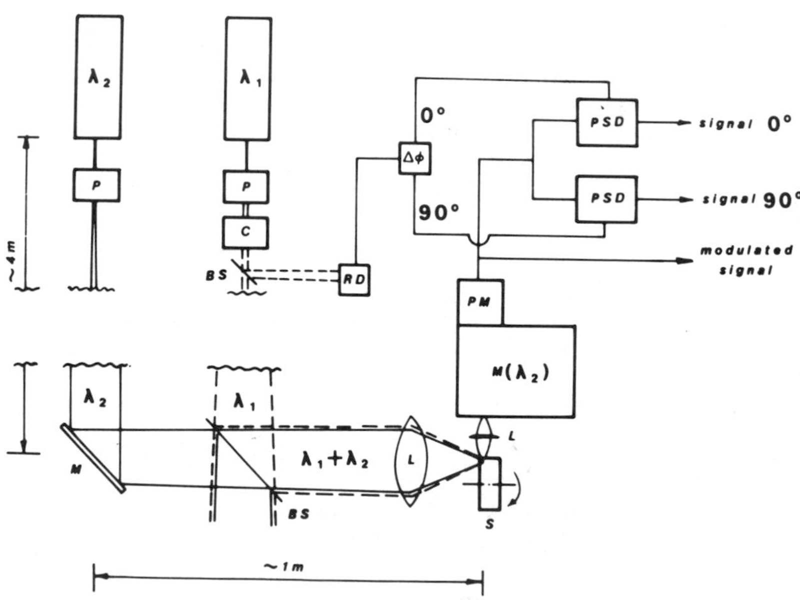

FIGURE 1. Apparatus for modulated excitation resonance Raman (MERR) spectroscopy (9). λ_1 and λ_2 are two lasers with different wavelengths, λ_1 being used to photolyse the sample and λ_2 to probe its Raman spectrum. P, plasma line filter; C, chopper; M, mirror; BS, beam splitter (Al on silica); L, silica lens; S, spinning sample cell; M(λ_2), monochromator set to measure Raman spectra excited by λ_2; PM, photomultiplier; RD, reference detector; $\Delta\phi$, phase shifter; PSD, phase-sensitive detector.

$$1.\quad [Ru^{II}(bpy)_3]^{2+} \xrightarrow{h\nu} [^1Ru^{III}(bpy)_2(bpy^{-\cdot})]^{2+*} \longrightarrow [^3Ru^{III}(bpy)_2(bpy^{-\cdot})]^{2+*}$$

$$2.\quad [^3Ru^{III}(bpy)_2(bpy^{-\cdot})]^{2+*} + MV^{2+} \longrightarrow [Ru^{III}(bpy)_3]^{3+} + MV^{+\cdot}$$

$$3.\quad [Ru^{III}(bpy)_3]^{3+} + MV^{+\cdot} \longrightarrow [Ru^{II}(bpy)_3]^{2+} + MV^{2+}$$

We developed a special apparatus for modulated excitation resonance Raman (MERR) spectroscopy, as shown in Fig. 1, for our studies of this reaction and used it also to probe the mechanism of a similar photoreaction using proflavin as the sensitizer (9). Apart from confirming Woodruff's (10) earlier conclusions about the asymmetric structure of the excited triplet state of the ruthenium-based photosensitizer, we also have accumulated extensive data on the $MV^{+\cdot}$ free radical cation species. In the presence of a suitable catalyst, e.g. colloidal platinum, $MV^{+\cdot}$ will reduce water to hydrogen and thus acts as a convenient electron relay in a solar cell based on this photochemistry. A set of Raman spectra from ca. 10^{-4} mol dm^{-3} aqueous solutions of the viologen radical, showing a variety of resonance effects, are shown in Fig. 2 (11).

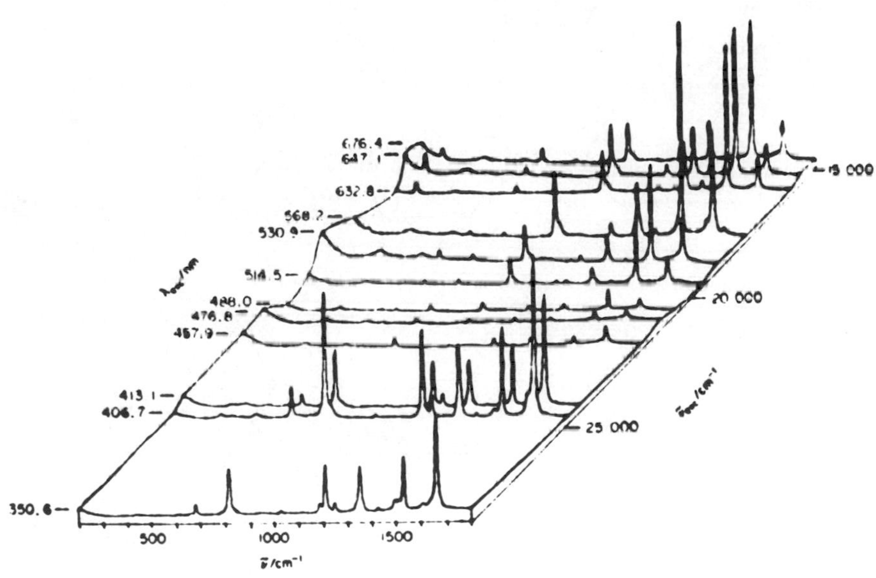

FIGURE 2. Raman spectra from ca. 10^{-4} mol dm^{-3} $MV^{+\cdot}$ in water with various excitation wavelengths, λ_{exc} (11).

We have performed detailed normal coordinate analyses (NCA) on the MV^{2+} and $MV^{+\cdot}$ species in order to help interpret their observed spectra and to try to understand the meaning of the shifts in vibrational band frequencies between these species (12). For molecules as complicated as these, the observed spectra represent elaborate coupling patterns involving simultaneously bond stretching and inter-bond angle deformation vibrations throughout the molecules. Such coupling patterns define the normal coordinates and a description of their form clearly is fundamental to any detailed analysis of vibrational spectra. However, only rarely are sufficient experimental data available on which to base such an elaborate calculation and it is commonplace to infer only very crudely approximate descriptions of vibrational modes. Our calculations for viologen species have been based on the Urey-Bradley type of force field for the rings and a local symmetry field for the methyl groups. The Shimanouchi set of computer programs was used. Although it is impractical as well as inappropriate to give the full results of this NCA analysis in this

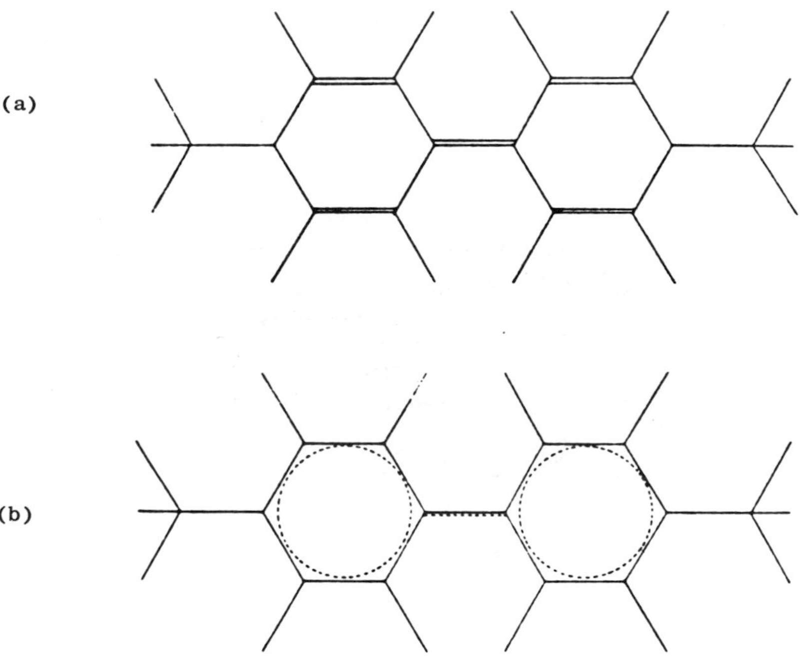

FIGURE 3. Structure and bonding in (a) MV^{2+} and (b) $MV^{+\cdot}$ as shown by normal coordinate analysis (12).

abstract, it is worth pointing to the fact that no single band in the spectra of either MV^{2+} or $MV^{+\cdot}$ appears to be associated solely with the stretching mode of the important inter-ring bond in these molecules. Nonetheless the NCA treatment does show clearly that this bond force constant increases from 4.1 to 5.4 mydyn A^{-1} in going from MV^{2+} to $MV^{+\cdot}$ and other changes found in the force constant set also are consistent with the change from a largely delocalized aromatic-type structure to a more localized quinonoid-type structure in the free radical form, as depicted in Fig. 3.

REFERENCES

1. E. E. Ernstbrunner, R. B. Girling, W. E. L. Grossman and R. E. Hester, J. Chem. Soc. Perkin Trans. 2, 177 (1978); J. Chem. Soc. Faraday Trans. 2, 74, 501 (1978); 74, 1540 (1978); J. Raman Spectrosc. 10, 161 (1981).
2. R. E. Hester and E. M. Nour, Indian J. Pure Appl. Phys. 16, 167 (1978); J. Mol. Struct. 62, 77 (1980); J. Raman Spectrosc. 11, 35, 39, 43, 49, 59, 64 (1981); J. Chem. Soc. Dalton, 939 (1981).
3. R. E. Hester and K. P. J. Williams, J. Chem. Soc. Faraday Trans. 2, 78, 573 (1982); J. Chem. Soc. Perkin Trans. 2, 561, 559 (1982), 852 (1981).
4. J. C. Merlin, J. L. Lorriaux and R. E. Hester, J. Raman Spectrosc. 11, 384 (1981).
5. R. Wilbrandt, N. H. Jensen, P. Pagsberg, A. H. Sillesen, K. B. Hansen and R. E. Hester, Chem. Phys. Lett. 80, 315 (1979); J. Chem. Phys. 71, 3326 (1979); J. Raman Spectrosc. 11, 24 (1981).
6. M. Forster and R. E. Hester, Chem. Phys. Lett. 81, 42 (1981).
7. M. Forster and R. E. Hester, J. Chem. Soc. Faraday Trans 2, 77, 1535 (1981); R. E. Hester and K. P. J. Williams, J. Chem. Soc. Faraday Trans. 2, 77, 541 (1981); J. Raman Spectrosc. 10, 169 (1981).
8. M. Forster, R. E. Hester, B. Cartling and R. Wilbrandt, Biophys. J. 38, 111 (1982).
9. M. Forster and R. E. Hester, Chem. Phys. Lett. 85, 287 (1982).
10. R. F. Dallinger, S. Farquharson, W. H. Woodruff and M. A. J. Rogers, J. Am. Chem. Soc., 103, 7433 (1981).
11. M. Forster, R. B. Girling and R. E. Hester, J. Raman Spectrosc. 12, 36 (1982).
12. R. E. Hester and S. Suzuki, J. Phys. Chem., accepted (1982).

PHOTOCHEMICALLY INDUCED ELECTRON TRANSFER REACTIONS OF trans - STILBENE

W.Hub, S.Schneider and F.Dörr*

Institut für Physikalische und Theoretische Chemie, Technische Universität München, Garching, Germany-W.

J.T.Simpson, J.D.Oxman and F.D.Lewis*

Department of Chemistry, Northwestern University, Evanston, Illinois

Time-resolved resonance raman spectroscopy (TRRR) so far has mainly been used to characterize and identify short-lived intermediates by their vibrational frequencies. But the method also allows to measure the kinetic of formation and decay of the transient species as well as the excitation profile of their resonance raman spectra. We will show, that these additional informations may be successfully used for the identification of an unknown intermediate in cases, where the identification is not possible from the TRRR spectrum alone.

We have studied the photochemical addition reaction of tertiary alkylamines to singlet excited trans-stilbene, where we could measure the TRRR spectrum and the decay of the solvated trans-stilbene anion radical ($TS^{\overline{\cdot}}$) even

* Travel expenses for this project were provided by NATO Research Grant 1911.

ISBN 0-12-066280-9

with amines, that do not form stable adducts with TS. We also identified the cation radical of TS in two different types of reactions: in the photoxidation of TS, sensitized by cyanoanthracenes, and in photolyzed solutions of TS and fumaronitrile. All reactions are carried out in acetonitrile solutions at room temperature. The apparature used is described in ref.1 .

I. FORMATION OF trans-STILBENE RADICAL ANION

Tertiary alkylamines generally quench the fluorescence of TS in non-polar and polar solvents. In non-polar solvents reversible exciplex formation is detected by fluorescence spectroscopy(2). In polar, aprotic solvents some of these amines form stable adducts with TS(3). The mechanism for this reaction was supposed to include three steps: first electron-transfer, followed by proton-transfer and a final free radical recombination step(3).

We observe one and the same TRRR spectrum after photolysis of solutions of TS with the following amines: triethylamine (I), ethyldiisopropylamine (II)(4), quinuclidene (III), 1,4-diazabicyclo 2.2.2 octane (IV)(4) and tetraethylethylenediamine (V), from which only I and II form stable adducts with TS. The RR frequencies and the relative intensities correlate well with those of the sodium salt of TS (see table). The shifts to lower frequencies may be explained by the influence of different solvents and/or counterions. The excitation spectra of the 1578/1555, 1251 and 1180 cm^{-1} frequencies in the wavelength range between 445 and 495 nm are very similar to the visible absorption spectrum of $Na^{+}TS^{-}$ (6). So the formation of $TS^{\overline{\cdot}}$ in this reaction is now established and it can be assumed, that the electron transfer process is responsible for the quenching of the TS fluorescence in polar solvents. The transient spectrum could be observed up to a microsecond after the photolysis pulse. The decay of the 1578/1555 cm^{-1} band of $TS^{\overline{\cdot}}$ follows a simple second order law (Fig.1), which shows, that the $TS^{-\cdot}$ molecule observed is not part of a singlet or triplet radical ion pair (which decays according to first order), but is the solvated radical ion, which disappears by recombination with the amine cation radical (see scheme 1). A thorough investigation of the decay of $TS^{-\cdot}$ as a function of increasing amine concentrationof I and II revealed, that the decay rate is decreasing in both cases and that the initial yield (t=0 ns) of $TS^{-\cdot}$ increases with II, but is independent of the concentration of I. The last observation can be explained by the quenching of a precursor of the solvated $TS^{-\cdot}$ by triethylamine (I), whereas this quenching is not observed with II, because the formation of a N-N bond between the amine and the amine cation radical is sterically hindered.

These results demonstrate for the first time the application of TRRR spectroscopy in the study of the kinetic of a short-lived species.

Table: RR frequencies of the anion and cation radicals

TS anion radical		t-stilbene	$TS^{+\cdot}$
this work 480nm/CH_3CN (ρ)	ref.5 488nm/THF	(ρ)	this work 488nm/CH_3CN
1578 s (0.24)	1584 s	1642 vs	
1555 s	1554 s	1596 s (0.27)	1610 s
	1531 w		1568 w
	1465 vw	1488 m	
1432 w	1432 w	1445 m (0.64)	
1326 w	1328 w	1337 s	
	1313 vw		
1251 m (0.31)	1264 m	1191 s (0.27)	1285 m
1180 m (0.43)	1185 m	1180 s	
	1168 w		
1145 w	1143 w		
978 m (0.09)	982 m	997 vs	995 w
848 m (0.25)	859 m	866 m	870 w
624 m	628 m	616 m	625 w
	614 vw		

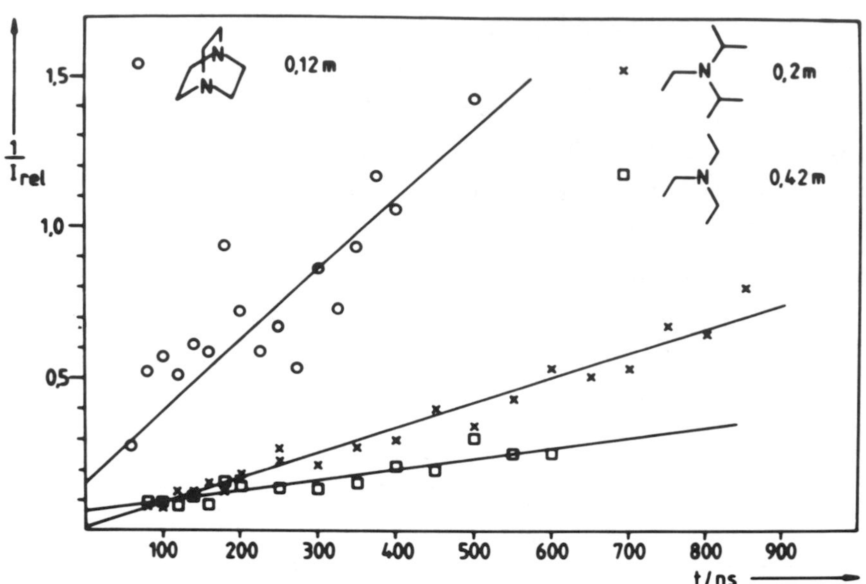

Fig.1 Second-order decay of the 1578/1555 cm^{-1} band of $TS^{\cdot-}$ in solutions of TS (0.01 m) with triethylamine (0.42 m), ethyldiisopropylamine (0.2m) and 1,4-diazabicyclo[2.2.2]octane (0.12 m). All amines quench 50 % of singlet excited TS at these concentrations.

$^1TS^*$ + A $\longrightarrow$ $^1[\text{Exciplex}]^*$ $\longrightarrow$ products

CS TS+hv $^1[TS^{\cdot-}\ A^{\cdot+}]$ $\longrightarrow$ $TS^{\cdot-}A^{\cdot+}A$ $\longrightarrow$ TS+2A

CS + A $\longleftarrow$ $^3[TS^{\cdot-}A^{\cdot+}]$ $\longleftarrow$ $TS^{\cdot-}$ + $A^{\cdot+}$

TS + A

Scheme 1: Mechanism of reaction of t-stilbene with alkylamines in polar solvent

II FORMATION OF trans-STILBENE CATION RADICAL

Irradiation of the ground-state complex between TS and fumaronitrile (VI) leads to the production of the isomers (7). The formation of radical ion pairs is highly exergonic (8) , as it is in the TS/amine case. Evidence for the intermediate formation of ion radicals is provided by a recent CIDNP experiment (7). The reaction mechanism is shown in scheme 2. It is supported by the observation, that the yield of cis-stilbene is increased by the presence of oxygen in the solution (9).

We measured a transient resonance raman spectrum (Fig.2), which disappeared within 600 ns after the photolysis pulse. There are two strong bands in the spectrum at 1610 and 1285 cm^{-1} and weaker ones at 995, 870 and 625 cm^{-1}. The spectrum may be assigned to trans-stilbene cation radical.

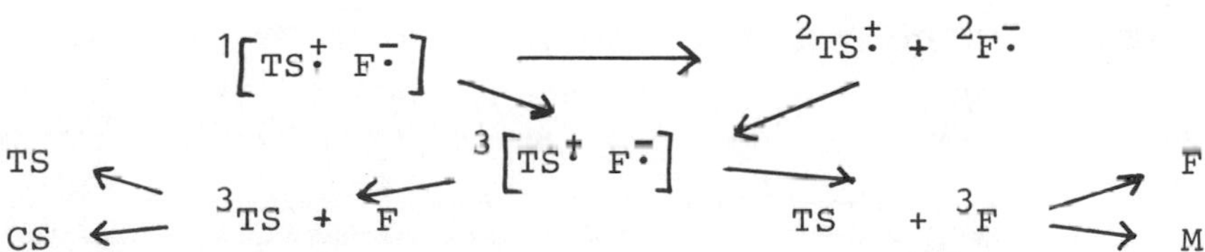

Scheme 2: Reaction between trans-stilbene (TS) and fumaronitrile (F).CS=cis-stilbene,M=maleonitrile.

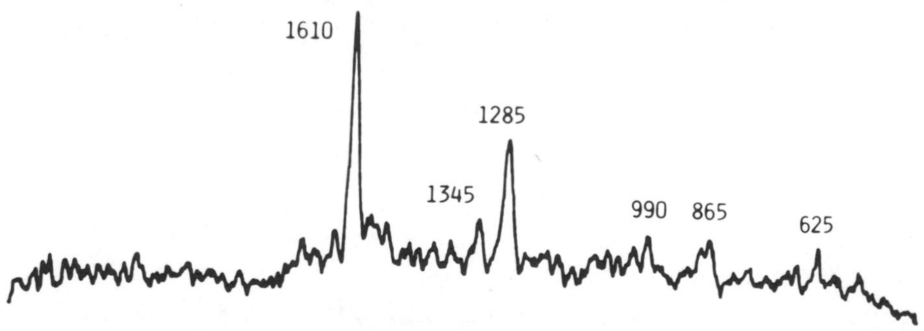

Fig.2: Transient RR spectrum in photolyzed TS/F solution

In order to experimentally identify the intermediate species, we investigated a different photochemical reaction, in which the production of $TS^{\dot{+}}$ is very likely. The irradiation of 9-cyanoanthracene (VII) or 9,10-dicyanoanthracene (VIII) in polar solvents containing O_2 and trans-stilbene results in the formation of oxidation products of TS like benzaldehyde, benzile and others(10) (scheme 3). By ESR spectroscopy the formation of the anion radical of VIII was measured (11), whereas no signal for a cation radical could be found. The transient absorption spectra of $VII^{\dot{-}}$, $VIII^{\dot{-}}$ and $TS^{\dot{+}}$ have been observed (10).

$$TS + {}^1DCA \longrightarrow TS^{\dot{+}} + DCA^{\dot{-}}$$

$$DCA^{\dot{-}} \xrightarrow{O_2} O_2^{\dot{-}} + DCA$$

$$TS^{\dot{+}} + O_2^{\dot{-}} \longrightarrow C_6H_5\text{-}CH\text{-}CH\text{-}C_6H_5 \text{ (with O—O bridge)}$$

$$C_6H_5\text{-}CH\text{-}CH\text{-}C_6H_5 \text{ (O—O)} \longrightarrow 2\ C_6H_5{=}O$$

Scheme 3: Sensitized photooxidation of TS with 9,10-dicyanoanthracene (DCA=VIII)

In our TRRR experiments we observed the same transient spectrum as with TS/fumaronitrile (fig.3) in solutions with both cyanoanthracenes. We therefore assign the transient spectrum to the $TS^{\dot{+}}$. The observability of the species up to a microsecond after the photolysis pulse can be explained by the assumption, that we are observing the solvated cation radical. The kinetic data taken so far do not allow a clear decision about the order of its decay reaction. The excitation spectrum of the RR frequencies at 1610 and 1285 cm^{-1} are very similar to the absorption spectrum of $TS^{\dot{+}}$ (12). RR frequencies of low intensity at 2211,1650 and 1172 cm^{-1}, which are only observable in nitrogen-saturated solutions, may be interpreted as being due to $VIII^{\dot{-}}$.

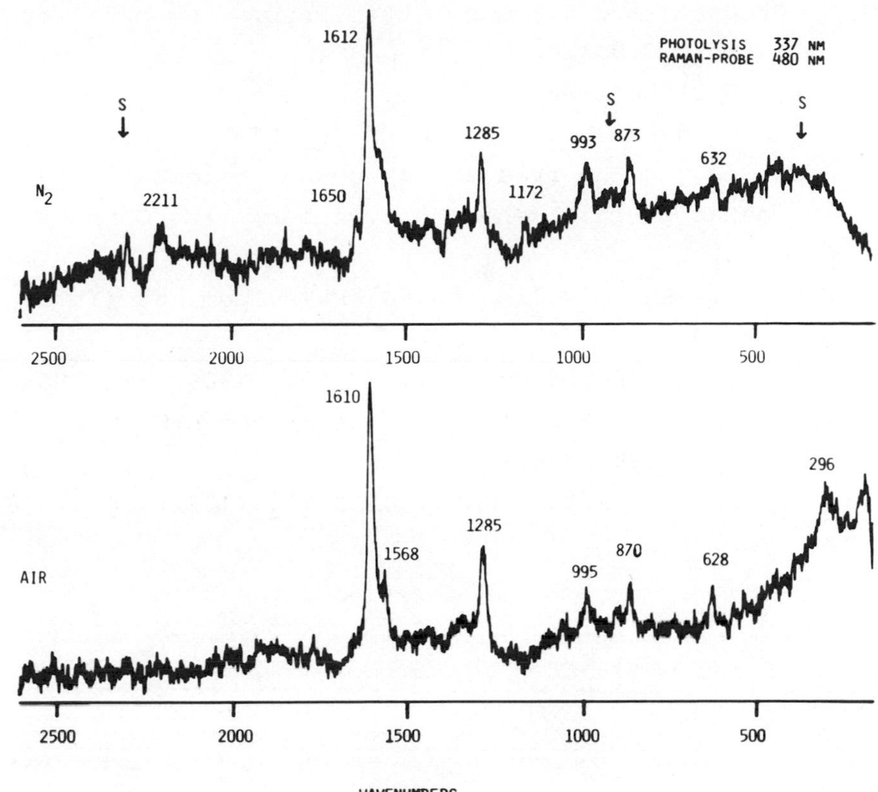

Fig.3: TRRR spectra of solutions of TS and VIII in CH_3CN at 100 ns after photolysis (λ_{exc}=480 nm)

REFERENCES

(1) F.Dörr,W.Hub and S.Schneider,J.Molec.Struct.60(1980) 233

(2) F.D.Lewis, Acc.Chem.Res. 12 (1979) 152

(3) F.D.Lewis,T.-I.Ho and J.T.Simpson, J. Org. Chem. 46 (1979) 1077-1082

(4) W.Hub,S.Schneider,F.Dörr,J.T.Simpson,J.D.Oxman and F.D.Lewis, JACS 104 (1982) 2044-2045

(5) C.Takahashi and S.Maeda, Chem.Phys.Letters 28 (1974) 22

(6) G.Levin, B.E.Holloway, C.R.Mao and M.Szwarc, JACS 100 (1978) 5841-5844

(7) D.R.Arnold and P.C.Wong, JACS 101 (1979) 1894-1895

(8) According to the free energy change calculated by the equation in D.Rehm and A.Weller, Isr. J.Chem. 8 (1970) 259-271

(9) F.D.Lewis and J.T.Simpson, J. Phys.Chem. 83 (1979) 2015-2019

(10) L.T.Spada anf C.S.Foote, JACS 102 (1980) 391-393

(11) A.P.Schaap, K.A.Zaklika, B.Kaskar and L.W.-M.Fung, JACS 102 (1980) 389-391

(12) R.Gooden and J.I.Brauman, JACS 104 (1982) 1483-1486

ELECTRON TRANSFER REACTIONS IN STILBENE ISOMERS

H. Shindo* and G. H. Atkinson

Department of Chemistry
Syracuse University
Syracuse, New York

INTRODUCTION

Electron transfer processes are considered to be of fundamental importance in an extremely wide range of chemical reactions including energy storage, photosynthesis, ion-pumping in biomembranes, and electron device fabrication at surfaces. In spite of their ubiquity, much of the molecular detail in even simple electron transfer reactions remains obscure, especially with respect to the structural characterization of ionic intermediates. This information is unavailable largely because of the experimental difficulties associated with obtaining structural data for short-lived, solution phase species. Data on the kinetic behavior of these transients are, of course, an equally important elements in constructing accurate mechanisms. Transient absorption spectroscopy provides kinetic information, but in general does not comment on structural questions. By contrast, time-resolved resonance Raman (TR^3)

*Permanant Address: National Chemical Laboratory for Industry Tsukuba Research Center, Yatabe, Ibaraki, Japan

ISBN 0-12-066280-9

spectroscopy has been shown in several electron-transfer studies to be an especially powerful method for recording both structural and kinetic data simultaneously (1,2).

In this paper, we address a question concerning the vibrational structure and kinetic behavior of anions of stilbene isomers. The radical anion of *trans*-stilbene ($T^{\overline{\cdot}}$) has been observed by TR^3 spectroscopy and its presence during the electron photodetachment reaction of the dianion of *trans*-stilbene (T^{2-}) has been confirmed (3). The radical anion of *cis*-stilbene ($C^{\overline{\cdot}}$) also has been proposed as an important intermediate in room temperature ionic reactions (4,5). Previous studies, however, have not provided a clear understanding of $C^{\overline{\cdot}}$ properties, especially with regard to stability. Optical absorption (6) and electron spin resonance (4) spectra of $C^{\overline{\cdot}}$ have been reported at low temperatures where the anion is relatively stable. At room temperature, transient absorption with a 100μs lifetime has been assigned to $C^{\overline{\cdot}}$(5). If correct, such an optical spectrum makes it experimentally feasible to record the TR^3 spectrum of $C^{\overline{\cdot}}$ and thereby to firmly establish its presence through its distinct vibrational structure.

TR^3 experiments designed for this purpose have been performed (7). They are based on the photodetachment of an electron from the radical anion of perylene ($Pe^{\overline{\cdot}}$) in the presence of an electron acceptor such as a neutral stilbene isomer (i.e., T or C). This method has been used to obtain TR^3 spectra of several ions in room temperature solutions including $T^{\overline{\cdot}}$ (7), but has been unsuccessful in recording the vibrational spectrum of $C^{\overline{\cdot}}$. Its failure in the $C^{\overline{\cdot}}$ case can be attributed to either the reaction of $C^{\overline{\cdot}}$ (presumedly to $T^{\overline{\cdot}}$) at a rate faster than the 10^{-8} time-resolution of these TR^3 experiments or the presence of an ionic intermediate which cannot be correctly described as either $T^{\overline{\cdot}}$ or $C^{\overline{\cdot}}$. While the first possibility remains understudy (7), resonance Raman spectra are reported here for an anion structurally-related to $C^{\overline{\cdot}}$, namely the 5H-dibenzo (a,d) cycloheptene radical anion ($DBCH^{\overline{\cdot}}$), which demonstrate that the second possibility is not a correct explanation.

EXPERIMENTAL

Perylene (Pe) was purified by sublimation. *Trans*-stilbene (T) also was purified by sublimation after being recrystallized twice from ethanol. *Cis*-stilbene (C) was recrystallized twice from n-hexane at lower temperatures (8) and then distilled. The purity of C was measured by infrared spectroscopy which showed no significant amount of T. DBCH was prepared by reduction of 5H-dibenzo (a,d) cyclohepten-t-one with aluminum isopropoxide (9) and was recrystallized from acetone (melting point of 131°C).

Radical anions ($Pe^{\overline{\cdot}}$ and $DBCH^{\overline{\cdot}}$) were prepared by exposing a THF solution of the parent molecule (Pe and DBCH) to a sodium mirror. In TR^3 experiments, a stilbene isomer (either T or C) was added to the $Pe^{\overline{\cdot}}$ solution to act as an electron acceptor.

Detailed descriptions of the TR^3 instrumentation are presented elsewhere (10). In this work, electron photodetachment from $Pe^{\overline{\cdot}}$ was initiated by 575 nm pulsed (15 ns FWHM) excitation while the resonantly enhanced Raman scattering from the anions was generated by pulsed (8ns FWHM) radiation between 508 nm and 495 nm.

The Raman spectra of DBCH and $DBCH^{\overline{\cdot}}$ were generated by 488nm CW excitation from an argon ion laser. Raman scattering was recorded with a double monochromator (Spex 1402) equipped with photon counting.

RESULTS and DISCUSSION

The TR^3 spectrum recorded 40 ns after electron photodetachment in a $Pe^{\overline{\cdot}}/T$ solution contained only bands characteristic of $T^{\overline{\cdot}}$ with the most prominent features appearing at 1185, 1246, 1554, and 1584 cm^{-1} (3,11). Only these same bands were observed when a $Pe^{\overline{\cdot}}/C$ solution was examined under identical experimental conditions. In separate TR^3 experiments on $Pe^{\overline{\cdot}}/C$ samples the probe laser wavelength used to generate resonance Raman scattering from the transient was tuned from 508 nm to 495 nm (the 40 ns time resoltuion remained unchanged). TR^3 bands were observed only at the red end of this tuning range and then only those assignable to $T^{\overline{\cdot}}$. No TR^3 band were found using the blue end of the tuning range where $C^{\overline{\cdot}}$ has been reported to exhibit transient absorption (6).

In order to establish firmly that the *cis*-isomeric form of the anion should have a resonance Raman spectra distinct from its *trans*-counterpart, the resonance Raman spectra of $DBCH^{\overline{\cdot}}$ and, for comparison, DBCH were recorded (Figure 1). The wavenumber displacements for the bands are given in Table I.

Although there are some similarities between the $DBCH^{\overline{\cdot}}$ and $T^{\overline{\cdot}}$ spectra, there are significant differences which should be noted. The strong $T^{\overline{\cdot}}$ bands at 1584 cm^{-1} and 1554 cm^{-1} assigned to the C=C ethylenic and ν_{8a}-type ring vibrational modes, respectively have their strong analogue bands in the $DBCH^{\overline{\cdot}}$ spectrum at 1584 cm^{-1} and 1530 cm^{-1}. The 24 cm^{-1} difference in the 1554/1530 cm^{-1} band positions serves partially to distinguish a *cis*-form of the anions from a *trans*-configuration. A more dramatic difference is found in the 1166 cm^{-1} $DBCH^{\overline{\cdot}}$ band which is 80 cm^{-1} lower in frequency than the 1246 cm^{-1} band from $T^{\overline{\cdot}}$ assigned to the C-C stretching mode. These single bonds apparently are strenghtened by the delocation caused in the addition of the extra electron when $DBCH^{\overline{\cdot}}$ is formed.

Although there are also differences in the less intense spectral features, the presence of these two district spectral regions provide that clearest support for the conclusion that the transient ionic form should be characterized as either $T^{\bar{\cdot}}$ or $C^{\bar{\cdot}}$. The bridging CH_2 group in DBCH, of course, guarantees that a *cis* or *cis*-like conformation is maintained during electron attachment. These results suggest that the absence of a TR^3 spectrum for $C^{\bar{\cdot}}$ should be interpreted in terms of its rapid conversion to $T^{\bar{\cdot}}$. This conclusion is consistent with recent results on the excited-state isomerization of T to C which placed the rate in the picosecond time regime (12).

TABLE I Vibrational frequencies (cm^{-1}) of Raman bands observed for the THF solutions of 5H-dibenzo (a,d) cycloheptene (DBCH) and its radical anion ($DBCH^{\bar{\cdot}}$). Spectra are presented in Figures 1A and 1B, respectively

DBCH				$DBCH^{\bar{\cdot}}$	
Band	Freq. (cm^{-1})	Band	Freq. (cm^{-1})	Band	Freq. (cm^{-1})
a	168	n	840	a	342
b	271	o	1041	b	697
c	304	p	1109	c	734
d	329	q	1159	d	1132
e	391	r	1200	e	1166
f	447	s	1217	f	1416
g	496	t	1310	g	1471
h	581	u	1413	h	1530
i	620	v	1489	i	1584
j	703	w	1564		
k	731	x	1601		
l	762	y	1620		
m	801				

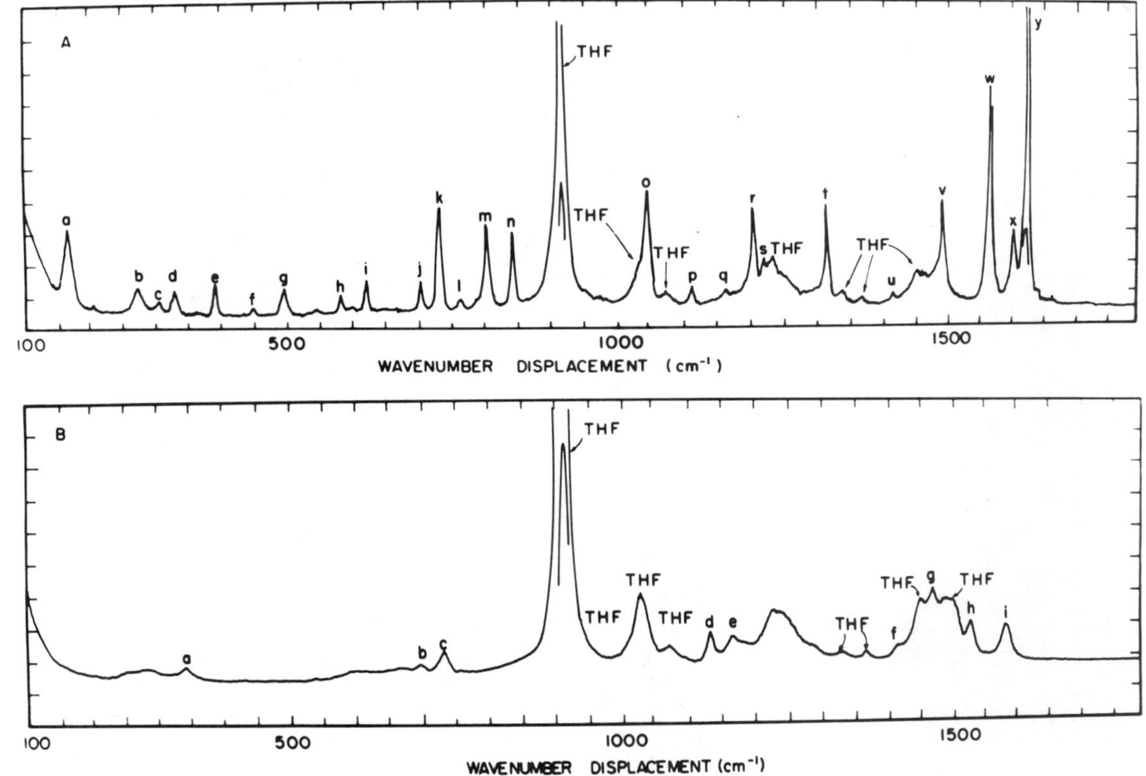

Figure 1 *Resonance Raman spectrum of (A) DBCH and (B) $DBCH^{\bar{\cdot}}$ obtained with CW argon ion laser excitation at 488 nm. Solvent was THF.*

The rate of ground-state isomerization of $C^{\bar{\cdot}}$ to $T^{\bar{\cdot}}$ may also occur on the picosecond time scale as a result of the weakened C=C bonding caused by the delocalization effects of the additional electron. TR^3 spectra of the ionic intermediates recorded with picosecond time resolution should contribute significantly to testing such a proposed mechanism.

Acknowledgement

The early instrumental contributions of Ms. J. B. Pallix and Mr. D. A. Gilmore to this work are gratefully acknowledged. This research was supported by grants from the National Science Foundation.

REFERENCE

1. G. H. Atkinson "Time-Resolved Raman Spectroscopy" in Advances in Infrared and Raman Spectroscopy, Vol. 9 (R. J. N. Clark and R. E. Hester, Eds.) Heyden, London 1982 pp 1-67.

2. G. H. Atkinson "Time-Resolved Raman Spectroscopy" in Advances in Laser Spectroscopy Vol. 1, (B. A. Garetz and J. R. Lombardi, Eds.) Heyden, London 1982, pp 155-165.

3. L. R. Dosser, J. B. Pallix, G. H. Atkinson, H. C. Wang, Levin and M. Szwarc, Chem. Phys. Lett., 62 (1979) 555.

4. F. Gerson, H. Ohya-Nishiguchi, M. Szwarc and G. Levin, Chem. Phys. Lett. 52 (1977) 587.

5. H. C. Wang, G. Levin and M. Szwarc, J. Amer. Chem. Soc., 99 (1977) 2642.

6. T. Shida and W. H. Hamill, J. Chem. Phys., 44 (1966) 4372.

7. G. H. Atkinson, H. Shindo, and J. B. Pallix (unpublished results).

8. D. S. Brackman and P. H. Plesch, J. Chem. Soc., (1952) 2188.

9. N. L. Wendler, D. Taub and R. D. Hoffsommer, Jr., Chem. Abstr., 65 (1966) 5428b.

10. G. H. Atkinson, D. A. Gilmore, L. R. Dosser and J. B. Pallix J. Phys. Chem. 86, (1982) 2305.

11. C. Takahaski and S. Maeda, Chem. Phys. Lett., 28 (1974) 22.

12. M. Sumitani and K. Yoshihara, Bul, Chem. Soc. Jpn., 55 (1982) 85.

TIME RESOLVED RAMAN STUDIES OF RADIATION PRODUCED RADICALS†

G. N. R. Tripathi
R. H. Schuler

Radiation Laboratory and Department of Chemistry
University of Notre Dame
Notre Dame, Indiana

The time resolved resonance Raman spectra of semiquinone and phenoxyl radicals produced by electron pulse radiolysis are obtained and analyzed. The kinetic behavior, spectral characteristics and nature of the resonant electronic transitions are briefly discussed.

I. INTRODUCTION

Electron pulse radiolysis is a convenient method for producing reactive free radicals in high concentration for short durations under controlled conditions. An apparatus for time resolved Raman studies of radiation produced radicals has been developed. The resonance Raman spectra of a number of semiquinone anions and p-aminophenoxyl radical are obtained and analysed. These radicals absorb moderately ($\varepsilon \sim 4\text{-}7 \times 10^3\ M^{-1}cm^{-1}$) between 400 - 450 nm and have mean lifetimes ranging between a few microseconds to a few seconds (1). They play a vital role as intermediates in many chemically and biologically important processes such as photosynthesis and mitochondrial electron transport. The observed Raman

†Research described herein was supported by the Office of Basic Energy Sciences of the Department of Energy. This is Document No. NDRL-2360 from the Notre Dame Radiation Laboratory.

ISBN 0-12-066280-9

spectra reveal structural and kinetic information about the radicals and have been useful in illucidating the nature of the electronic transitions in resonance.

II. EXPERIMENTAL

The radicals are produced in the Raman cell by radiation chemically oxidizing the hydroquinones or phenols in N_2O saturated aqueous solutions using 10 ns to 5 μs electron pulses of 2.3 MeV at current ∿ 1 A. The Raman scattering is generated by a N_2 laser pumped dye laser (FWHM = 10 ns), tunable between 337-900 nm. A Spex 0.85m double monochromater (1800 g/mm gratings) has been used for the spectral dispersion and a gated optical multi-channel analyser (OMA) (reticon, gate width, 5-20 ns) or a photomultiplier tube (PMT) in conjunction with a sampling oscilloscope (gate width, 350 ps) as detectors. The experiments were performed at a rate of 7.5 to 50 per second and data collected by an LSI 11 microcomputer for signal averaging purposes.

III. RESULTS AND INTERPRETATION

The resonance Raman spectra of p-benzosemiquinone and 15 of its derivatives were obtained using PMT detection (2). Fig. 1 shows the typical Raman spectra of p-benzosemiquinone and its fluorinated derivative ($C_6H_4\bar{O}_2^{\bullet}$ and $C_6F_4\bar{O}_2^{\bullet}$). Fig. 2 depicts the kinetic behavior of $C_6F_4\bar{O}_2^{\bullet}$ (tetra-fluoro-p-benzosemiquinone) (3). This radical exhibits an exponential decay ($k = 0.55 \times 10^3 s^{-1}$; half life = 1.25 ms) in the initial stages of its production (< 2 ms). At longer times the observed radical decay is slowed down considerably, apparently due to secondary reactions producing $C_6F_4\bar{O}_2^{\bullet}$. The nonfluorinated p-benzosemiquinones are comparatively less reactive and decay on the seconds time scale or longer depending on the concentration. Since the radiation chemical system used is primarily oxidizing, p-benzosemiquinone is essentially the only radical present and decay should initially take place by disproportionation process. Once the equilibrium is reached between hydroquinone, semiquinone and benzoquinone, further decay of the semiquinone occurs due to slow reactions yielding coupled products.

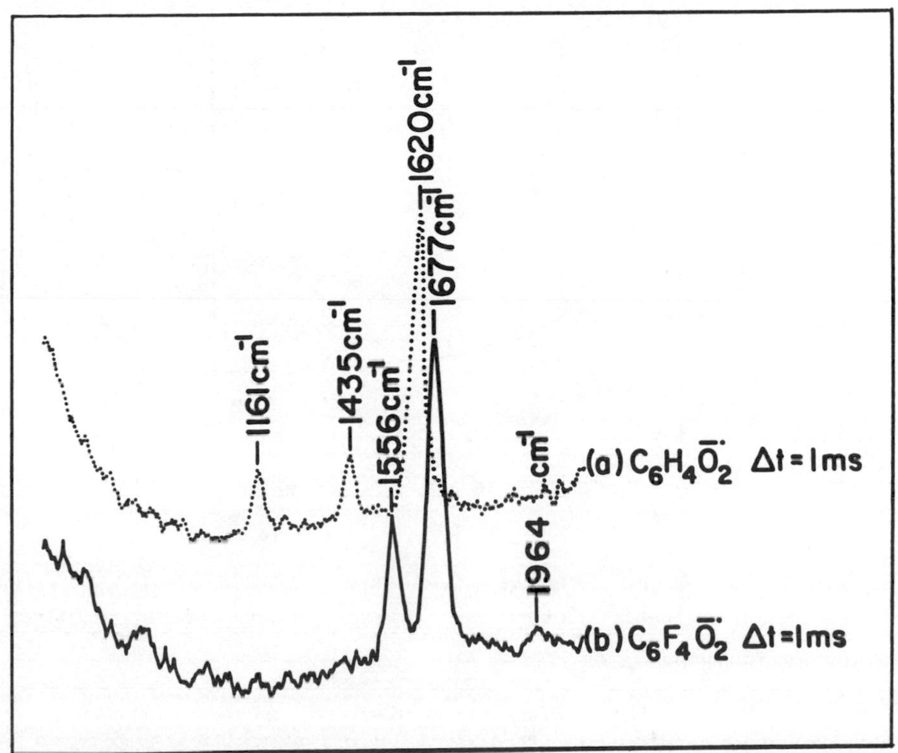

FIGURE 1. Resonance Raman spectrum (solid) of tetrafluoro-p-benzosemiquinone anion observed 1 ms (Δt) after 1 μs electron-pulse radiolysis of tetrafluorohydroquinone (2×10^{-3}M; pH ∿ 10.5, N_2O saturated) in water. Probe laser wavelength = 437 nm. Spectrum is averaged over 25 scans. The dotted line shows the Raman Spectrum of p-benzosemiquinone anion under similar conditions.

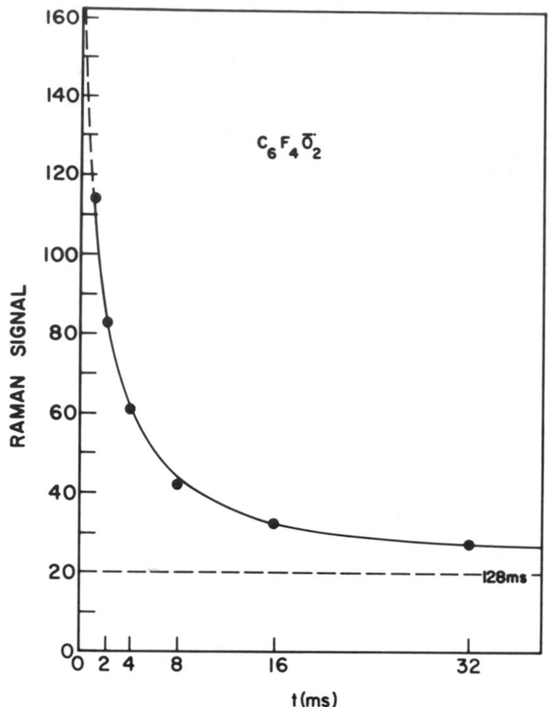

FIGURE 2. Kinetic plot of the 1677cm^{-1} signal of tetrafluoro-p-benzosemiquinone anion.

The p-benzosemiquinone radical ($C_6H_4\bar{O}_2^{\bullet}$) exhibits 4 polarized Raman bands corresponding to the totally symmetric (a_g) modes at 1620cm^{-1} (CC stretch, 8a), 1435cm^{-1} (CO stretch, 7a), 1161cm^{-1} (CH planar bend, 9a) and 481cm^{-1} (CCC bend, 6a) and some weaker bands due to combination modes (2). The a_g ring breathing (Wilson mode 1) and CH stretch (Wilson mode 2) vibrations are not observed due to lack of sufficient resonance enhancement. The CO stretch frequency indicates weaker CO bonds (bond order $\sim$ 1.5) than p-benzoquinone. In the substituted p-benzosemiquinones, additional Raman bands characteristic of the nature, number and positions of the substituent groups appear. For derivatives with moderately interacting substituents, the ring CC stretch (8a) frequency (ν_{CC}) increases with the (0,0) band energy ($\nu_{0,0}$). ($\partial\nu_{CC}/\partial\nu_{0,0}$ = 2.06×10^{-2}). The CO stretch frequency shows an opposite trend ($\partial\nu_{CO}/\partial\nu_{0,0}$ = -3.43×10^{-2}). In derivatives with highly interacting substituents like tetrafluoro-p-benzosemiquinone anion, the CC as well as CO stretch bands (1677 and 1556cm^{-1}) are considerably shifted towards higher frequencies with respect to the protonated counterpart (1620 and 1435cm^{-1}).

The resonance Raman spectroscopy is a sensitive probe for illucidating the nature of the electronic transitions in resonance. The fundamental vibrations most resonance enhanced are those which are responsible for converting the molecule from its ground to excited state geometry. For the 400-450 nm electronic absorption in p-benzosemiquinone anions, conflicting assignments have been proposed ($^2B_{3g} \longrightarrow {}^2B_{1u}$ and $^2B_{3g} \longrightarrow {}^2A_u$) by the quantum chemists (4). Fig. 3 shows the calculated bond orders in the ground $^2B_{3g}$ and the excited $^2B_{1u}$ and ^2A_u electronic states.

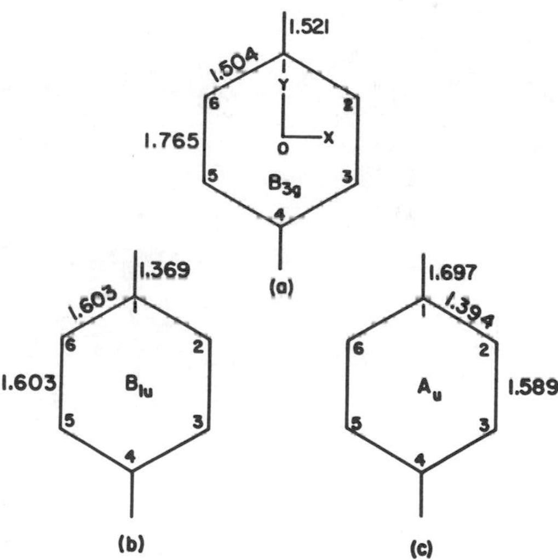

FIGURE 3. Calculated bond orders in the $^2B_{3g}$, $^2B_{1u}$ and ^2A_u states of p-benzosemiquinone anion.

Fig. 4 depicts the normal modes 8a and 1. The transition $^2B_{3g} \longrightarrow {}^2B_{1u}$ favors the resonance Raman enhancement of the 8a phenyl mode and $^2B_{3g} \longrightarrow {}^2A_u$ of 1. The 8a mode is strongly resonance enhanced in p-benzosemiquinone while the ring breathing mode 1 is not observed. This clearly supports the assignment, $^2B_{3g} \longrightarrow {}^2B_{1u}$.

The phenoxyl radicals are very much more reactive and are observed to disappear in a process second order in radical concentration on the microsecond time scale.

$$\phi O^{\cdot} + \phi O^{\cdot} \xrightarrow{k_2} \text{products}$$

where second order rates are limiting. The maximum radical concentration available for observation at any given time is C(t), which can be expressed as

$$C(t) = \frac{1}{1/C(0)+2k_2t} < \frac{1}{2k_2t}$$

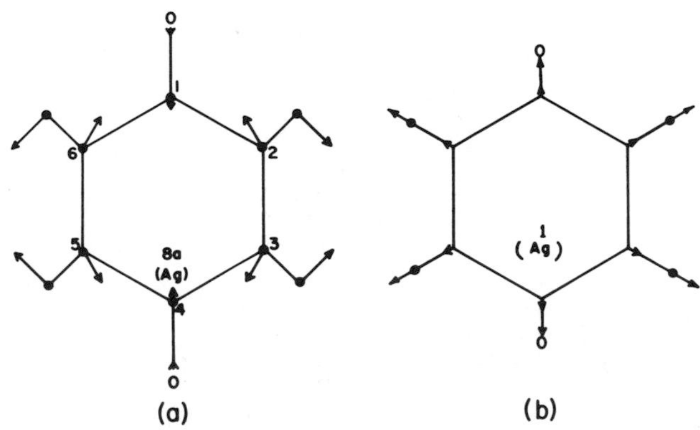

FIGURE 4. CC stretch (8a) and ring breathing (1) phenyl modes.

Since k_2 for diffusion controlled reaction is > $10^9 M^{-1} S^{-1}$, concentration will necessarily be less than micromolar at the millisecond times and the measurements must be performed at μs time scale (5,6). Since Cerenkov radiation is more intense than the Raman signal some form of a gated detector must be used. We have obtained the Raman spectrum (Fig. 5) of p-aminophenoxyl radical ($\lambda max = 440$ nm, $\varepsilon \sim 7000\ m^{-1} cm^{-1}$, $2k_2 = 2.5 \times 10^9 M^{-1} S^{-1}$) 150 ns after the termination of a 1 μs electron pulse using OMA detection (initial concentration, $\sim 3 \times 10^{-5}$M). The spectrum was generated by excitation at 437 nm. The radical exhibits two

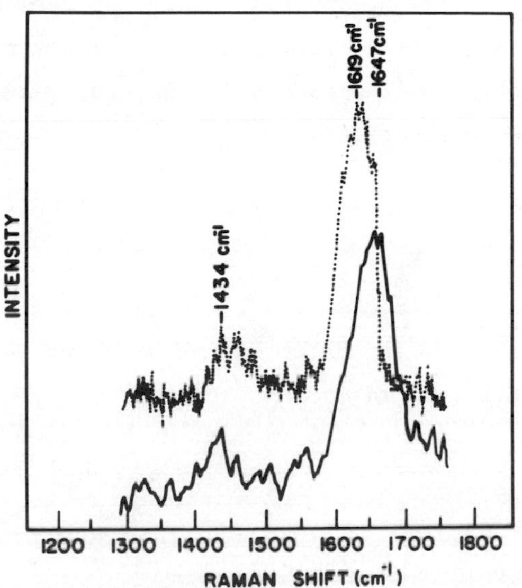

FIGURE 5. Resonance Raman spectrum of p-aminophenoxyl radical observed in the pulse radiolysis of a p-aminophenol solution (5×10^{-3}M, N_2O saturated ph ∿ 10) 150 ns after a 1 μs electron pulse. Spectrum represents the data from 3500 experiments which have been smoothed using a 25 point parabolic smoothing routine. Background recorded with a 200 μs delay between the laser pulse (10 ns) and detector gate pulse (20 ns) has been subtracted. Raman spectrum (dotted) observed for tert-butyl p-benzosemiquinone under similar conditions is given for comparison purposes (700 experiments).

resonance enhanced Raman bands at 1434 and 1647cm^{-1}. The radical concentration is reduced to ∿ 60% at 10 μs, ∿ 10% at 100 μs and ∿ 1% at 1000 μs, as measured by Raman band intensities, corresponding to the dependence expected from a second order rate constant of ~$10^9 M^{-1} S^{-1}$. The 8a phenyl mode (1647cm^{-1}) is strongly resonance enhanced. The electronic transition (λmax = 440 nm) is, therefore, analogous to the 440-450 nm band system in p-benzosemiquinone and is assigned as $^2A_1 \longrightarrow {}^2A_1$ (C_{2v} point group).

These studies have demonstrated the power of the kinetic Raman spectroscopy in identifying the reactive species in the radiation induced reactions, revealing their structure as manifested in their vibrational spectra, following the reaction kinetics, and assigning the electronic transitions in the poorly resolved absorption spectra in solution.

REFERENCES

1. Schuler, R. H., Radiat. Res. 69, 417 (1977). This paper describes the experimental technique used for observing the kinetic optical spectra in our laboratory.
2. Tripathi, G. N. R., and Schuler, R. H., J. Chem. Phys. 76, 2139 (1982); Tripathi, G. N. R., J. Chem. Phys. 74, 6044 (1981).
3. Tripathi, G. N. R. and Schuler, R. H., J. Phys. Chem., in press.
4. Harada, Y., Mol. Phys. 8, 273 (1964); Chang, H. M., Jaffe, H. H. and Masmanidis, C. A., J. Phys. Chem. 78, 1118 (1975).
5. Tripathi, G. N. R. and Schuler, R. H., J. Chem. Phys., 76, 4289 (1982).
6. Tripathi, G. N. R. and Schuler, R. H., Chem. Phys. Lett. 88, 253 (1982).

RAMAN SPECTRA OF PHENOXY RADICALS

Hitoshi Shindo

National Chemical Laboratory for Industry
Tsukuba Research Center
Yatabe, Ibaraki, Japan

I. INTRODUCTION

Time resolved resonance Raman spectroscopy was applied to the study of structure and kinetics of reaction intermediates during oxidation of phenols(1,2). A stopped-flow reaction apparatus was combined with a Raman spectrometer which is equipped with a multi-channel detection system(PAR; OMA). The frequency range of about 1000cm^{-1} is covered at a time. Computer control facilitates repetition of measurement and accumulation of signal.

Acidic(0.5M H_2SO_4) aqueous solutions of phenols(10mM) were used as reactants and cerium(IV) solution(5-20mM) was used as an oxidant. Two solutions were mixed rapidly(dead time 5ms) and the Raman spectrum of the mixture was observed after certain time intervals. Reaction process was followed with a time resolution of 20ms.

ISBN 0-12-066280-9

II. RESULTS AND DISCUSSION

In Fig.1(A) is shown the Raman spectra observed during oxidation of phenol-h_6. Relative Raman intensities, against solvent bands, of the Raman bands with asterisks indicated similar time dependence, so that they are attributable to a single chemical species. Similar spectral patterns were observed for various monosubstituted phenols. This characteristic pattern seems to reflect a common structure of the reaction intermediates. By reference to the esr measurement(3) with the same reaction system, the spectra were assigned to those of phenoxy radicals.

Nine Raman bands were observed for $C_6H_5O\cdot$, of which five stronger bands are polarized. The band at 1615cm^{-1} is probably polarized, too. In order to help the assignment of the Raman bands to particular vibrational modes, the same experiment was performed with phenol-d_6. The result is shown in Fig.1(B).

Compared to other monosubstituted benzenes, the Raman spectrum of $C_6H_5O\cdot$ has very different features such as; 1) missing of the ring breathing mode; 2) high intensity and small deuteration shift of the band at 1485cm^{-1}; 3) appearance of a polarized band at 1615cm^{-1} and its small d-shift; 4) disappearance of bands at 1374, 1262 and 1202cm^{-1} by deuteration. These differences seem to come from the peculiar structure of the radical. The unpaired electron on the oxygen atom is highly delocalized all over the benzene ring (3), giving the structure shown in Fig.2, in which $r_{23} < r_{34} < r_{12}$. The C-O bond has a considerable double bond character.

The assignment of the Raman bands of the radical is given below considering frequencies, intensities and depolarization ratios. The 1615cm^{-1} band with the smallest d-shift is assigned to C=O stretching mode in accordance with ir data for stable phenoxy radicals. For monosubstituted benzenes, seven($4A_1+3B_2$) ring vibrations are expected above 700cm^{-1}. The two strong bands of the radical at 1596 and 1485cm^{-1} are clearly polarized, thus,

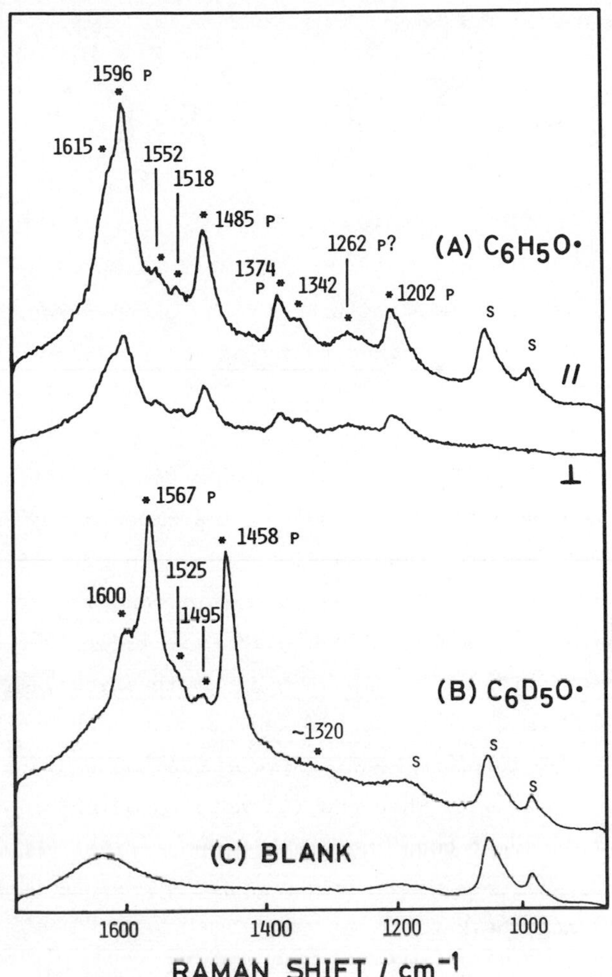

FIGURE 1. Resonance Raman spectra of the phenoxy radical-h_5 and -d_5 observed by 488nm excitation. (A); $C_6H_5O\cdot$ with parallel and perpendicular polarization, (B); $C_6D_5O\cdot$ with parallel polarization, (C); without phenol. The letters P and S denote polarized bands and solvent bands, respectively.

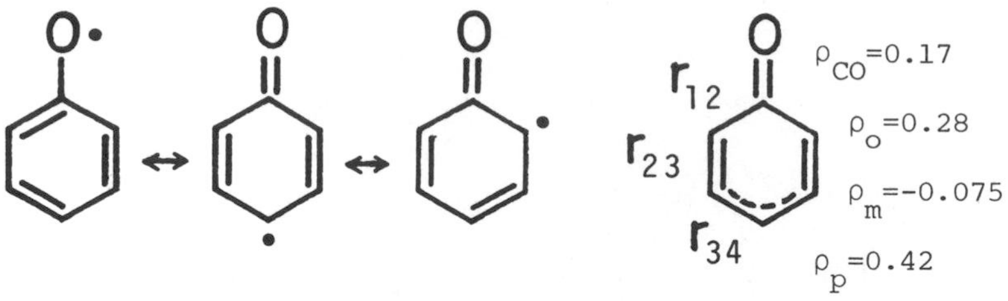

ρ_{CO}=0.17
r_{12}
ρ_o=0.28
r_{23}
ρ_m=-0.075
r_{34}
ρ_p=0.42

FIGURE 2. Resonance structures of the phenoxy radical (ρ denotes spin density of the odd electron).

assigned to ν_{8a} and ν_{19a} vibrations, respectively. It should be noted that these two modes are roughly described as symmetric stretching of r_{23} and r_{34} in Fig.2. Three bands at 1552, 1518 and 1342cm^{-1} are assigned to B_2 modes, ν_{8b}, ν_{19b} and ν_{14}, respectively.

The remaining three bands in Fig.1(A) at 1374, 1262 and 1202 cm^{-1} are assigned to C-H in-plane bending vibrations. Two of them at 1374 and 1202cm^{-1} are polarized and belong to A_1 species. The fact that C-D in-plane bending modes are too weak to be observed seems to indicate that the intensities of these bands mainly come from their coupling with the ring vibrations.

Similar experiments were performed using various monosubstituted phenols and the results are shown in Fig.3. Ortho- and para- substituted species indicated different spectral patterns in accordance with the monomeric structure of the species.

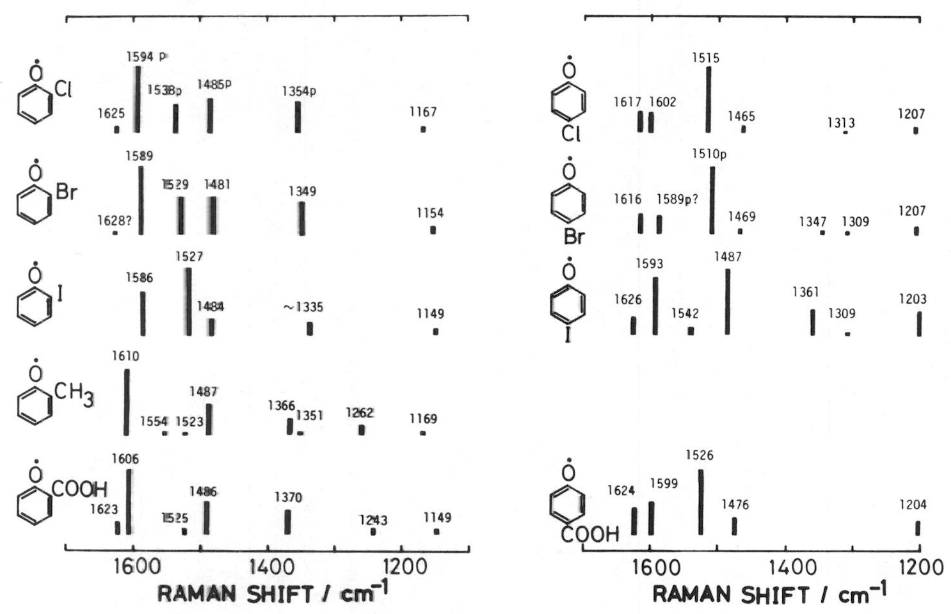

FIGURE 3. Raman bands of monosubstituted phenoxy radicals observed by 488nm excitation.

REFERENCES

1. Shindo, H. and Hiraishi, J., Chem. Phys. Lett., 80, 238 (1981).
2. Shindo, H. and Hiraishi, J., J. Raman Spectry., 12, 194 (1982).
3. Stone, T. J. and Waters, W. A., J. Chem. Soc., 213 (1964).

THE ELECTRONIC AND RESONANCE RAMAN SPECTROSCOPY OF THE INORGANIC TRANSIENTS S_2^-, S_3^- AND Se_2^- STABILIZED IN ULTRAMARINE

Robin J.H. Clark
David P. Fairclough
Mohamedally Kurmoo

Christopher Ingold Laboratories
University College London
London, U.K.

Electronic and resonance Raman spectral studies have established the nature of the key chromophores in ultramarine blue, green, violet (various shades thereof) and pink, as well as in ultramarine selenium. They are S_3^-, S_2^-, S_4 or S_4^- and Se_2^-, as demonstrated by excitation profile studies. Spectroscopic constants for the radical anions are presented. Franck-Condon fits to the Raman-band excitation profiles allow the calculation of bond length changes for S_2^- and Se_2^- on excitation to the resonant electronic states.

ISBN 0-12-066280-9

INTRODUCTION

The species responsible for the blue colour formed when alkali metal polysulphides are dissolved in highly polar solvents such as dimethylformamide and hexamethylphosphoramide has, until recently, been very uncertain. However, resonance Raman (RR) studies have now shown that this species (λ_{max} = 600 nm, $\varepsilon_{max} \simeq 10^4$ $M^{-1}cm^{-1}$) is the radical anion S_3^- (1). It has not proved possible to prepare salts of this ion since, when normally effective counter-ions are added to its solutions in these solvents, the S_3^- ion dimerizes to S_6^{2-} and other species. The rate of this dimerization process and of similar processes in other polar solvents, even in the absence of counter-ions, is very rapid. A study of the kinetics of these processes, and of the nature of other possible short-lived intermediates formed, could be important to our understanding of the many reactions (e.g. vulcanisation of rubber, bioinorganic processes, sulphur-cell battery processes) in which sulphur radical anions may play key roles.

Earlier e.s.r. and resonance Raman studies have shown that the S_3^- ion is trapped at cubic sites in the semi-precious mineral lazurite (approximately $Na_8[Al_6Si_6O_{24}]S_4$) (2,3). Indeed this mineral, and its synthetic equivalent ultramarine blue, has been through the ages an important pigment owing to the richness and durability of its royal blue colour. Evidence has also been presented for small amounts of S_2^- (λ_{max} = 380-400 nm) in ultramarine blue (1).

By modifications to the preparative conditions it is possible to prepare further forms of ultramarine which are green, various shades of violet, and pink (or red). Moreover, it is also possible to substitute selenium for sulphur in such preparations, and to obtain a brick-red material referred to as ultramarine selenium. By a combination of electronic and RR spectroscopy, it is possible to investigate the nature of the chromophores present in each of these materials and to establish values for various spectroscopic constants of the chromophores.

ULTRAMARINE GREEN

Previous studies of ultramarine green had been confined, on the shortwavelength side, to 454.5 nm (1). This exciting line only falls within the low energy wing of the lowest allowed electronic band of the S_2^- ion, and much higher quality RR spectra could be obtained by use of violet (413.1 and 406.7 nm) and (to a lesser extent) ultra-violet (356.7, 350.7 and 337.5 nm) lines. In particular, with 406.7 nm excitation the RR spectrum of the S_2^- ion chromophore is very intense and consists of a progression in $\nu(SS)$ to 9ν. These observations are similar to those made for S_2^- doped in alkali halide lattices (4,5). The derived spectroscopic constants ω_e and $\omega_e x_e$ have the values 597.0 and 2.5 cm^{-1}, respectively.

The excitation profiles of the ν, 2ν and 3ν bands of S_2^-, as well as of $\nu_1(a_1)$ of S_3^-, as derived from studies of ultramarine green, clearly demonstrate that the 380 nm band arises from S_2^- and the 600 nm band from S_3^-. The observed excitation profiles for ν, 2ν and 3ν of S_2^- can be fitted on the Franck-Condon scattering scheme (6-8) by taking ν_{00} for the $^2\Pi_u \leftarrow {}^2\Pi_g$ transition to be 20,035 cm^{-1} and the excited-state value for ν to be 360 cm^{-1} (5); the best fit yields a shift parameter (change in SS bond length on excitation to the $^2\Pi_u$ state) of 0.29 Å and a Γ value (damping factor) of 500 cm^{-1}. Thus the SS bond length change on $^2\Pi_u \leftarrow {}^2\Pi_g$ excitation is similar to that for the I-I bond of I_2 on $^3\Pi^+_{ou} \leftarrow {}^1\Sigma^+_g$ excitation (0.35 Å).

ULTRAMARINE VIOLET AND PINK

Extensive studies of a wide range of differently shaded ultramarine violets and also of ultramarine pink (sometimes referred to as ultramarine red) lead to the conclusion that all these species contain S_3^-, S_2^- and a further species (which may be either S_4 or S_4^-). The proportions of the first two ions diminish while that of the last increases, as the shade of the ultramarine tends towards pink. The nature of this third species has not been identified for certain, but it is characterised by Raman bands at 355, 653 and 674 cm^{-1}, all of whose excitation profiles maximise at ca. 520 nm, which corresponds to the absorption maximum of the band responsible for the pink colour of ultramarine pink.

ULTRAMARINE SELENIUM

Ultramarine selenium, which is brick red, possesses two absorption maxima at 490 and 350 nm. Irradiation within the contour of the 490 nm band yields a RR spectrum which is dominated by a long progression (to 13ν with 488.0 nm excitation) in a single mode for which ν = 327.9 cm^{-1}. This mode can readily be identified with the SeSe stretching mode of the Se_2^- ion, by analogy with its value when doped into alkali halide lattices (325 cm^{-1} in KI from Raman measurements (2), ca. 327 cm^{-1} in KI from fluorescence measurements (9-11)). The assignment is confirmed by the lack of infrared activity for the 327.9 cm^{-1} band, and by the absence of any bands which might be attributed to asymmetric stretching or bending modes of other possible ions present in the lattice, such as Se_3^-. The RR spectrum of Se_2^- in ultramarine selenium is much longer and more spectacular than that observed for the ion in alkali halide lattices (to 3ν only) (2). Derived spectroscopic constants for the Se_2^- ion are $\omega_e = 329.6$ cm^{-1} and $\omega_e x_e = 0.70$ cm^{-1}. The resonant transition is, in this case also, considered to be $^2\Pi_u \leftarrow {}^2\Pi_g$ (10-12).

Franck-Condon simulation of the excitation profiles of ν, 2ν and 3ν of Se_2^-, using the experimental value for ν_{00} of 16,100 cm^{-1} (from fluorescence measurements) and the value for ν in the excited state (220 cm^{-1}) (11,12), leads to a band-shift parameter of ca. 0.27 Å on excitation to the resonant $^2\Pi_u$ state ($\Gamma \simeq 300$ cm^{-1}).

CONCLUSION

The results indicate that the ultramarine lattice is an excellent one in which to trap otherwise short-lived sulphur and selenium radical anions. RR studies enable the chromophores to be identified, their spectroscopic constants to be calculated from the observed Raman-band progressions (13), and their bond-length changes on excitation to the resonant excited state to be deduced from Franck-Condon simulation of the Raman band excitation profiles.

ACKNOWLEDGMENT

The authors thank the S.E.R.C. for financial support and Reckitt's Colours Ltd. for some of the samples.

REFERENCES

1. R.J.H. Clark and D.G. Cobbold, Inorg. Chem. 1978, 17, 3169.
2. W. Holzer, W.F. Murphy and H.J. Bernstein, J. Mol. Spectrosc. 1969, 32, 13.
3. R.J.H. Clark and M.L. Franks, Chem. Phys. Letters, 1975, 34, 69.
4. L. Vannotti and J.R. Morton, Phys. Rev., 1967, 161, 282.
5. C.A. Sawicki and D.B. Fitchen, J. Chem. Phys., 1976, 65, 4497.
6. R.J.H. Clark and B. Stewart, J. Amer. Chem. Soc. 1981, 103, 6593.
7. M. Samoc, W. Siebrand, D.F. Williams, E.G. Woolgar and M.Z. Zgierski, J. Raman Spectrosc., 1981, 11, 369.
8. R.J.H. Clark, T.J. Dines and M.L. Wolf, J. Chem. Soc. (Faraday II), 1982, 78, 679.
9. J. Rolfe, J. Chem. Phys., 1968, 49, 4193.
10. A.C. Boccara, J. Duran, B. Briat and P.J. Stephens, Chem. Phys. Letters, 1973, 19, 187.
11. M. Ikezawa and J. Rolfe, J. Chem. Phys., 1973, 58, 2024.
12. G.J. Vella and J. Rolfe, J. Chem. Phys., 1974, 61, 41.
13. R.J.H. Clark in Advances in Infrared and Raman Spectroscopy (R.J.H. Clark and R.E. Hester, eds.) Vol. 1, p. 143. Heyden, London, (1975).

TIME-RESOLVED RESONANCE RAMAN SPECTROSCOPY OF THE K_{610} AND O_{640} PHOTOINTERMEDIATES OF BACTERIORHODOPSIN[1]

Steven O. Smith
Mark Braiman[2]
Richard Mathies

Department of Chemistry
University of California
Berkeley, California

I. INTRODUCTION

Photon absorption by the retinal chromophore in bacteriorhodopsin (BR), the purple membrane protein in *Halobacterium halobium*, initiates a cyclic photochemical reaction which pumps protons across the bacterial cell membrane [1,2]. The sequence of intermediates in the BR photocycle (Figure 1) was determined by time-resolved absorption spectroscopy [3]. Resonance Raman and chromophore extraction studies have demonstrated that the retinal chromophore in the light-adapted pigment, BR_{570}, has an all-*trans* configuration and that the retinal-lysine Schiff base linkage is protonated [4-6]. Photolysis of BR_{570} forms a red-absorbing intermediate, K_{610}, in ~10 picoseconds [7]. K_{610} thermally decays to a blue-absorbing intermediate, M_{412}, that contains an unprotonated 13-*cis* Schiff base chromophore [4-6,8]. Conversion of M_{412} back to the parent BR_{570} proceeds through

[1]*This research was supported by grants from the NSF (CHE-8116042) and the NIH (EY-02051). A cavity dumper was borrowed from the San Francisco Laser Center, supported by the National Science Foundation under Grant #CHE-16250, awarded to the University of California at Berkeley in collaboration with Stanford University.*

[2]*Present address: Departments of Chemistry and Biology, Massachusetts Institute of Technology, Cambridge, MA 02139.*

ISBN 0-12-066280-9

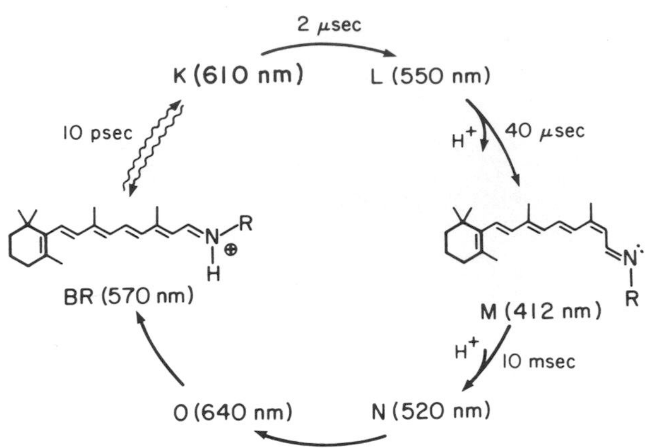

Figure 1. Bacteriorhodopsin photocycle: Absorption maxima of intermediates are given in parentheses.

another red-shifted intermediate, O_{640}. We are developing new time-resolved resonance Raman techniques to probe the K_{610} and O_{640} structures, which are "transition states" in the rapid photoisomerization of BR_{570} to M_{412} and in the slower thermal isomerization of M_{412} back to BR_{570}.

To obtain resonance Raman spectra of these intermediates, a strong "pump" laser beam is used to initiate photocycling of the BR_{570} pigment. Maximum photoconversion is achieved by selecting a photolysis laser wavelength (514 nm) which is strongly absorbed by BR_{570}. Then, by employing a Raman "probe" laser wavelength within the absorption band of the photointermediate of interest, we can resonantly enhance its Raman spectrum selectively. Optimal resonance enhancement of the red-absorbing intermediates, K_{610} and O_{640}, is obtained using laser excitation wavelengths above 630 nm. The best experimental design for obtaining resonance Raman spectra of K_{610} and O_{640}, then, is a green-pump/red-probe experiment with the probe beam temporally delayed from the pump beam.

Room temperature Raman spectra of K_{610} with 60-nsec time resolution were obtained with cavity-dumped pump and probe lasers. These spectra are very similar to red-probe spectra of K previously obtained at 77°K [9], except for a shift of the C_{15}-H hydrogen out-of-plane (HOOP) vibration from 973 cm^{-1} to 987 cm^{-1}. Raman spectra of O_{640} with msec time resolution were obtained using a flow system with spatially separated CW laser beams. Strong HOOP vibrations are also observed in the Raman spectrum of O_{640}, at

945, 959 and 977 cm^{-1}. The intense HOOP vibrations in K_{610} and O_{640}, the only intermediates in the BR photocycle in which they have been observed, are characteristic of twisted chromophore structures, as might be expected for transition states in photochemical or thermal isomerizations.

II. THE K_{610} INTERMEDIATE

A. *Experimental Design*

The experimental apparatus used for the dual-pulse K Raman experiment is shown in Fig. 2. Two cavity-dumped lasers, each generating 25-nsec pulses at 1 MHz, sequentially illuminate a flowing suspension of purple membrane. The time delay between the 514-nm photolysis and 647-nm probe pulses is set at 60 nsec. Because the 514-nm photolysis pulse would generate a prohibitively large fluorescence background, it is necessary to gate the detector on only during the probe pulses.

As in the earlier 77°K steady-state experiments [10], the photoalteration parameter F can be used to describe the approach to a steady state between BR and K. For a short pulse, F is expressed most conveniently in terms of the pulse energy E (photons/pulse), the focused beam area A, and the absorption

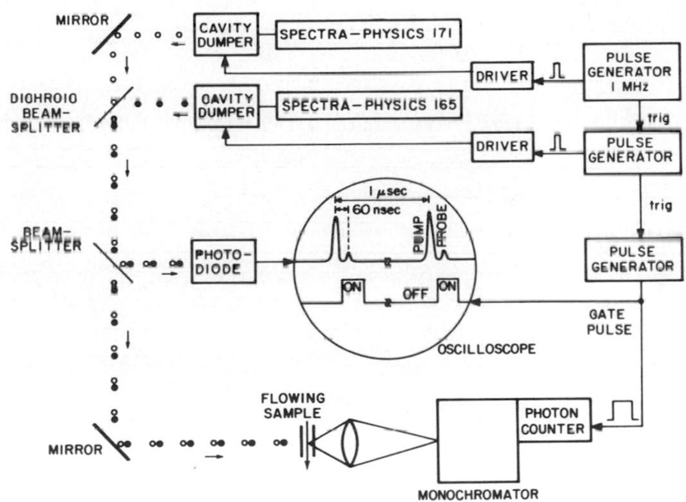

Figure 2. Schematic of nsec time-resolved Raman apparatus.

cross sections (σ_{BR},σ_K) and quantum yields (Φ_{BR},Φ_K) for the forward and reverse photoreactions:

$$F = \frac{E}{A} (\sigma_{BR} \Phi_{BR} + \sigma_K \Phi_K).$$

Using published values for the absorption cross sections and quantum yields [11], we calculate that F is 6 for the 100-nJ 514-nm photolysis pulses and 1 for the 30-nJ 647-nm probe pulses (A=100μm^2). Since the pump beam photoalteration is greater than unity, the pump pulse creates a quasi-steady-state mixture of K and BR. The probe pulse also has a large F and thus reconverts most of the K formed by the photolysis pulse back to BR while the Raman spectrum is generated.

The sample, flowing through a 1-mm diameter capillary at a speed of 300 cm/sec, traverses the coaxially focused laser beams in 4 μsec. This is slow enough to ensure that the sample is effectively stationary between the pump and probe pulses. While 4 μsec is also sufficiently long for the K produced to decay to L, it is unlikely that L makes a large contribution to the spectra. The concentration of L and its relative scattering cross section can be estimated for our experimental conditions. The L scattering contribution to our spectra should be $\lesssim$20% of the K contribution. The intensity and frequency of the ~1535 cm^{-1} shoulder on the K ethylenic line in Fig. 3C are consistent with this estimate. It would theoretically have been possible to keep the L contribution down to essentially 0%, by decreasing the repetition rate of the lasers to 250 kHz, thus ensuring that the sample experienced only one set of pulses per pass through the laser beam. This, however, was undesirable because it would have substantially decreased our signal levels and resulted in longer data collection times.

B. *Results and Discussion*

Figure 3 presents our time-resolved K_{610} Raman data. With both the 514-nm pump and the 647-nm probe beams, the spectrum contains scattering from both K and BR (Fig. 3A). The probe-only spectrum, corresponding to pure BR (Fig. 3B), was subtracted to produce the spectrum in Fig. 3C. This K_{610} spectrum is similar to those previously obtained at room temperature by single-beam, time-resolved techniques [12,13]. However, the earlier spectra, obtained with bluer Raman probe wavelengths, exhibited additional Raman scattering which could be due to the presence of small amounts of other photointermediates. Because we use a red probe wavelength which selectively enhances the K_{610} photointermediate, our Raman spectra do not exhibit these additional lines.

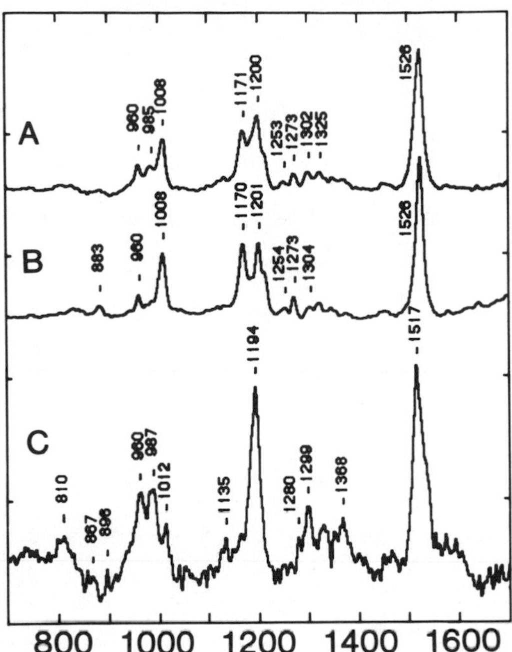

Figure 3. Time-resolved Raman spectra of BR_{570} and K_{610} at 4°C. (A) 514-nm pump and 647-nm probe pulses, separated by 60 nsec. (B) 647-nm probe pulses only. (C) Raman spectrum of K_{610} obtained by subtracting ~80% of (B) from (A). Spectra obtained with 4 cm^{-1} resolution and 2 cm^{-1} bins.

In Fig. 4A,B we compare our 60-nsec time-resolved spectrum of K_{610} with a red-probe K spectrum obtained at liquid-N_2 temperature using a spinning cell [9]. We will denote the K species trapped at liquid-N_2 temperature as K_{625}. All of the major lines in the spectra are identical under the two sets of conditions except for a line at 987 cm^{-1} in the K_{610} spectrum. Figures 4C,D similarly compare K molecules which contain a 15-deuterio retinal chromophore. These spectra show that the 1100-1300 cm^{-1} fingerprint regions of the 15-H and 15-D chromophores are the same in the frozen (K_{625}) and 60-nsec time-resolved (K_{610}) species. Our 15-deuterio fingerprint comparison method was used previously to provide evidence that K_{625} contains a 13-*cis* chromophore [9]. On the basis of the very close similarity of the fingerprint regions shown in Fig. 4, we conclude that the configuration of retinal in the room temperature 60-nsec K_{610} species is also 13-*cis*.

The 15-deuterio spectra also demonstrate that the line which is observed at 987 cm^{-1} in the 60-nsec room temperature species (Fig. 4A) is due to the C_{15}-H HOOP vibration, since it shifts to 794 cm^{-1} in the 15-D molecule (Fig. 4C). This line corresponds to the 973 cm^{-1} vibration in K_{625} (Fig. 4B), which shifts to 786 cm^{-1} in the 15-D derivative (Fig. 4D). The 15-H and 15-D vibrations are thus both ~10 cm^{-1} higher in K_{610} than in K_{625}. The difference in the relative intensity of this line and the line at 960 cm^{-1} in Fig. 4A,B probably results from a change in the mixing of these nearly degenerate vibrations. The 960 cm^{-1} line and the shoulder at 945 cm^{-1} are due to the HC_7=C_8H and HC_{11}=$C_{12}H$ "A_u" HOOP vibrations [14,15].

Thus, the only observed structural differences between K at 77°K and room temperature appear to be localized at the Schiff base end of the molecule. It seems likely that the protein conformation in K_{625} is frozen, i.e., unable to relax to accommodate

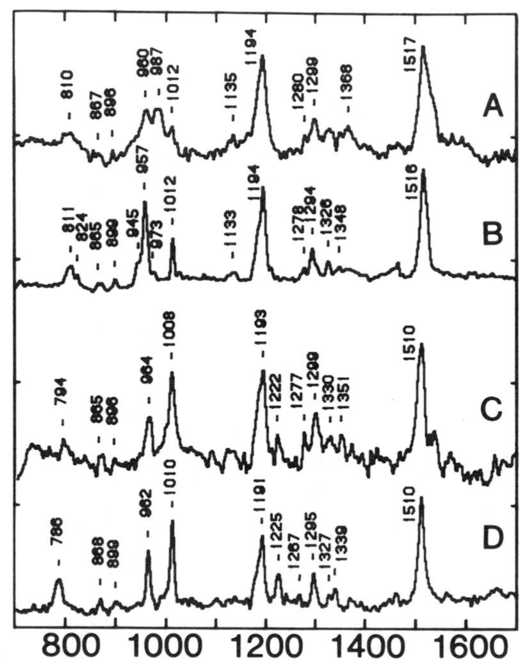

Figure 4. Comparison of 60-nsec time-resolved K_{610} spectra (A,C) with spectra of K_{625} obtained at 77°K using a spinning cell (B,D) [9]. Spectra in (A) and (B) are of native (15-H) K; spectra in (C) and (D) are of 15-D isotopically substituted K.

the newly isomerized chromophore. At room temperature, however, the protein residues near the chromophore should be able to move rapidly to a new free energy minimum. Our comparison of the two sets of K spectra suggests that a protein relaxation occurs near the Schiff base ($C{=}N^{+}H$) bond within 60 nsec after isomerization at room temperature.

III. THE O_{640} INTERMEDIATE

A. *Experimental Design*

O_{640} is a key intermediate in the transition between the 13-*cis* M_{412} intermediate and all-*trans* BR_{570}. To obtain resonance Raman spectra of the O_{640} intermediate we used the dual-beam flow system in Fig. 5. Bacteriorhodopsin was circulated at 200 cm/sec through a long section of glass capillary (diameter=1.4 mm). Raman spectra were taken with a 200-mW 752-nm probe laser beam spatially displaced from a 500-mW 514-nm pump beam. The time separation between the beams was controlled by translating the pump beam along the length of the

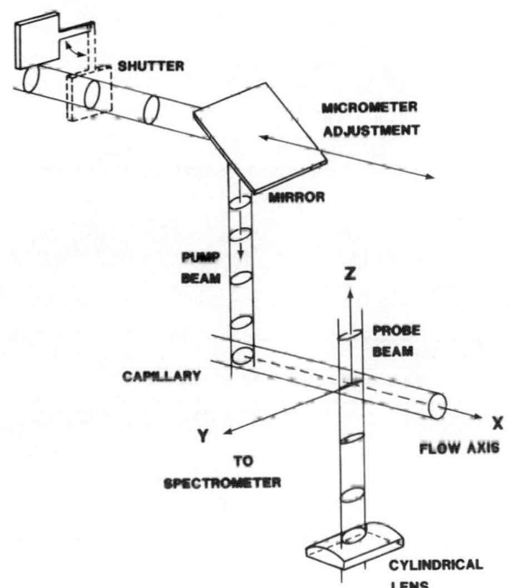

Figure 5. Dual-beam flow apparatus used to obtain resonance Raman spectra of O_{640}. A micrometer sets the time delay by displacing the pump beam upstream from the probe.

capillary. The useful time range of this apparatus was 0.2 to 20 msec. The optimal spatial separation for observing Raman scattering from O_{640} (4-6 msec) was determined by monitoring the laser-induced fluorescence from O_{640}.

The probability that the O_{640} chromophore will photoreact during passage through the beam is given by the photoalteration parameter [16]. Using extinction coefficients at 752 nm from ref. [3] and assuming a quantum yield of unity, we adjusted the probe laser power to keep its photoalteration less than 0.1. The probe beam was cylindrically focused along the flow axis, to permit a higher probe laser power without increasing the photoalteration parameter [4]. The pump beam was spherically focused to irradiate the entire diameter of the flow capillary, thereby maximizing the conversion of BR_{570} to O_{640}. The photoalteration of the pump beam was 1.3. A shutter system (Uniblitz #26L) was interfaced to a PDP-11/23 computer and programmed to take alternate pump-and-probe and probe-only data points at each wavelength setting of the monochromator.

B. *Results and Discussion*

Figure 6 presents the resonance Raman spectra of purple membrane taken with a 752-nm probe beam in the presence (Fig. 6A) and absence (Fig. 6B) of a spatially separated 514-nm pump beam. Subtraction of the probe-only (BR_{570}) spectrum from the pump-plus-probe (BR_{570} and O_{640}) spectrum yields the Raman spectrum of the O_{640} intermediate. Distinctive lines in the O_{640} spectrum are an ethylenic stretching frequency at 1509 cm^{-1}, HOOP modes at 945, 959 and 977 cm^{-1}, and fingerprint lines at 1168 and 1198 cm^{-1}. Raising the temperature to 40°C serves to increase the proportion of O_{640} formed [3], thus allowing for a more reliable subtraction of the BR_{570} component. Our red-probe flow spectrum of O_{640} is distinctly different from Raman spectra obtained previously using a stationary sample and a bluer probe [17].

The isotopic fingerprint comparison method has been used to determine the configuration of the chromophore in BR_{570}, M_{412}, and in K_{610} [4,9]. The 15-D-induced changes in the Raman spectra are also characteristic of the configuration about the 13=14 double bond in O_{640}. Comparison with deuteration-induced changes in all-*trans* and 13-*cis* pigments shows that the chromophore in O_{640} has an all-*trans* configuration [18]. Therefore, the conversion of M_{412} to O_{640} involves a protein-catalyzed dark isomerization. Furthermore, strong HOOP modes are observed in the Raman spectrum of O_{640}. Deuterium substitution at the C_{15} position has been used to assign the 977 cm^{-1} line to the C_{15}-H HOOP mode [18]. The two remaining intense HOOP lines at 945 and 959 cm^{-1} likely result from the $HC_7{=}C_8H$ and $HC_{11}{=}C_{12}H$ A_u HOOP modes [15].

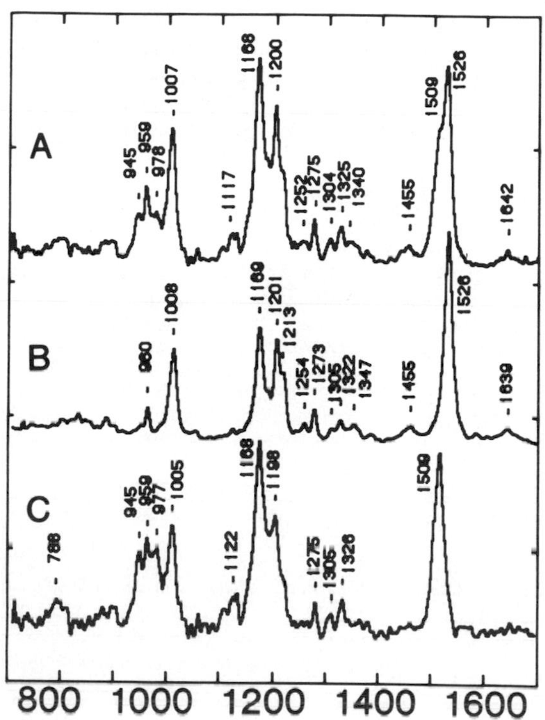

Figure 6. Resonance Raman spectra of purple membrane taken with 752-nm probe beam in the presence (A) and absence (B) of a temporally separated (4 msec) 514-nm pump beam. (C) Raman spectrum of O_{640} obtained by subtracting ∿60% of (B) from (A).

IV. COMPARISON OF K_{610} AND O_{640} - STRUCTURAL CONCLUSIONS

Both K_{610} and O_{640} are products of protein-catalyzed chromophore isomerizations. Although one isomerization requires a photon while the other proceeds thermally, it is not too surprising that these intermediates share some common structural features, as evidenced by similar spectroscopic behavior. The most obvious of these structural similarities are:

1. *An increase in* π*-electron delocalization relative to the parent* BR_{570} *molecule.* Both K_{610} and O_{640} have red-shifted absorption maxima, which in both molecules is accompanied by a decrease in frequency of the strong ethylenic line near 1520 cm^{-1}. Rimai *et al.* [19] and Aton *et al.* [8] noted the strong correlation between shifts in λ_{max} and $\nu_{C=C}$ in a series of retinyl chromophores, and pointed out that shifts in both spectroscopic parameters are a measure of changes in π-electron delocalization in the polyene chain.
2. *The unusual enhancement of the HOOP vibrations between 950-1000* cm^{-1} *provides strong evidence for a twisted ground state chromophore structure in both K and O.* In both molecules the same HOOP vibrations appear to be enhanced (the C_{15}-H HOOP and the $HC_7{=}C_8H$ and $HC_{11}{=}C_{12}H$ A_u HOOP modes) and they appear at remarkably similar frequencies in the K_{625} spectrum (973, 957, and 945 cm^{-1}) and the O_{640} spectrum (977, 959, and 945 cm^{-1}). Eyring *et al.* [20] and Warshel and Barboy [21] have demonstrated that HOOP modes can be strongly enhanced by out-of-plane deformation of the polyene chain.

It seems likely that the protein is perturbing these two intermediates similarly. Based on our 15-deuterio isotopic fingerprint comparison method, K has a 13-*cis* configuration, whereas O has an all-*trans* configuration. Thus, the same bond is isomerized when both K_{610} and O_{640} are formed, although in opposite directions. The same HOOPs are enhanced in both species because the isomerization causes torsional deformations in the same regions of the chromophore.

The increased π-electron delocalization and the torsional deformations may be related to the presence of negatively-charged protein residues in the chromophore binding pocket. An experimental model for these charges in bacteriorhodopsin [22] places one near the protonated Schiff base linkage and the other near the ionone ring. This model could explain the red-shift in the absorption maxima of K_{610} and O_{640} as a natural consequence of enhanced Coulombic interaction between the negative charges and the more delocalized π-electrons of K and O.

These charges could also stabilize twists about the same chromophore bonds in both K_{610} and O_{640}, giving rise to similar HOOP enhancements.

Bathorhodopsin, the primary photoproduct in the 11-*cis* to all-*trans* photoisomerization of visual pigments, shares a number of key features with K_{610} and O_{640}: a red-shifted absorption maximum, a decreased C=C stretching frequency, and strong HOOP enhancements [23]. We might expect all three of these "transition states" of retinal isomerization to require weakened double bonds, corresponding to an increase in π-electron delocalization, and out-of-plane distortions of the polyene chain which occur because the isomerized chromophore has not yet relaxed to a planar geometry within the protein. These three pigment intermediates are the only ones which are known to be direct products of chromophore isomerization. Thus, the spectroscopic features observed in K_{610}, bathorhodopsin, and O_{640} are indicative of structural perturbations which are common elements in protein-bound isomerizations of retinal.

REFERENCES

1. Stoeckenius, W., and Bogomolni, R.A. *Ann. Rev. Biochem. 51*, 587 (1982).
2. Ottolenghi, M. *Adv. Photochem. 12*, 97 (1980).
3. Lozier, R.H., Bogomolni, R.A., and Stoeckenius, W. *Biophys. J. 15*, 955 (1975).
4. Braiman, M., and Mathies, R. *Biochemistry 19*, 5421 (1980).
5. Lewis, A., Spoonhower, J., Bogomolni, R.A., Lozier, R.H., and Stoeckenius, W. *Proc. Natl. Acad. Sci. USA 71*, 4462 (1974).
6. Tsuda, M., Glaccum, M., Nelson, B., and Ebrey, T.G. *Nature (London) 287*, 351 (1980).
7. Applebury, M.L., Peters, K.S., and Rentzepis, P.M. *Biophys. J. 23*, 375 (1978).
8. Aton, B., Doukas, A.G., Callender, R.H., Becher, B., and Ebrey, T.G. *Biochemistry 16*, 2995 (1977).
9. Braiman, M., and Mathies, R. *Proc. Natl. Acad. Sci. USA 79*, 403 (1982).
10. Braiman, M., and Mathies, R. *Meth. Enzymol. 88*, 648 (1982).
11. Goldschmidt, C.R., Ottolenghi, M., and Korenstein, R. *Biophys. J. 16*, 839 (1976).
12. Terner, J., Hsieh, C.-L., Burns, A.R., and El-Sayed, M.A. *Proc. Natl. Acad. Sci. USA 76*, 3046 (1979).
13. Hsieh, C.-L., Nagumo, M., Nicol, M., and El-Sayed, M.A. *J. Phys. Chem. 85*, 2174 (1981).
14. Braiman, M. Ph.D. Thesis, University of California, Berkeley, in preparation.

15. Curry, B., Broek, A., Lugtenburg, J., and Mathies, R. *J. Am. Chem. Soc. 104*, 000 (1982).
16. Mathies, R., Oseroff, A.R., and Stryer, L. *Proc. Natl. Acad. Sci. USA 73*, 1 (1976).
17. Terner, J., Hsieh, C.-L., Burns, A.R., and El-Sayed, M.A. *Biochemistry 18*, 3629 (1979).
18. Smith, S., and Mathies, R. To be submitted for publication in *Biochemistry*.
19. Rimai, L., Heyde, M.E., and Gill, D. *J. Am. Chem. Soc. 95*, 4493 (1973).
20. Eyring, G., Curry, B., Mathies, R., Fransen, R., Palings, I., and Lugtenburg, J. *Biochemistry 19*, 2410 (1980).
21. Warshel, A., and Barboy, N. *J. Am. Chem. Soc. 104*, 1469 (1982).
22. Nakanishi, K., Balogh-Nair, V., Arnaboldi, M., Tsujimoto, K., and Honig, B. *J. Am. Chem. Soc. 102*, 7945 (1980).
23. Eyring, G., Curry, B., Broek, A., Lugtenburg, J., and Mathies, R. *Biochemistry 21*, 384 (1982).

TIME RESOLVED RESONANCE RAMAN STUDIES OF THE PHOTOCHEMICAL CYCLE OF BACTERIORHODOPSIN

Thomas Alshuth
Iris Grieger
Manfred Stockburger

Max-Planck-Institut
für biophysikalische Chemie
Göttingen, Germany

I. INTRODUCTION

Bacteriorhodopsin, a retinal-protein chromophore in the purple membrane of Halobacteria (Fig. 1), controls a light-driven proton pump which is used by the cell to drive metabolic processes (1-4). Following light absorption the BR chromophore

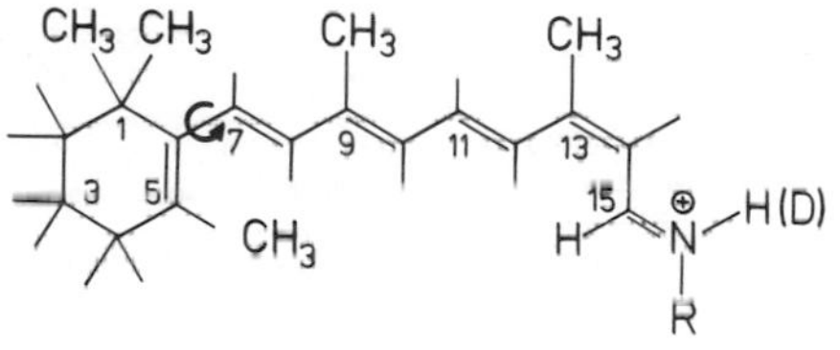

FIGURE 1. BR chromophore with 13-cis retinal and a Schiff base linkage to the protein.

runs through a series of short-lived intermediate states and is reconstituted within a few milliseconds (Fig. 2).

ISBN 0-12-066280-9

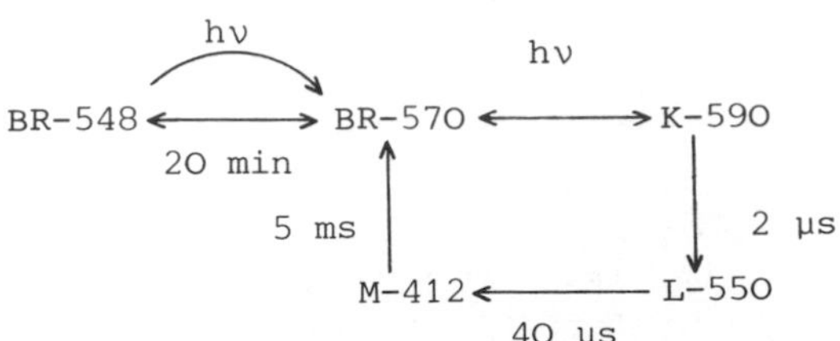

FIGURE 2. Photochemical cycle of BR. Time constants refer to PM suspensions at room temperature and neutral pH, the numbers to absorption maxima in nm (3,4).

Resonance Raman spectroscopy is an ideal method to study the different chromophores of the photochemical cycle, since it selectively probes the chromophoric site (5-11). It could be concluded from RR spectra that in BR-570 retinal is bound to the protein via a protonated Schiff base linkage (Fig. 1). It could further be shown that in the intermediate state M-412 the proton is removed. It was proposed, therefore, that translocation of the Schiff base proton is an essential step during the light-driven proton pump mechanism (12,13).

In the dark the parent chromophore, BR-570, in which the retinal chain is in the all-trans form, equilibrates to a 1:1 mixture with BR-548 (Fig. 2) in which the chain is found in the 13-cis configuration (14). This suggests that transitions between the two forms in the course of the cyclic process could play an important role (12-13).

This paper is concerned with the identification of the intermediate state L-550 and the interpretation of its most characteristic vibrational RR bands.

II. EXPERIMENTAL

Time resolution was achieved in the following way. - Two CW laser beams of different wavelengths, serving for photolysis and RR probing, were focussed into a rotating cell which contained the purple membrane suspension with an optical density of one at 514 nm. RR scattering was monitored perpendicular to the laser probe beam. The time which a molecule needs to flow from the photolytic to the probe beam provides a scale on which the temporal evolution of intermediates can be measured. This flow time is given by

$$\delta = \Delta s \ \upsilon$$

where Δs is the lateral distance between the foci of the two beams in the cell and υ is the flow velocity of the suspension. In our experiments υ was kept constant (6 ms^{-1}) and δ was varied by changing Δs. The focal diameters of the two beams were 160 μm and 60 μm for the photolytic and probe beam, respectively. A lower limit δ_{min} is given by Δs_{min} for which the two beams no longer overlap. Measurements were carried out from δ_{min} = 20 μs up to 500 μs.

In order to initiate the cyclic process at least one photochemical event has to take place during the residence time, Δt_p, of a molecule in the photolysis beam. When l_o denotes the rate constant for the primary photochemical process this means that

$$l_o \Delta t_p > 1$$

Since l_o is proportional to the laser power this condition can be easily fulfilled. On the other hand l_o has to be kept low enough to avoid secondary reactions of the intermediate L-550 by the photolytic radiation. These restrictions could be taken into account by using the 647 nm line of a krypton ion laser with a power of less than 500 mW for photolysis. For probing L-550 the 514 nm line of an argon ion laser was used. To avoid secondary photoreactions by this radiation the power was kept below 8 mW.

By irradiation at 647 nm the parent chromophore can only partially be converted into L-550. The reason for this lies in the photo-induced back reaction from the first intermediate K-590 which at higher power competes with thermal relaxation to L-550 (Fig. 2). This implies that in any case a mixture of BR-570 and L-550 is probed. To obtain "pure" spectra of L-550 the contribution of BR-570 of about seventy percent had to be subtracted. In order to minimize systematic errors in this procedure the spectra of the mixture and of the parent BR-570 chromophore were recorded simultaneously. This was achieved by chopping the photolysis beam with a period which is large compared with the 500 μs time scale of the flow experiment but smaller than the stepping period of the monochromator. For each wavenumber set, therefore, two signals were obtained which refer to the open and closed photolysis beam giving rise to the two spectra.

III. RESULTS

A new vibrational band appears in the mixed spectrum at 1550 cm^{-1}. Its intensity decreases in the same time range as the optically detected intermediate L-550. In particular its intensity decrease is correlated with the appearance of the well-known C=C stretch of M-412 at 1567 cm^{-1}. On the basis of this kinetic behaviour the 1550 cm^{-1} band can be assigned to the intermediate L-550.

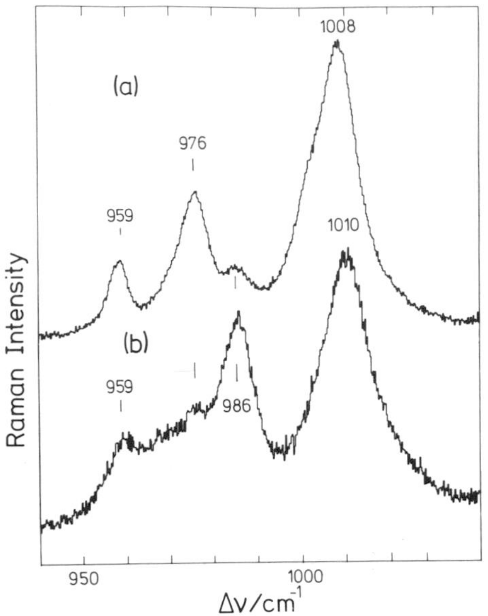

FIGURE 3. RR spectra in D_2O, (a) BR-570, (b) L-550.

RR spectra of the mixture were recorded for δ = 20 μs in the region between 700 and 1700 cm^{-1} and the spectrum of L-550 was deduced in the way described. The spectrum differs in many details from that of the parent chromophore indicating important structural changes. This will be demonstrated for two cases.

In Fig. 3 spectra of BR-570 and L-550 are shown in the region between 950 and 1050 cm^{-1}. They were obtained from a D_2O suspension of the purple membrane where the exchangable proton at the Schiff base nitrogen is replaced by a deuteron (Fig. 1). A characteristic band of BR-570 lies at 976 cm^{-1} which was assigned as in-plane bending vibration of the deuterium atom (11). In the spectrum of L-550 this band has disappeared and a fairly strong band arises at 986 cm^{-1}. When the hydrogen at the carbon C(15), which is adjacent to the Schiff base nitrogen, is substituted by deuterium, this band moves apart and therefore can be assigned as C(15)-H out-of-plane bending vibration of the hydrogen atom. It was found that such a band is also characteristic of the 13-cis chromophore BR-548 (10,11). We therefore conclude that the appearance of a fairly strong RR band close to 980 cm^{-1} reflects the 13-cis form of the retinal moiety in a chromophore and that in L-550 this moiety is in a "13-cis like" configuration.

This conclusion is confirmed by the fact that the spectra of BR-548 (13-cis) and L-550 also closely resemble each other in the "fingerprint" region (1120-1220 cm^{-1}) which is characteristic of a certain configuration. On the other hand a number of differences with respect to hydrogen bending vibrations do occur over the whole spectral range. This implies that a notion like "13-cis" describes the chromophoric structure only in a crude way. Each chromophore probably has a conformational finestructure which depends on its specific interaction with the protein environment.

An important peculiarity in the spectra of L-550 is observed in the C=C stretching region. It is demonstrated in Fig. 4 that the C=C stretch at 1550 cm^{-1} is composed of two overlapping bands of comparable intensity. This is in contrast

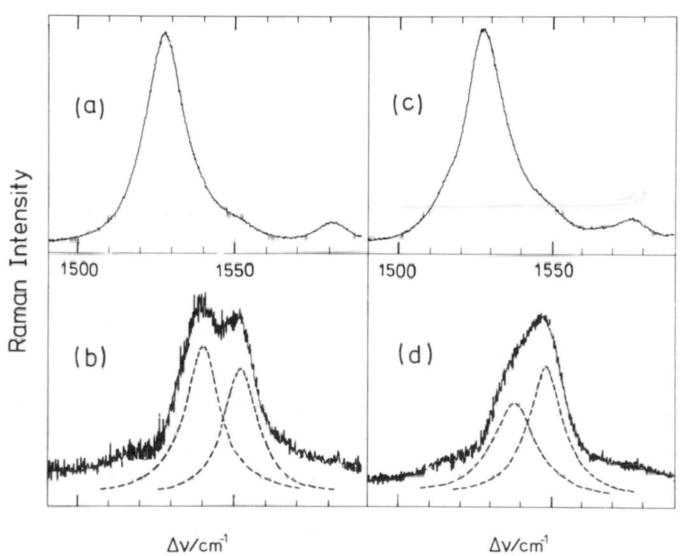

FIGURE 4. RR spectra of (a) BR-570, H_2O (b) L-550, H_2O (c) BR-570, D_2O (d) L-550, D_2O. The dotted bands were obtained from fits with Lorentzian lineshape functions.

to BR-570 but also to other chromophores like BR-548 (13-cis) and M-412 which are characterized by a single strong band in this region. One interpretation, therefore, would be that two chromophores of slightly different structure contribute to the spectrum of L-550. This interpretation obtains some support from the observed deuterium effect on the relative intensity of the two bands (Fig. 4). However, it was found that during the conversion from L-550 to M-412 the two bands show an identical time dependence. If we, therefore, accept the existence of two different chromophores they would be coupled to each other by a fast dark-reaction. From the fact that the RR spectrum of L-550 exhibits predominantly 13-cis vibrational features it had to be concluded that the two L-components would differ not in configuration but in another structural feature.

Finally it must be noted that it cannot be excluded that the two C=C bands are characteristic of a single L-chromophore. A resolution of this problem by future experimental and theoretical work would greatly assist a better understanding of the mechanism of the photochemical cycle of bacteriorhodopsin.

ACKNOWLEDGEMENTS

We thank Prof. D. Oesterhelt and Dr. W. Gaertner for leaving us purple membrane suspension and preparations with the C(15)-D modified chromophore.

REFERENCES

1. Oesterhelt, D., and Stoeckenius, W., Proc. Natl. Acad. Sci. USA 70, 2853 (1973).
2. Henderson, R., Ann. Rev. Biophys. Bioeng. 6, 87 (1977).
3. Stoeckenius, W., Lozier, R.H., and Bogomolni, R.A., Biochim. Biophys. Acta 505, 215 (1979).
4. Ottolenghi, M., Adv. Photochem. 12, 97 (1980).
5. Lewis, A. Spoonhower, J., Bogomolni, A., Lozier, R.H., and Stoeckenius, W., Proc. Natl. Acad. Sci. USA 71, 4462 (1974).
6. Aton, B., Doukas, A.G., Callender, R.A., Becher, R.H., and Ebrey, T.G., Biochemistry 16, 2995 (1977).
7. Terner, J., Hsieh, C.L., and El-Sayed, M.A., Biophys. J. 26 527 (1979).
8. Braiman, M., and Mathies, R., Biochemistry 19, 5421 (1980).
9. Stockburger, M., Klusmann, W., Gattermann, H., Massig, G., and Peters, R., Biochemistry 18, 4886 (1979).
10. Alshuth, Th., and Stockburger, M., Ber. Bunsenges. Phys. Chem. 85, 484 (1981).

11. Massig, G., Stockburger, M., Gaertner, W., Oesterhelt, D., and Towner, P., J. Raman Spectroscopy 12, 287 (1982).
12. Schulten, K., and Tavan, P., Nature 272, 85 (1978).
13. Orlandi, G., and Schulten, K., Chem. Phys. Lett. 64, 370 (1979).
14. Sperling, W., Carl, P., Rafferty, D.N., and Dencher, N.A., Biophys. Struct. Mech. 3, 79 (1977).

RESONANCE COHERENT ANTI-STOKES RAMAN SPECTROSCOPY OF THE K INTERMEDIATE IN THE BACTERIORHODOPSIN PHOTOCYCLE

Aaron Lewis*

School of Applied and Engineering Physics
Cornell University
Ithaca, New York 14853

I. INTRODUCTION

The retinylidene membrane proteins rhodopsin and bacteriorhodopsin have been studied extensively using resonance Raman spectroscopy(1). Both of these proteins store a significant fraction of the photon energy. In rhodopsin, the primary transducing entity in visual excitation, ≈60% of the photon energy(2) is stored in < 6 psec and this high energy state eventually stimulates in the photoreceptor cell the production of the chemical species which are essential in generating a neural response. In bacteriorhodopsin, which is found in the membrane of the bacterium _Halobacterium halobium_, light also generates a high energy state and this process eventually drives a proton across the bacterial cell membrane. The molecular structure of these high energy states in both rhodopsin and bacteriorhodopsin is a fundamental question that is of interest both from a structural-chemical and from a biological point of view. At a

* The National Aeronautics and Space Administration is gratefully acknowledged for supporting this research.

ISBN 0-12-066280-9

structural-chemical level it is important to understand in molecular detail how these systems within a few psecs after photon absorption are capable of storing as much as 60% of the photon energy. At a biological level very specific interactions between the light absorbing retinylidene chromophore and the covalently attached protein must be involved both in generating the chemical signals for visual transduction and in creating in bacteriorhodopsin the molecular alterations that convert stored light energy into a proton gradient.

II. PREVIOUS ATTEMPTS TO OBTAIN VIBRATIONAL SPECTRA OF THE K INTERMEDIATE

Resonance Raman spectroscopy is one of the most selective and sensitive probes of chromophore structure that could be used to determine the nature of the light induced molecular alterations which result when the high energy (batho) state is generated. In the case of rhodopsin, there is complete agreement on the major bands observed in the resonance Raman spectrum of the batho intermediate(3-5). In addition, there is agreement that the spectra at 77°K and room temperature are similar for this intermediate(4). Furthermore, it has been shown definitively in three independent investigations that the C=N stretch is completely unaltered in going from rhodopsin to bathorhodopsin(3-5). This strongly suggests that the protonated nature of the Schiff base is unaltered in going from rhodopsin to bathorhodopsin(3-5). Thus, even though questions still remain about the nature of the chromophore conformations and configurations that result in the bathorhodopsin resonance Raman spectrum, the data on which future experiments and discussions will be based is firmly in hand.

In bacteriorhodopsin the situation is far less clear, even though, four resonance Raman studies have recently been published on the batho (K) intermediate(6-9). All of the above resonance Raman investigations of K arrive at different conclusions as to which bands are indeed present in the K spectrum. The principal problem in obtaining a reproducible resonance Raman spectrum of K is the observation of fluorescence from bacteriorhodopsin in a frequency regime which seriously perturbs attempts at selectively exciting K chromophore Raman spectra without stimulating bacteriorhodopsin fluorescence(10). To circumvent these problems a low temperature rotating cell technique was used to slow down the decay of K and thus allow the use of a spatially separated resonance Raman probe laser from the pump laser which photochemically generates K from bacteriorhodopsin and stimulates the fluorescence(9). When this investigation is compared to previous resonance Raman spectra of K, little agreement is found. On the other hand, a comparison of this low temperature data with recent low temperature infrared measurements(11,12) provides convincing similarities in the observed vibrational modes.

In the infrared measurements, light versus dark, difference fourier transform infrared spectra are recorded at 77°K. These spectra appear to contain mainly vibrational modes of the chromophore and hence the comparison with the 77°K resonance Raman spectrum of K using spinning cell techniques. In view of the data on rhodopsin suggesting that the high energy batho state has the same spectrum at low temperature and room temperature it may seem reasonable to assume that in bacteriorhodopsin the low temperature data is a good representation of the room temperature K spectrum. However, there are several major differences between the rhodopsin and bacteriorhodopsin active sites and thus, such comparisons may not be valid.

III. RHODOPSIN AND BACTERIORHODOPSIN ACTIVE SITE STRUCTURAL COMPARISONS

In addition to the different initial configurations of the chromophore in these two systems [rhodopsin (11-cis) and bacteriorhodopsin (all trans)], there is a considerable difference, as we have noted(13,14) in the nature of the Schiff base linkage as probed by resonance Raman spectroscopy. Even though the Schiff base is protonated in both these systems(1), the frequency of the C=N-H+ vibrational mode in rhodopsin and in bacteriorhodopsin are not the same. In rhodopsin, the C=N-H+ stretching frequency is strikingly similar to values obtained

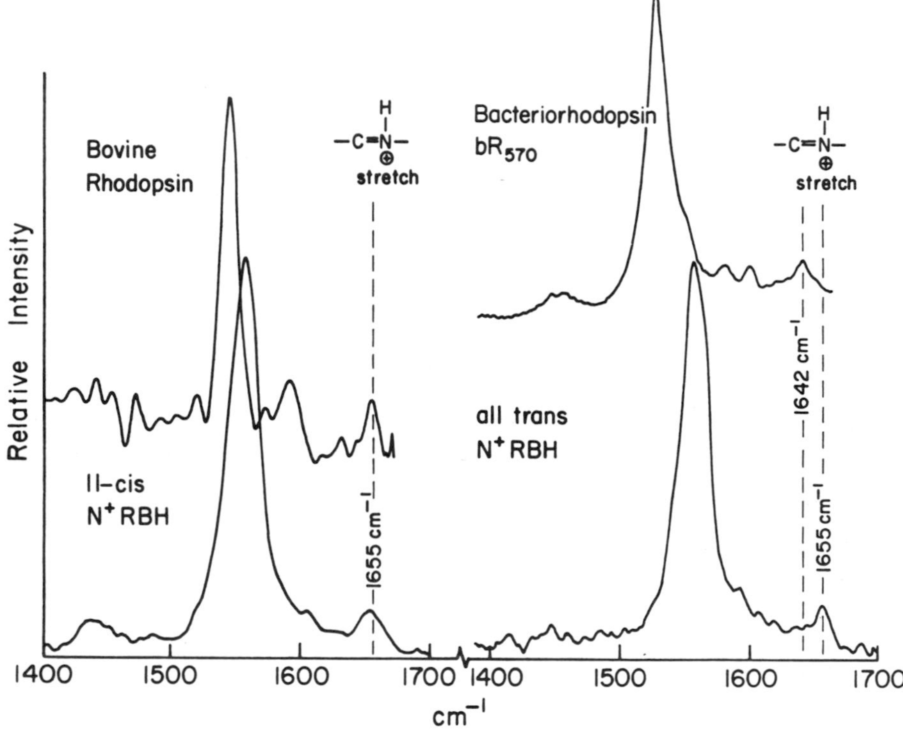

FIGURE 1. A Comparison of the Resonance Raman Spectra of Rhodopsin and Bacteriorhodopsin with Model 11-Cis and All Trans Retinylidene Chromophores in the Region of the Schiff Base Frequency.

for free protonated Schiff bases (see Figure 1) and furthermore, the C=N-H+ frequency is completely unperturbed when the high energy batho state is generated(3-5). In bacteriorhodopsin on the other hand there is a considerable perturbation of this frequency in the membrane relative to a free protonated Schiff base and the rhodopsin system(13,14) (see Figure 1).

IV. ROOM AND LOW TEMPERATURE BACTERIORHODOPSIN SPECTROSCOPY

In addition to these active site structural differences in rhodopsin and bacteriorhodopsin there are more direct indications suggesting that the low temperature resonance Raman spectrum of K may be different from the species generated at room temperature. Kalisky and Ottolenghi(15) have demonstrated that at low temperatures there are a number of additional "K type" species present when bacteriorhodopsin is illuminated. This is deduced from absorption measurements but these measurements give no information as to whether the chromophore structures in the "K type" species generated at low temperature are different to the relaxed room temperature form. Nagle et. al.(16) have been investigating procedures to deconvolute the K and bacteriorhodopsin spectra and have demonstrated that there is a large perturbation in the absorption spectrum of K when room temperature and 77°K spectra are compared. Their data indicate that, in spite of the K absorption peaking at 610 nm at 77°K, the room temperature maximum occurs at ≈582 nm which is very close to the absorption of bacteriorhodopsin at 568 nm before the action of light.

V. RESONANCE COHERENT ANTI-STOKES RAMAN SPECTROSCOPY

In view of all the differences noted above it is important to develop a selective and structurally sensitive technique that can monitor both the low temperature and room temperature spectra of the high energy species generated by light. For several Raman spectroscopy (CARS) as a selective probe of the vibrational spectrum of the bacteriorhodopsin chromophore(17,18). In this technique two pulsed lasers are focused simultaneously onto a sample. One laser is at a fixed frequency and the other laser is tunable. Both lasers are within or near the chromophore absorption. In the non-linear CARS process a coherent beam is generated, in the anti-Stokes region and at an angle that satisfies momentum conservation, whenever the energy difference between the fixed frequency and tunable laser impinging on the sample match a vibrational mode in the selectively excited chromophore. This allows the structurally sensitive chromophore vibrational spectrum to be detected in the anti-Stokes region where fluorescence from the lasers is not a problem. Although this is important, of even greater significance is the coherent nature of the CARS signal as compared to the emission process which occurs in all directions. This allows the experimenter to readily discriminate fluorescent emission and the anti-Stokes Raman spectrum.

VI. THE K CARS EXPERIMENT

In terms of bacteriorhodopsin this CARS technique provides us with a probe of the chromophore vibrational spectrum which is not complicated by fluorescence problems. In addition, the use of a pulsed Nd:YAG laser with a pulse width of ≈ 7 nsec and a repetition rate of ≈ 20 Hz gives us the ability to kinetically detect at room temperature, the K spectrum and then allow the system to regenerate the original pigment state in 10 msec before the next laser pulses interrogate the sample. The lack of fluorescence problems is particulrarly important in our K investigations because only 30% of excited bacteriorhodopsin molecules go on to K. Thus, in the photostationary state created by the pulsed lasers impinging on the sample, there is only a small percentage of K molecules and therefore it is important to be able to probe the K spectrum with lasers to the red of the bacteriorhodopsin absorption where the K intermediate preferentially absorbs light. This is precisely the wavelength region where bacteriorhodopsin fluoresces and so the utility of the CARS technique to this problem is obvious. Furthermore, freed from these fluorescence problems the sample can also be irradiated with a third laser beam in the green. This laser is preferentially absorbed by bacteriorhodopsin and maintains a high photostationary concentration of K in the pulsed beams while the sample is being interrogated with the red lasers to stimulate the K spectrum. The desireability of using the green laser arises from the fact that the red (CARS) beams also regenerate, photochemically, the original bacteriorhodopsin species as shown in the reaction sequence below:

$$\text{bacteriorhodopsin } (\lambda_{max} = 568 \text{ nm}) \underset{\lambda_{red}}{\overset{\lambda_{green}}{\rightleftharpoons}} \text{K} \quad \begin{matrix} \lambda_{max}{=}610\text{nm};{-}196\,^{\circ}\text{C} \\ \lambda_{max}{\approx}582\text{nm};25\,^{\circ}\text{C} \end{matrix}$$

To obtain the K cars spectrum the three beams were focused onto the sample: one in the green to excite photochemistry at 530 nm and two in the red to record the CARS spectrum. One of the two beams required to generate the CARS spectrum was at a fixed wavelength of 650 nm and the other was tunable between ≈ 690 nm and 720 nm. The photochemical pump beam at 530 nm was made co-linear with the fixed frequency CARS pump beam at 650 nm. The samples used to obtain the CARS spectra had an O.D. of 4/mm so that the non-resonance background from water was negligable. This reduced but did not eliminate the dispersive lineshapes normally detected in anti-Stokes Raman spectra of diluted samples. Even at these high concentrations the membranes were still in suspension. For the thickness of the sample used the combination of the green and red photons assured establishment of the photostationary equilibrium within the 7 nsec pulse.

An example of the spectra obtained in the structurally sensitive fingerprint region is seen in Figure 2.

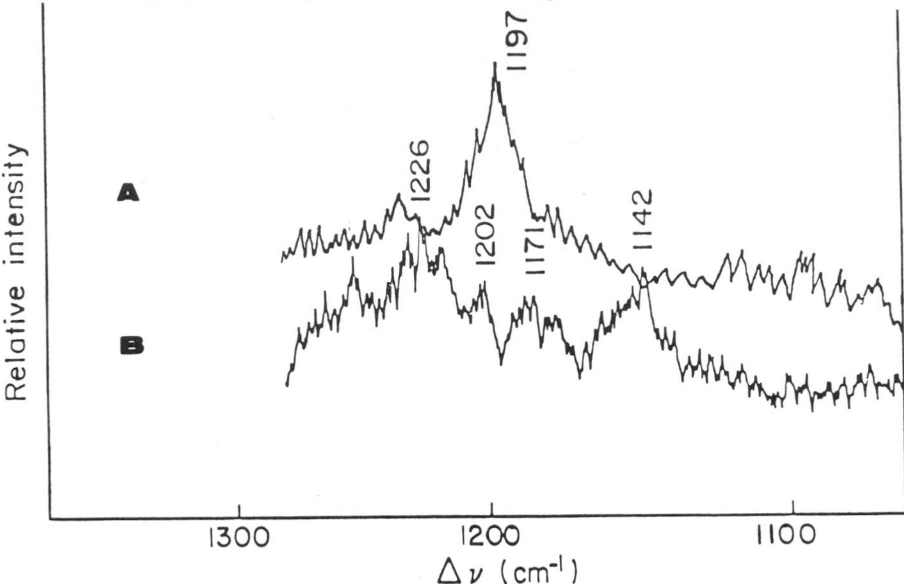

FIGURE 2. Resonance Coherent Anti-Stokes Raman Spectra Obtained with a 650 nm Pump Laser at (A) 77°K and (B) 273°K.

The spectrum of the fingerprint region shown in Figure 2 was recorded at 77°K with all three laser beams illuminating the sample. This spectrum is identical to the data obtained in the low temperature spinning cell Raman study(9) and compares well with the results of the two infrared investigations(11,12). If the temperature is raised from 77°K to room temperature and all other conditions are kept the same there is a dramatic alteration in the spectrum. In spite of the considerable absorption blue shift upon warming the sample(16), it appears that this alteration in the spectrum is indicative of a structural difference between the species at 77°K and the species at room temperature that is generated and exists for 7 nsec after the absorption of light. We reach this conclusion based on the essentially non-dispersive lineshapes we have obtained with similar concentrations under a variety of conditions(17,18). However, the lineshapes we obtained with the above excitation wavelengthsdo appear somewhat dispersive upon warming the sample. Thus, we are presently generating a series of spectra with different combinations of excitation wavelengths to establish whether the observed room temperature alterations are related to real structural perturbations of the K chromophore at 77°K or whether the 77°K and room temperature spectral differences may be related to the absorption blue shift relative to the excitation wavelengths. Such shifts could cause perturbations in the spectra by causing alterations in the dispersive character of the observed bands.

VII. SUMMARY

In summary we have demonstrated that CARS is a technique that can be used to probe selectively the K chromophore vibrational spectrum at a variety of temperatures without encountering the fluorescence problems that have plagued conventional resonance Raman measurements. It certainly appears that this nonlinear technique is the most flexible method with which to probe the K spectrum both under photostationary state and kinetic conditions. Our results clearly indicate that the resonance CARS technique will play an important and complimentary role in helping elucidate the critical structural constraints which allow bacteriorhodopsin to store the photon energy in a few picoseconds and then use that energy to drive a proton across the bacterial cell membrane.

ACKNOWLEDGMENTS

Dr. A. Hochberg is acknowledged for his participation in some of these experiments.

REFERENCES

1. Lewis, A. in Methods in Enzymology, Vol. 88 (L. Packer, ed.), p. 561 Academic Press, New York, (1982).
2. Cooper, A., Nature (London) 282, 531 (1979).
3. Sulkes, M., Lewis, A., and Marcus, M. A., Biochemistry 17, 4712 (1978).
4. Eyring, G., and Mathies, R., Proc. Natl. Acad. Sci. (U.S.A.) 76, 33 (1979).
5. Aton, B., Doukas, A., Narva, D., Callender, R. H., Dinur, U. and Honig, B., Biophys. J. 29, 79 (1980).
6. Terner, J., Hsieh, C. -L., Burns, A. R. and El-Sayed, M. A., Proc. Natl. Acad. Sci. (U.S.A.) 76, 3046 (1979).
7. Pande, J., Callender, R. H. and Ebrey, T. G., Proc. Natl. Acad. Sci. (U.S.A.) 78, 7379 (1981).
8. Hsieh, C., Nagumo, M., Nicol, M. and El-Sayed, M. A., J. Phys. Chem. 85, 2714 (1982).
9. Braiman, M. and Mathies, R., Proc. Natl. Acad. Sci. (U.S.A.) 79, 403 (1982).
10. Lewis, A. and Perreault, G. J. in Methods in Enzymology, Vol. 88 (L. Packer, ed.), p. 217 Academic Press, New York (1982).
11. Bagley, K., Dollinger, G., Eisenstein, L., Singh, A. K. and Zimanyi, L., Proc. Natl. Acad. Sci. (U.S.A.) 79, 4972 (1982).
12. Rothschild, K. J. and Marrero, H., Proc. Natl. Acad. Sci. (U.S.A.) 79, 4045 (1982).
13. Lewis, A., Marcus, M. A., Ehrenberg, B. and Crespi, H., Proc. Natl. Acad. Sci. (U.S.A.) 17, 4722 (1978).
14. Marcus, M. A. and Lewis, A., Biochemistry 17, 4722 (1978).
15. Kalisky, O. and Ottolenghi, M., Photochem. Photobiol. 35, 109 (1982).

16. Nagle, J., Parodi, L. A. and Lozier, R. H., Biophys. J. 38, 161 (1982).

17. Nelson, E. T., Lewis, A., and MacFarlane, R., Biophys. Soc. Abs. 25, 79a (1979).

18. Nelson, E. T., Ph.D. Dissertation, Cornell University (1980).

RESONANCE RAMAN SPECTRA OF PHOTOCHEMICAL PICOSECOND TRANSIENTS: METHOD AND APPLICATION TO STUDY BACTERIORHODOPSIN PRIMARY PROCESSES

M. A. El-Sayed[*1]

Institut fur Physikalische Chemie
Technische Universitat Munchen
Garching, West Germany

Chung-Lu Hsieh[2]
Malcolm Nicol

Department of Chemistry
University of California
Los Angeles, California

I. TIME RESOLVED RESONANCE RAMAN SPECTROSCOPY, INTRODUCTION

The use of lasers in determining the Raman and resonance Raman spectra of stable chemical and biological systems has produced many important and useful structural results over the past decade. The use of lasers for determining the Raman spectrum of transients has been developing over the same period, but at a much slower rate. The use of picosecond lasers and vidicon detection in Raman spectroscopy was first discussed in 1977 (1). Undoubtedly, the use of vidicon detection has greatly assisted in the development of resolved techniques. The high intensity of

*Alexander von Humboldt Awardee, January - August 1982

[1]Permanent address: Department of Chemistry, University of California, Los Angeles

[2]Permanent address: Technical Center, The Chlorox Company, Pleasonton, California

The financial support of the Department of Energy is acknowledged.

ISBN 0-12-066280-9

the laser made it possible to obtain Raman spectra with the laser frequencies within the absorption band of the substance studied. This gives rise to large enhancements thus reducing the high concentration requirement for conventional (nonresonance) Raman spectroscopy. In spite of this fact, the number of papers published on the resonance Raman spectroscopy of picosecond photochemical transients (i.e. transients that are chemically different from the parent molecule) remains very small (2). A computer search* of chemical abstracts (and physics abstracts) during the 1972 - mid-1982 period reveals that of a total of 16,700 (11,200) papers appeared on Raman spectroscopy and 1360 (2237) appeared on picosecond research, only 76 (140) are on picosecond Raman and of these, only 10 (10) are on Raman of picosecond transients.

The use of picosecond Raman spectroscopy in determining the kinetics of vibrational relaxation of vibrationally excited liquids was among the first applications of Raman picosecond techniques (3). However, as indicated by the library search results given above, the number of reported spontaneous spectra of picosecond photochemical or photobiological transients remains very small (10 of the 76 or 100 papers) (2). The number of reports on the resonance Raman spectra of photochemical and photobiological transients using nanosecond pulsed lasers (4) is larger, for both the spontaneous and the coherent Raman modes. A number of techniques have been developed that employ c.w. lasers as a source together with either choppers of variable slits (5) or sample flow techniques (6,7) to determine the resonance Raman spectra of microsecond transients. A qualitative review (8) of some time-resolved

There is a good overlap between the references cited in chemical and in physics (given in parentheses) abstracts. The larger number of conventional Raman papers in chemical abstracts is due to papers on photobiology and inorganic complexes. On the other hand, for picosecond work, more work is cited in physics abstracts on the technology of this field.

Raman techniques in the ml-micro and nanosecond time scales was published in 1979 (8).

II. RESONANCE RAMAN OF PICOSECOND TRANSIENTS

Picosecond pulses have been used extensively in determining the optical (mostly the visible) spectra of photochemical picosecond transients. These spectra have so far given limited structural information about these picosecond transients. This is because of the fact that most of the systems studied with the available lasers are too large with broad absorption spectra. Most small molecules, with well-resolved spectra, absorb in the deep U.V. for which useful picosecond continua for absorption measurement are not yet available.

Vibration spectroscopy has proven to be very useful over the years for the synthetic organic chemist in determining the structure of new compounds. Both infrared and Raman spectroscopy have been used for this purpose. For this reason, time-resolved resonance Raman techniques are very essential to develop if picosecond lasers are to be used to identify the mechanisms of rapid photochemical and photobiological primary processes. In spite of this need, and the fact that picosecond lasers have already been used to determine the spontaneous and coherent Raman spectra of stable liquids and of their coherently vibrationally excited picosecond transients (3) (photophysical transients), as was pointed out in the previous section, only a handful of papers appeared in the literature which reports on the spontaneous Raman spectra of picosecond photochemical transients (2). The reason for this is undoubtedly the low Raman signal expected to be observed. Raman scattering, even after resonance enhancement, has a cross section which is far smaller than that for optical absorption or fluorescence (if the latter has non-zero quantum yield). This, together with doing the monitoring on the picosecond time scale as well as the low repetition rate of the pulsed picosecond lasers makes the number of collected spontaneously Raman-scattered photons by picosecond transients very small. Of course one might increase the laser intensity, in particular if the photolysis laser is also the Raman laser source, as in the case for all our systems studied so far. In this case, barring saturation effects, the Raman signal should increase quadratically with the laser intensity. However, the other nonlinear processes taking place in solution, e.g., filamentation, stimulated Brillion and Raman processes, multiphoton ionization and dissocation, etc., would compete (and probably win) if one tries to greatly increase the intensity of pulsed lasers. In order to solve this problem, we have used the synchronously pumped, cavity dumped dye laser system (9). Each pulse has only a few nanojoules in energy but it produces a <u>million pulses per second</u>. The scattering radiation is collected with a spectrometer fitted with vidicon detection at such a high duty cycle, to make up for the low energy that each pulse possesses (which might minimize the danger of sample destruction).

The above system, however, introduces another problem if the picosecond transient changes chemically or photochemically to another species, unless the sample is changed in-between pulses. Using a large volume of the sample with continuous stirring (a method used with high power, low-repetition rate pulsed lasers) might not assure a complete change in the sample within the laser focal volume during the microsecond period in-between the pulses of the laser used here. Electronic gating is of no use in solving this problem.

Sample-flow techniques (7,10), when combined with laser microbeam (tight focusing) techniques (11), can be used to secure that fresh sample is present for each picosecond laser pulse when using the synch-pump system. Using available pumps, the solution sample can be flown through a syringe needle with speeds of up to 40 meters per sec. If the laser is focused on the sample as it comes out of the syringe needle to a spot of few microns (using microscope lense), the sample residence time in the laser beam

$= \frac{4 \times 10^{-4} \text{ cm}}{40 \times 10^{2} \text{ cm/sec}} = 0.10$ μsec. This is ten times shorter than the time in-between pulses, which insures that by the time a new laser pulse arrives at the flowing sample, a complete new fresh supply of the sample would have indeed replaced the sample exposed by the previous pulse. A detailed application of this technique is given elsewhere (12)

III. RESONANCE RAMAN SPECTRA OF NANOSECOND-μSECOND TRANSIENTS BY FLOW TECHNIQUES (7,10)

If the exposed sample by one laser pulse is not removed prior to the arrival of the following pulse, transients formed during the residence time of the flowing sample in the laser focused volume (in the micro-nanosecond time scale) might cause Raman scattering. Thus the time resolution of the experiment becomes the sample residence time in the laser focus rather than the pulse width of the picosecond laser used. In this case the picosecond laser might well be replaced by the much cheaper c.w. laser. The details of this technique are given elsewhere (10).

A. C.W. laser initiates the photochemistry, as well as acting as a source for the Raman scattering. The time resolution in this case is determined by the residence time of the photolabile sample in the laser beam, which under tight focus and rapid flow could be as short as 80 nanoseconds (10). For slow flow and more diffuse focus, scattering from transients appearing at longer time scales can be detected and studied. Two lasers, photolysis-probe, techniques focused at different spots of the flowing sample to give a predetermined time delay might be used in the μsecond time scale. The two laser techniques have the advantage of using different frequencies for the photolysis (to maximize photolysis) and the probe (to maximize the resonance Raman enhancement for the daughter) lasers.

IV. THE SYSTEM AND THE RESULTS

The picosecond technique discussed in Section II can be used for any photolabile system such as:

$$A \xrightarrow{h\nu} B \rightsquigarrow C$$

where A and B both have strong overlapping absorption bands, as is the case for many photobiological systems. The Raman spectrum is obtained at low laser powers (negligible photolysis) and then at high laser powers (efficient photolysis). Computer subtraction techniques are used to determine the characteristic bands of the picosecond transient B. It helps greatly if the photochemical yield of B is high, its resonance enhancement is large and neither A nor B is fluorescent. To avoid Raman photon losses from the absorption by A and B, the concentration of A, the laser wavelength as well as the optical arrangement have to be carefully adjusted to give a maximum resonance Raman signal of B at high laser powers.

The system discussed here is bacteriorhodopsin (bR) (13), the second photosynthetic system (14) in nature (the first being chlorophyll). It is a retinal-protein complex of molecular weight of about 28,000 a.m.u. The primary processes resemble those for rhodopsin (13). Using optical flash techniques (15), a scheme is derived showing the kinetic behavior of the different intermediates from the changes of the retinal absorption in the visible region. It is believed (16) that 11-cis to all-trans isomerization is the first step in vision, occurring on the picosecond time scale (17). Some kind of isomerization was also believed to occur for bR on a similar time scale (18). Recently, it was observed that the rate of the first step in these two systems was slower (17) in D_2O than in H_2O. These results raised questions (17) whether cis-trans isomerization can even occur on this time scale. Instead, they suggested that the first step involves a proton translocation, e.g., the Schiff base proton on the nitrogen connecting the retinal system with the protein or other protons.

In the present work we present the resonance Raman spectra of bacteriorhodopsin transients in the 30-50 psec time scale of two vibrational regions of retinal, one (1100-1400 cm^{-1}, the fingerprint region) is sensitive to retinal conformational changes and the other (~1646 cm^{-1}) is of the protonated C=N stretching vibration of the Schiff base.

The results in the fingerprint region are shown in Figure 1.

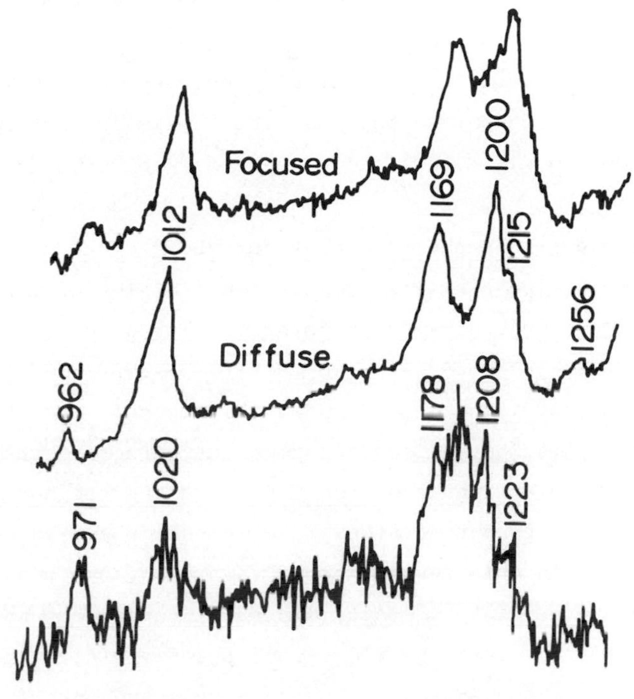

FIGURE 1. The effect of photolysis of 50 psec, 15 nanojoules, mega Hz modelocked synch. pump dye laser at 587 nm on the resonance Raman spectrum of bacteriorhodopsin in the fingerprint region of retinal (1000-1400 cm^{-1}). The third spectrum shows the best computer difference between the "unphotolyzed" spectrum (using diffuse laser focus) and the photolyzed one (focused laser). The result suggests that retinal isomerization indeed takes place in less than 50 psec within the protein pocket.

A comparison of the bottom two spectra shows that the spectrum of the parent (containing all-trans retinal) and that for the picosecond transient are indeed different. This suggests a change in the retinal conformation during the 30-50 picosecond pulse-width of the laser used, unless a proton transfer process is taking place which greatly affects the retinal spectrum in this region. If this were indeed the case, this region should be sensitive to deuterium effects. The observed spectrum in this region in H_2O and in D_2O of the parent (bR 570) is found to be similar (19). This might suggest that retinal conformation changes might be responsible for most of the observed changes in this region. Furthermore, the qualitative similarity between the transient spectrum in this region and that for the 13-cis in dark adapted bacteriorhodopsin might (20) add support for conformational changes.

One should stress the fact that our above conclusion does not eliminate the possibilities of proton translocation prior, during or after the conformation changes. The only proton that we can follow in our studies is the Schiff base proton. The $C{=}N^{+}H$ frequency at 1640 cm^{-1} is known (21) to shift to 1620 cm^{-1} upon proton ionization. Figure 2 shows the effect of the photolysis (focused laser condition) on the $-C=N^{+}H-$ region (at 1646 cm^{-1}) of the parent (diffuse laser condition). While the signal to noise in this region is not very good, the results suggest a change in the intensity distribution in this region upon transformation into the picosecond transient(s). Broadening and/or shift is observed for the band in this region. The broadening might be due to unrelaxed protein environment on the picosecond time scale of the experiment. The shift might result from a change of the degree or the strength of the protonation of the Shiff base nitrogen. This might be a result of the retinal conformation changes. For details of this work, see reference (19).

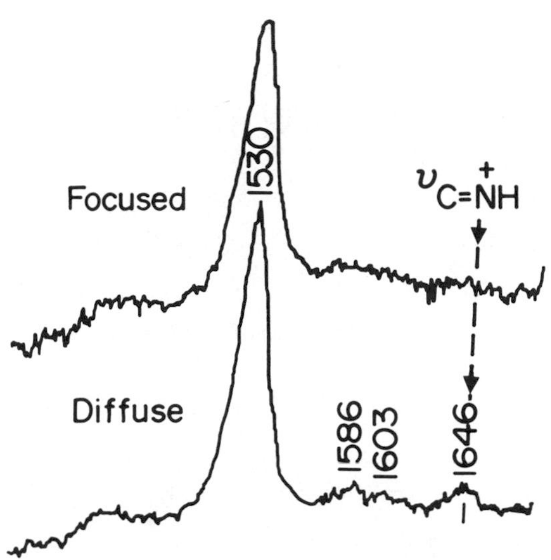

FIGURE 2. The effect of photolysis of the same laser described under Figure 1 on the $-C=NH^+$ *stretching region of the Schiff base of retinal observed at 1646* cm^{-1}. *Increasing the laser intensity (by focusing the laser) is found to decrease the relative scattering in this region. This could be a result of the inhomogeneous broadening of this vibrational transition for the picosecond transient or due to a shift in the frequency of this vibration for the psec transient. In any case, this might suggest a change in the electronic structure of this group upon picosecond photolysis.*

ACKNOWLEDGMENTS

M. A. El-Sayed would like to thank the Alexander von Humboldt Foundation for the Senior Scientist Award given him during the period February-August 1982. He would also like to thank Professor W. E. Schlag for his hospitality during this period and for stimulating scientific interactions with him and his group.

REFERENCES

1. See e.g.: Bridoux, M., and Delhaye, M., in "Advances in Infrared and Raman Spectroscopy" *2* (R. J. H. Clark and R. E. Hester, eds.) p. 140. Heyden (1976).
2. a. Terner, T., Spiro, T., Nagumo, M., Nicol, M. F., and El-Sayed, M. A., *J. Am. Chem. Soc. 102*, 3238 (1980).
   b. Coppey, M., Tourbez, H., Valat, P., and Alpert, B., *Nature* (London) *284*, 568 (1980).
   c. Hayward, G., Carlsen, W., Siegman, A., and Stryer, L., *Science, 211*, 942 (1981).
3. a. Laubereau, A., Kehl, G., and Kaiser, W., *Opt. Commun. 11*, 74 (1974).
   b. Alfano, R. R., and Zawadzkas, G. A., *Phys. Rev. A. 9*, 822 (1974).
   c. Clerc, M., Jones, R., and Rentzepis, P. M., *Chem. Phys. Lett. 26*, 167 (1974).
   d. For a recent discussion see: Laubereau, A., *Philos. Trans. R. Soc. London, Ser. A., 293*, 441 (1979).
4. a. Campion, A., Terner, J., and El-Sayed, M. A., *Nature 265*, 659 (1977).
   b. Dallinger, R. F., Nestor, J. R., and Spiro, T. G., *J. Am. Chem. Soc. 100*, 6251 (1978).
   c. Woodruff, W. H., and Farquharson, S., *Science 201, 831 (1978)*.
   d. Lyons, K. B., Friedman, J. M., and Fleury, P. A., *Nature 275*, 565 (1978).
   e. Scrivastava, R. B., Schuyler, M. W., Dosser, L. R., Purcell, F. J., and Atkinson, G. H., *Chem. Phys. Lett. 56*, 595 (1978).
5. Terner, J., Campion, A., and El-Sayed, M. A., *Proc. Natl. Acad. Sci. U.S.A. 74*, 5212 (1977).
6. Campion, A., El-Sayed, M. A., and Terner, J., *Biophysical J. 20*, 369 (1977).

7. Marcus, M. A., and Lewis, A., *Science 195,* 1328 (1977).
8. El-Sayed, M. A., "Time Resolved Resonance Raman Spectroscopy in Photochemistry and Photobiology," in *Multichannel Image Detectors in Chemistry, ACS Symposium Series Bk. 102,* Chap. 10, pp. 215-227 (1979).
9. Nicol, M., Hara, Y., Wiget, J., and Anton, M. F., *J. Mol. Struct. 47,* 371 (1980).
10. Terner, J., Hsieh, C.-L, Burns, A. R., and El-Sayed, M. A., *Proc. Natl. Acad. Sci. U.S.A. 76,* 3046 (1979).
11. Berns, M., *Exp. Cell. Res. 65,* 470 (1971).
12. Terner, J., Strong, J. D., Spiro, T., Nagumo, M., Nicol, M., and El-Sayed, M. A., *Proc. Natl. Acad. Sci. U.S.A. 78,* 1313 (1981).
13. For relevant reviews see:
a. Henderson, R., *Ann. Rev. Biophys. Bioeng. 6,* 87 (1977).
b. Stoeckenius, W., Lozier, R. H., and Bogomolni, R. R., *Biochem. Biophys. Acta 505,* 215 (1979).
c. Stoeckenius, W., *Acc. Chem. Res. 13,* 337 (1980).
d. Ottolenghi, M., in "Advances in Photochemistry" (J. N. Pitts, G. S. Hammond, K. Gollnick, and D. Grosjean, eds.) p. 97. John Wiley and Sons, New York (1980).
14. Oesterhelt, D., and Stoeckenius, W., *Proc. Natl. Acad. Sci. 70,* 2853 (1973).
15. Lozier, R. H., Bogomolni, R. A., and Stoeckenius, W., *Biophys. J. 15,* 955 (1975).
16. a. Mathies, R., Freedman, T. B., and Stryer, L., *J. Mol. Biol. 109,* 367 (1977).
b. Doukas, A. G., Aton, B., Callender, R. H., and Ebrey, T. G., *Biochemistry 16,* 2430 (1978).
17. Applebury, M. L., Peters, K. S., and Rentzepis, P. M., *Biophys. J. 23,* 375 (1978).
18. Mowery, P. C., and Stoeckenius, W., *Biochemistry 20,* 2302 (1981).

19. Hsieh, C.-L., El-Sayed, M. A., Nicol, M., Nagumo, M., and Lee, J. H., "Time Resolved Bacteriorhodopsin in the Picosecond and Nanosecond Time Scales," *Photochem. Photobiol.*, sent for publication.
20. For more details of the conclusions, see Hsieh, C.-L., Nagumo, M., Nicol, M., and El-Sayed, M. A., *J. Phys. Chem.* *85*, 2714 (1981).
21. Marcus, M. A., Lewis, A., *Science 195*, 1328 (1977).

PICOSECOND EVENTS IN BIOLOGICAL PROCESSES

E. F. Hilinski
P. M. Rentzepis

Bell Laboratories
Murray Hill, New Jersey

I. INTRODUCTION

Direct observation of the ultrafast primary events of many biological and chemical reactions is possible by means of picosecond spectroscopy. Technological advances in lasers and detection systems have resulted in the continuing growth of picosecond emission and absorption spectroscopy and development of new areas such as picosecond Raman spectroscopy and coherent anti-Stokes Raman spectroscopy. Judicious utilization of these various techniques permits, in principal, elucidation of the detailed mechanisms of many photoinitiated reactions. Currently various picosecond laser systems with temporal resolution ranging from less than one to several picoseconds exist which can be used to observe the spectra of the transient intermediates involved in a dynamic process and to measure their formation and decay kinetics. In this paper, we will briefly describe a picosecond laser system which is currently being used and then focus upon picosecond studies of the primary photochemical event involved in the visual transduction process which were performed in our laboratory.

II. EXPERIMENTAL SYSTEM

Picosecond laser systems in use today may vary widely in specific detail, although they most commonly rely upon solid-state oscillators or synchronously pumped dye lasers for the initial generation of picosecond light pulses. In the past,

ISBN 0-12-066280-9

solid-state oscillators, such as Nd^{3+}:YAG or Nd^{3+}:silicate glass, generated the picosecond pulses used to probe ultrafast reactions. In fact, the studies on the primary event in the visual transduction process to be discussed later were all performed with laser systems based upon solid-state oscillators. More recently, synchronously pumped dye laser systems have become popular as a result of their wavelength tunability and ability to generate shorter pulses. Because the manipulation of picosecond pulses for the purposes of spectroscopy and kinetics measurements are basically the same after pulse generation and amplification have been done, regardless of the means of pulse generation, the description of the synchronously pumped dye laser system presented in Fig. 1 is essentially the same as the one for solid-state lasers, of course, with minor variations, depending upon the specific experiment to be performed.

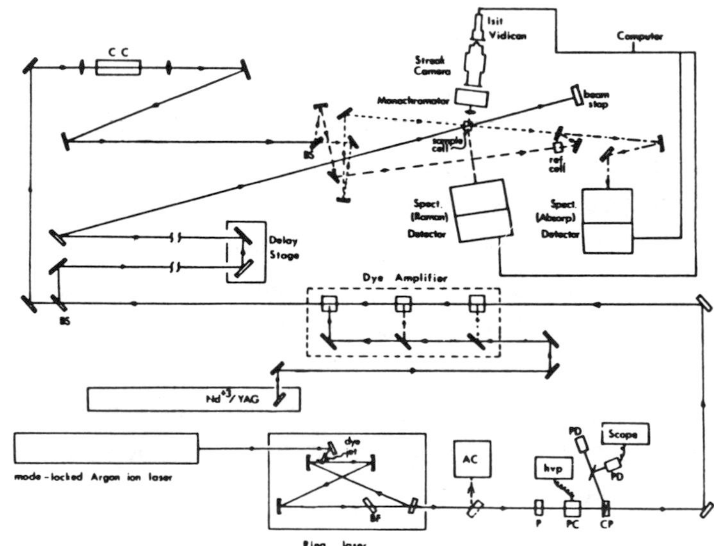

FIGURE 1. Experimental arrangement for picosecond absorption, emission, and Raman spectroscopy. AC: autocorrelator; P: polarizer; PC: Pockels cell; CP: crossed polarizer; hvp: high-voltage pulser; PD: photodiode; SHG: second harmonic generating crystal; BS: beam splitter; CC: continuum cell.

The laser system illustrated in Fig. 1 is composed of an actively mode-locked Argon ion laser which produces a train of 150-ps, 514.5-nm pulses at a rate of 123 MHz. These pulses pump a ring dye laser to generate low-energy pulses (~100 mW average power) and ~1-ps FWHM. In this particular system, the ps pulses are emitted by the ring laser at a rate of 246 MHz and are tunable in wavelength from ~570 to 670-nm with rhodamine 6G (R6G) as the lasing medium. The quality of these pulses are monitored by means of a background-free autocorrelator.

Spectroscopic studies of most chemical and biological processes of interest require the use of a single pulse possessing 10 μJ or greater energy or a train of pulses generated at a high repetition rate with lower individual energy content. When amplification is necessary, a single pulse first is selected by means of a Pockels cell situated between two crossed Glan polarizers and then amplified by a factor of ~10^6 by passing the pulse through a series of flowing R6G dye cells which are pumped by a 10-ns pulse of the second harmonic, 532-nm, of a Nd^{3+}:YAG laser. The amplified pulse has sufficient energy to be used directly for excitation or converted, by means of frequency doubling and/or stimulated Raman scattering, to wavelengths which are not emitted by the dye laser.

Appropriate direction and splitting of this ps pulse, as illustrated in Fig. 1, permits the detection and recording of the spectra of transient intermediates and measurements of their lifetimes by means of emission, absorption, or Raman spectroscopy. For time-resolved emission studies, the emission of light is measured by means of a streak camera/vidicon/optical multichannel analyzer (OMA)/minicomputer arrangement. The emission spectra of a transient may be obtained by means of a spectrometer/vidicon/OMA/minicomputer assembly. For absorption and Raman spectroscopy, the pulse is split into two parts. The first is used for excitation of the sample. The second is used either to generate a probe continuum for absorption spectroscopy by passing it through a cell containing D_2O/H_2O or for light scattering in the case of Raman spectroscopy.

III. INTERMEDIATES IN THE VISUAL TRANSDUCTION PROCESS

Of the various types of picosecond spectroscopy described above, picosecond absorption spectroscopy has been a tool that recently has revealed, in detail, the nature of the primary event involved in sensory process of visual transduction. Although the sense of sight plays perhaps the greatest role in the process by which we learn about and understand our environment, the cascade of events involved in the visual transduction process and triggered by the absorption of a photon by the visual chromophore, rhodopsin, is not well established. Rhodopsin is composed of a protein, opsin, and a polyene aldehyde, 11-*cis*-retinal, that are joined together via a Schiff base linkage between an ε-amino group of a lysine residue within the protein and the carbonyl group of the aldehyde. Much effort has been put forth to gain an understanding of the initial event and the nature of the first intermediate. However, in spite of the wealth of information which has been obtained by means of absorption and emission spectroscopy, controversy still exists with regard to the nature of the first intermediate, generally accepted to be bathorhodopsin, and the process by which it is formed. Using picosecond absorption spectroscopy, Busch, et al. for the first time measured the room temperature formation and decay kinetics of bathorhodopsin and found that it was formed within 6 picoseconds of excitation and decayed to the second intermediate, lumirhodopsin, with a time constant of ~30-ns (1). The remarkably fast formation of bathorhodopsin prompted the low-temperature studies of Peters, et al. in which the risetime of the appearance of bathorhodopsin was found to be extremely fast, 36-ps, even at 4K (2) (Figure 2).

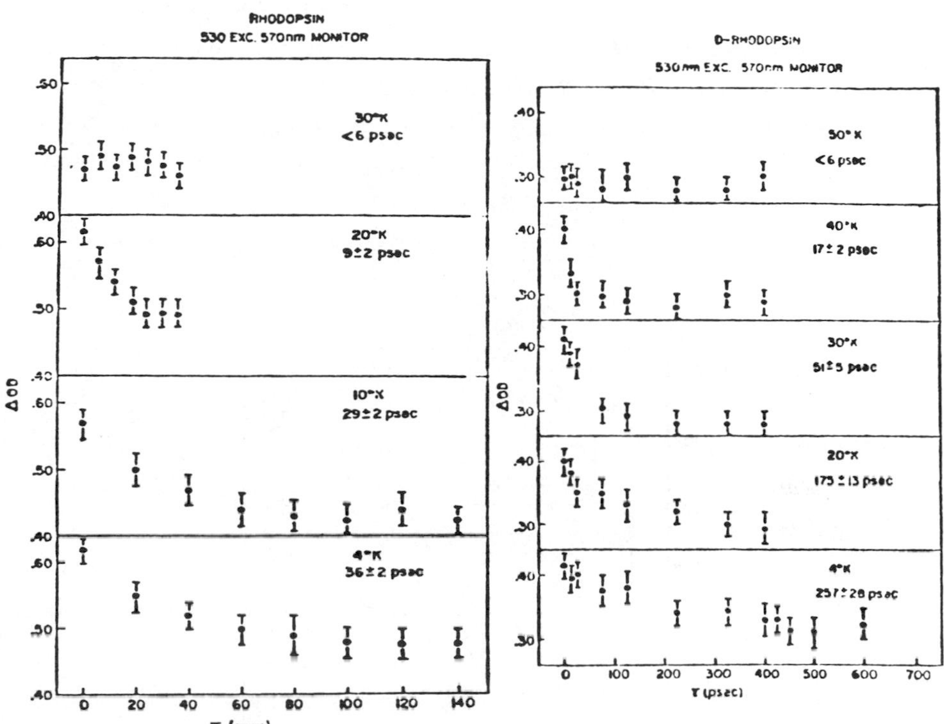

FIGURE 2. Formation kinetics of 11-cis-bathorhodopsin at various temperatures (530-nm excitation; 570-nm probe). (a) rhodopsin; (b) deuterium-exchanged rhodopsin.

Prior to the measurement of the bathorhodopsin formation rate, the primary event which forms bathorhodopsin was considered to be only cis-trans isomerication of the 11-cis-retinal moiety of rhodopsin to an all-trans form (3,4,5). The rapid rate at which bathorhodopsin is formed, even at low temperature, led

Peters, et al., (2) to postulate an alternate mechanism - proton translocation. To test this hypothesis, Peters, et al. measured the low-temperature kinetics of bathorhodopsin generated from a sample of deuterium-exchanged rhodopsin prepared by shaking the rhodopsin preparation with D_2O (2). A pronounced deuterium isotope effect of $k_H/k_D = 7$ was observed. Furthermore, the rate of bathorhodopsin formation was found to exhibit non-Arrhenius behavior and, at low temperature, to become practically independent of temperature. Proton translocation was occurring via quantum mechanical tunneling at low temperature!

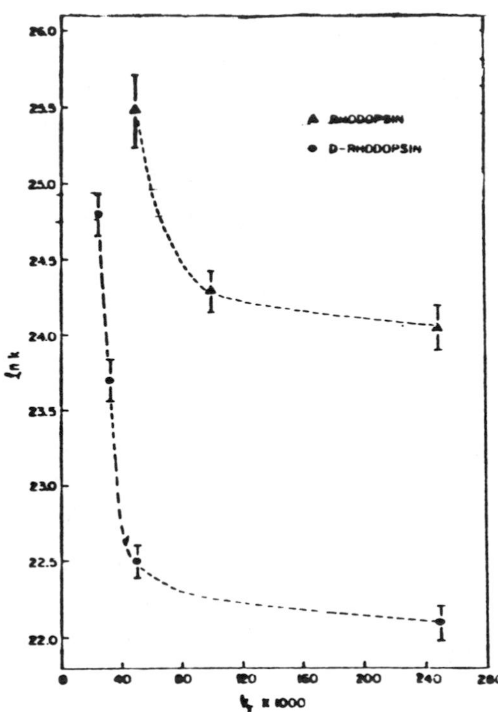

FIGURE 3. Arrhenius plot of the formation rate of bathorhodopsin and deuterium-exchanged bathorhodopsin (D-bathorhodopsin).

Analysis of the kinetic data revealed that a tunneling distance of ~0.5Å was present. Since the Schiff base proton is, by this method, the only deuterium-exchangeable proton within the retinal moiety, it was concluded that proton translocation plays a prominent role in the first event of visual transduction.

The belief that cis-trans isomerization is the primary event has been supported by the commonly occurring statement that both 11-cis-rhodopsin and 9-cis-rhodopsin lead to a common bathorhodopsin intermediate (3). In their recent picosecond absorption study, Spalink, et al. obtained evidence which demonstrated that the batho-intermediates formed by controlled excitation of carefully prepared samples of 9-cis-rhodopsin and 11-cis-rhodopsin indeed are not the same (6). Difference absorption spectra were recorded 85-ps after excitation to insure that only bathorhodopsin was observed since decay of excited-state rhodopsin occurs within 6-ps and bathorhodopsin decays to lumirhodopsin, the next intermediate in the cycle, with a time constant of ~30-ns.

Figure 4 illustrates the difference spectra obtained 85-ps after excitation 9-cis- and 11-cis-rhodopsin at 290K. An observation that had gone unnoticed previously is the shift of the transient absorption band, bleaching band, and isosbestic point resulting from excitation of 9-cis-rhodopsin to longer wavelengths, by ~10-nm, relative to the corresponding positions in the 11-cis difference spectrum. This spectral difference persists to a time of 8-ns after excitation of the sample. These data clearly show that a common intermediate is not formed when 9-cis-rhodopsin and 11-cis-rhodopsin are excited with 532-nm light!

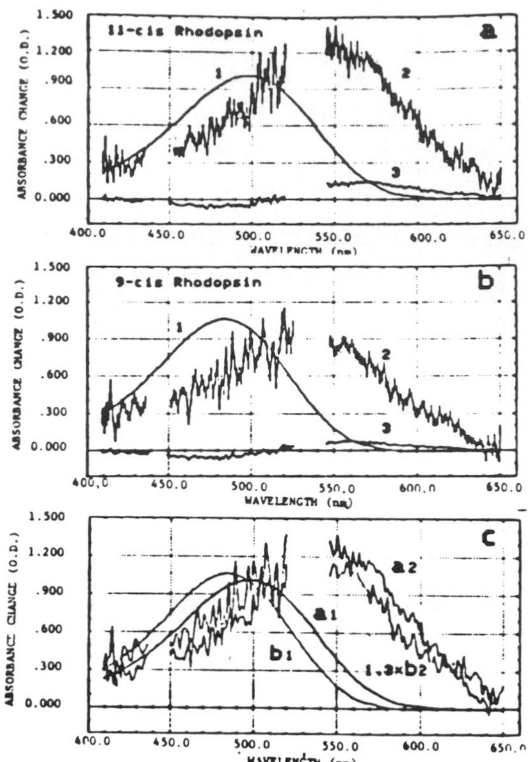

FIGURE 4. Room-temperature absorption spectra of (a) 11-*cis*-rhodopsin static spectrum a1; difference spectrum a3, estimated 11-*cis*-bathorhodopsin spectrum a2; (b) same as (a) but starting material is 9-*cis*-rhodopsin; (c) superposition of curves a1, a2, b1, and b2 normalized to have the same absorbance maximum.

The picosecond absorption studies described above indicate that, while *cis*-trans isomerization may be an important event in the visual transduction process, proton translocation is a prominent process which occurs in the primary event of visual transduction. *Cis*-*trans* isomerization, long-considered to be the primary event on the basis of common intermediate formation from 9-*cis*- and 11-*cis*-rhodopsin, can no longer be considered as a criterion for the classic explanation of the primary event in the visual transduction process in light of the recent picosecond study of Spalink, et al. The power of the technique of picosecond spectroscopy to reveal mechanistic detail of a chemical or biological process, <u>as it evolves</u>, is clearly illustrated in the examples given above.

REFERENCES

1. Busch, G. E., Applebury, M. L., Lamola, A. A., and Rentzepis, P. M., Proc. Natl. Acad. Sci. USA 69, 2802 (1972).
2. Peters, K., Applebury, M. L., and Rentzepis, P. M., Proc. Natl. Acad. Sci. USA 74, 3119 (1977).
3. Ottolenghi, M., Adv. in Photochem. 12, 97 (1980).
4. Wald, G., Nature 219, 800 (1968).
5. Rosenfeld, T., Honig, B., Ottolenghi, M., Hurley, J., and Ebrey, T. G., Pure Appl. Chem. 49, 341 (1977).
6. Spalink, J. D., Reynolds, A. H., Rentzepis, P. M., Applebury, M. L., and Sperling, W., submitted for publication.

APPLICATIONS OF TIME-RESOLVED RESONANCE RAMAN SPECTROSCOPY IN RADIATION CHEMISTRY AND PHOTOBIOLOGY: STRUCTURE AND CHEMISTRY OF CAROTENOIDS IN THE EXCITED TRIPLET STATE

Robert Wilbrandt
Niels-Henrik Jensen

Risø National Laboratory
Roskilde, Denmark

I. INTRODUCTION

Pulse radiolysis provides a means of investigating fast chemical reactions in solution irradiated by ionizing radiation such as γ-rays or high energy electrons. Different primary short-lived species are generated dependent on the choice of solvent. In polar solvents ionic processes predominate, whereas nonpolar solvents in general lead to the production of excited species.

The primary species generated in aqueous solutions are the solvated electron e^-_{aq}, OH-radicals and, with a lower yield, H-atoms. Each of these species can be selectively scavenged under appropriate conditions. The solvated electron e.g. is converted to OH-radicals by the reaction with nitrous oxide. In such a way well defined conditions for the reaction of one of these primary species with a given solute can be obtained and reduction and

ISBN 0-12-066280-9

oxidation processes can be studied.

In nonpolar organic solvents such as benzene or cyclohexane excited singlet and triplet states of these molecules are predominantly formed by a rapid recombination of the primarily produced cation-electron pairs, and this energy is then available for energy transfer processes to solute molecules leading to the production of solute excited states.

In comparing pulse radiolysis with its "bigger brother" flash photolysis, certain differences should be noticed: in pulse radiolysis the primary excitation energy is deposited in the solvent whereas solutes can be excited directly and selectively in photolysis. In pulse radiolysis relatively large volumes can be excited homogeneously, whereas the homogeneity of excitation in photolysis is dependent on solute concentration. The primary species generated by the two techniques are basically different: free electrons and ions, radicals and excited states of solvents in pulse radiolysis and excited states of solutes in photolysis. With these differences in mind, it is obvious that comparative and often complimentary experiments can be performed by the two techniques.

The present discussion is restricted to pulse radiolysis and deals with the application of time-resolved resonance Raman spectroscopy (RRS) in combination with it. Our group has worked in this area of research since 1974, when the first spectrum of a transient species was published (1). In 1978 it was shown for the first time that this method was applicable to the study of short-lived excited states as well (2). Since then a number of different chemical systems and short-lived species was studied with the main interest in excited triplet states of biological polyenes like β-carotene and retinal (3). The reduction of cytochrome c by solvated electrons (4) and radical cations and anions of certain disulfides (5) were studied as well. Other groups have published work on carotenoids (6), semiquinone radical anions (7), the galvinoxyl free radical (8), the aminophenoxyl radical (9) and viologen radicals (10).

Among the different work mentioned above we shall here concentrate on a discussion of recent results concerning the sensitized photoisomerization of carotenoids and the application of time-resolved resonance Raman spectroscopy to the study of their lowest excited triplet states involved in the isomerization processes.

II. SENSITIZED PHOTOISOMERIZATION

The photochemistry and photophysics of carotenoids is a field which still is full of puzzling features although numerous studies have been published. Carotenoids have very strong $S_o - S_1$ absorptions, but the excited singlet states do not fluoresce or undergo intersystem crossing to the triplet manifold. Hence nonradiative internal conversion of the excited singlet state back to the ground state must be extremely rapid. Recent experiments (6) indicate a lifetime of S_1 shorter than 1 ps. In spite of this carotenoids act as accessory pigments in photosynthetic organisms (11), a function which involves energy transfer from excited singlet carotenoids to the chlorophylls. The detailed mechanism of this process is not yet known, but the elegant work of Moore and coworkers (12) suggest that the spatial arrangement of the carotenoids relative to the chlorophylls is very important.

Because of the low $S_1 - T_1$ intersystem crossing efficiency the triplet manifold of the carotenoids can only be populated by sensitization. The triplet energy appears to be low, as triplet chlorophyll <u>a</u> (13) and singlet oxygen (14) are efficient sensitizers of e.g. the triplet state of β-carotene. In intact photosynthetic organisms the triplet states of carotenoids are important intermediates in the protection of the photosynthetic apparatus against photodestruction (15), and the sensitizers are believed to be chlorophyll and/or singlet oxygen.

Studies of antenna complexes and reaction centers from several photosynthetic bacteria by RRS have strongly suggested that the carotenoids (not β-carotene) associated with the reaction centers exist in a cis-conformation which is strikingly similar with a central mono-cis conformation (16-18). The carotenoids associated with the antenna complexes appear as all-trans. Whether the conformation of the reaction center carotenoids is a strained s-cis conformation imposed by the protein or a true geometric double-bond isomer is still an open question.

Recently, triplet states of carotenoids have been studied by time-resolved RRS by us (3) and by Woodruff and coworkers (6). The experimental results obtained by the two groups are similar, but the interpretations of the triplet state spectra are different. Woodruff and coworkers suggested that the spectral changes in RRS were due to changes in signs of the interaction constants connecting the C-C and C=C vibrational modes when going from the ground to the excited triplet state. Within this model, major changes in geometry upon electronic excitation are considered to be unimportant. We have, on the other hand, suggested that the triplet Raman spectra reflect two major changes: i) decreased C=C double bond order and ii) substantial twisting around the central 15,15'- double bond in the excited triplet state. Our model implies that at least some trans-cis isomerization in the triplet state of all-trans carotenes is to be expected. However, several reports in the literature claim that only cis-trans and <u>not</u> trans-cis isomerization occurs upon triplet sensitization(19).

Faced with these conflicting reports and suggestions we have initiated a study of the triplet state chemistry of carotenoids with special emphasis on the sensitized photoisomerization of β-carotene (20). As triplet sensitizers we have used chlorophyll <u>a</u> and chlorophyll <u>a</u> plus oxygen because these systems are biologically important, and because the S_o - S_1 absorption of Chl <u>a</u> is spectrally separated from the absorption of β-carotene. Analysis by HPLC shows that illumination with red light (λ>610 nm) of all-trans β-carotene in the presence of Chl <u>a</u> causes trans-cis isomerization of β-carotene. The primary products are 9-cis, 13-cis and 15-cis-β-carotene. By prolonged illumination a photostationary mixture is obtained. Illumination of 15-cis-β-carotene with red light in the presence of Chl a yields all-trans as the primary product, but the photostationary mixture from 15-cis is similar to that from all-trans. Chromatograms illustrating these results are shown in figure 1.

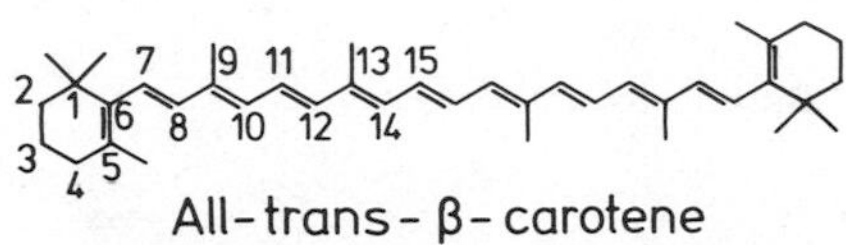

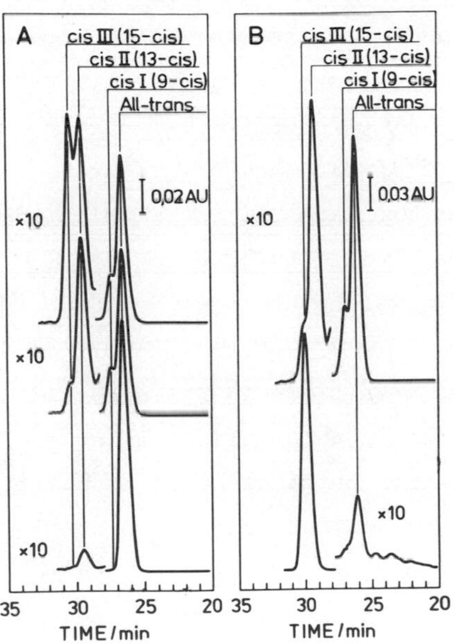

FIGURE 1. Chromatograms (reversed-phase HPLC) of β-carotene before and after Chl a sensitized photoisomerization. The Ar-sat solutions all contained $2x10^{-5}$ M Chl a. Analysis wavelength, 440 nm; solvent, benzene/acetone 1/1. A) Lower trace: 1 mM all-trans before illumination. Middle trace: The same after illumination for 15min with λ>610 nm. Upper trace: Illuminated solution mixed with 15-cis. Samples diluted 10 times before HPLC. B) Lower trace: 1 mM 15-cis before illumination. Upper trace: After illumination for 15 min with λ>610 nm. Samples diluted 5 times before HPLC.

Assignment of the fractions is based upon the absorption spectra of the fractions.

The relative quantum yields of isomerization show that the isomerization of 15-cis is an order of magnitude more efficient than that of all-trans. In the presence of oxygen isomerization is still observed for all-trans and 15-cis, but the quantum yields are lower than in the absence of oxygen. The photostationary mixtures contain more all-trans in the presence of oxygen. Finally we have observed that the composition of the photostationary mixture obtained from all-trans depends on the concentration of β-carotene.

III. TRIPLET STATE RESONANCE RAMAN SPECTRA

RRS provides detailed information on the conformational structure of carotenoids. Recently Koyama et al. (18) have recorded the spectra of different isomers of β-carotene in the ground state and shown that each isomer has its own characteristic RRS. We present here a comparison of the spectra of two isomers in their triplet states, namely that of 15-cis- and of all-trans-β-carotene. As these triplet states are expected to exist as intermediates in the above mentioned isomerization processes, time-resolved RRS should provide a very useful means of investigating such processes in detail.

The short-lived triplet states were produced in pulse radiolysis of solutions of the carotenoids in a nonpolar solvent (p-xylene) by triplet energy transfer from naphthalene as a sensitizer. The transient triplet-triplet optical absorption spectra were determined and showed peaks at approx. 520 nm for both isomers. The lifetime of the transients was approx. 3 μs. Raman spectra were excited by the second harmonic of a Q-switched Nd:YAG laser at 532 nm and detected by an optical multichannel analyzer.

Figure 2 shows a comparison of the obtained RRS of the two isomers in their triplet states (A,B) and their ground state(C,D).

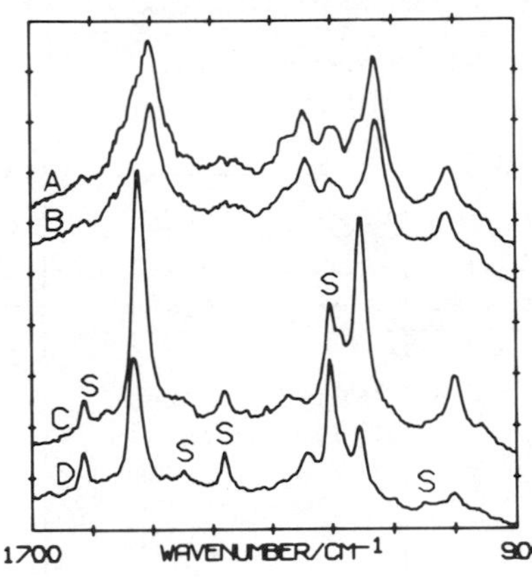

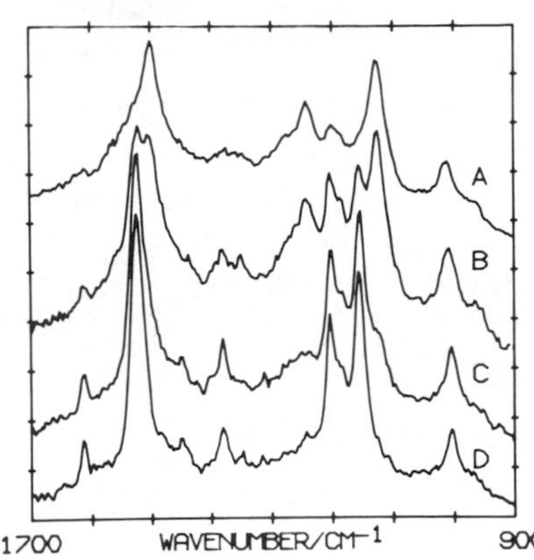

FIGURE 2. FIGURE 3.

Resonance Raman spectra from 10^{-4} M β-carotene in Ar-sat. p-xylene containing 0.01 M naphthalene. Fig.2: A) all-trans 2 μs after electron irradiation, B) 15-cis 2 μs after electron irradiation, C) all-trans without D) 15-cis without electron irradiation. Fig.3: A) 2 μs, B) 6 μs, C) 12 μs, D) 40 ms after electron irradiation of 15-cis. Dose, 22 krad; excitation, 532 nm; 2-4 laser pulses accumulated in each spectrum, spectra not smoothed; S: solvent bands.

In figure 3 the spectra obtained at different time delays with respect to the electron pulse are shown.

From figure 2 it can be seen that the RRS of the two isomers in their triplet states are identical within limits of error, whereas the ground state spectra are markedly different. With respect to the above mentioned different interpretations of the RRS of the triplet state of all-trans, the present results favour our own suggestion, which would explain the strong similarity or identity of the triplet states in the two isomers.

Wavenumbers are tabulated in table I. The shift in wavenumber from 1532 to 1526 cm^{-1} when going from the unirradiated (Fig.2D) to the irradiated solution (Fig.3D) and the relative

TABLE I. Vibrational wavenumbers/cm^{-1} of 15-cis and all-trans β-carotene in their ground state and lowest excited triplet state[a)]

15-cis				all-trans	
ground state Lutz[b)]	ground state this work	triplet state	after irradiation	ground state	triplet state
		969w			965w
1006	1001	1014s	1005	1001s	1014s
1160	1156	1127vs	1156	1156vs	1129vs
1197	1195[c)]	1188m		1190w	1188m
1237/1245	1242	1244s	1242	1213vw	1247s
		1351vw			1358vw
1540	1532	1500vs	1526	1526vs	1501vs

a) in p-xylene, calibrated with indene, b) in cyclohexane at 30 K (16), c) in benzene

intensities of this and the band at 1156 cm^{-1} indicate a 15-cis → all-trans isomerization via the triplet state with a very high quantum efficiency which is in agreement with the results obtained by HPLC.

We have also studied the triplet state of all-trans lycopene, the major pigment in red tomatoes. This pigment does not contain the cyclohexene end groups found in e.g. β-carotene and it may thus provide information about the importance of the rings with respect to the vibrational properties of the triplet state. The RRS of the triplet state of all-trans lycopene were excited at two different wavelengths: 532 nm close to the maximum of triplet-triplet absorption (535 nm) and 572.3 nm. Spectra are shown in Fig. 4 and 5 and wavenumbers are tabulated in Table II. The major features observed are i) a downshift of the strong bands at 1153 and 1511 cm^{-1} to 1129 and 1495 cm^{-1}, ii) a shift of the band at 1004 cm^{-1} to 1014 cm^{-1} and iii) a new band at 1254 cm^{-1} when going from the ground to the triplet state. The main differences compared to other carotenoids are seen in the C=C stretching region, where lycopene shows a complex of bands around 1500 cm^{-1} in the triplet state.

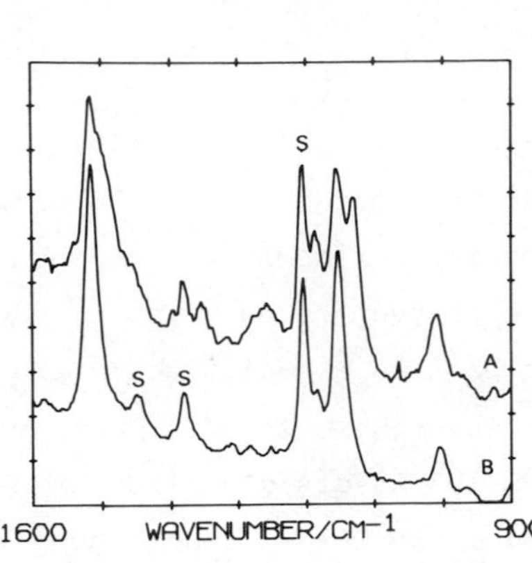

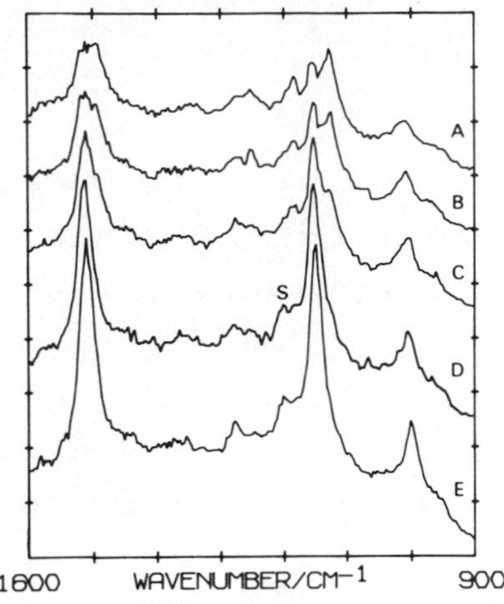

FIGURE 4. FIGURE 5.

Resonance Raman spectra from 10^{-4} M all-trans lycopene in Ar-sat. p-xylene containing 10^{-2} M naphthalene. Fig.4: A) 2 μs after electron pulse, B) without electron pulse. Excitation, 572.3 nm; dose, 22 krad. Fig.5: A) 1 μs, B) 3 μs, C) 5 μs, D) 12 μs after and E) without electron irradiation. Dose, 22 krad. 2-4 laser pulses accumulated in each spectrum. S: solvent bands.

TABLE II. Vibrational wavenumbers/cm^{-1} of all-trans lycopene in the ground state and lowest excited triplet state

ground state		triplet state	
572.3 nm	532 nm	572.3 nm	532 nm
965 w	965 w	975 w	971 w
1003 m	1004 m	1010 m	1014 m
1152 s	1153 s	1130 s	1129 s
			1137 sh
1184 w	1184 w	1185 m	1183 m
		1262 m,br	1254 m
	1278 w		1272 m
		1355 m	
1514 s	1511 s	1497 s,sh	1495 s
			1505 s
			1520 s
			1570 sh

a) wavenumbers are calibrated with indene

IV. DISCUSSION

The results obtained from the experiments on the sensitized photoisomerization of all-trans β-carotene suggest that the triplet state produced from all-trans consists of a mixture of several noninterconvertible species, each of them being twisted around the 9-, 13- or 15- C=C double bond as exemplified in the tentative reaction scheme in Fig. 6. The similarity of the triplet Raman spectra obtained from all-trans and 15-cis could indicate that the triplet species twisted around the 15-C=C double bond is the predominant one in the mixture of twisted triplet species produced from all-trans. However, we cannot yet exclude the possibilities that the 9- and 13-twisted species have weak Raman spectra or that all of the twisted species have similar resonance Raman spectra in the excited triplet. In both cases RRS would only give little information about the structure of the relaxed triplet state. The distinctly different resonance Raman spectra of the ground states of various cis-isomers of β-carotene reported by Koyama et al. (18), however, give hope that the Raman spectra of the triplet states will show differences too. Work on the other mono-cis-isomers of β-carotene is presently in progress in our laboratory.

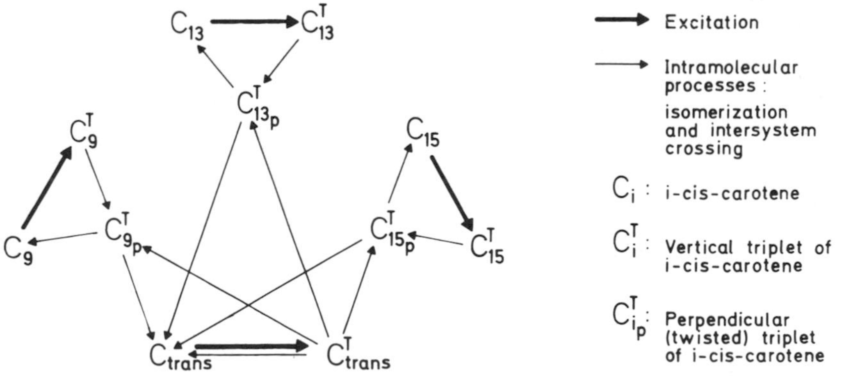

FIGURE 6. Reaction scheme for β-carotene isomerization

The present results are also of interest in connection with the recent reports concerning the apparent occurrence of cis-carotenoids in the reaction centers of several photosynthetic bacteria (16-18). Lutz and coworkers suggested that the reaction center carotenoids exist in a di-cis configuration (16), while Koyama et al. have pointed out that the reaction center carotenoids with respect to resonance Raman are more like 15-cis-β-carotene than any of the other cis-isomers included in their study. Taken together with our demonstration that 15-cis-β-carotene is indeed formed in the chlorophyll sensitized trans-cis photisomerization of all-trans a feasible mechanism for the formation of cis-carotenoids in the reaction centers would be photoisomerization sensitized by the bacteriochlorophylls. However, it should be stressed that this hypothesis cannot explain the apparent thermal lability of the supposed cis-isomers. Lutz et al. (16) reported that the reaction center carotenoids upon extraction in the dark isomerized very rapidly from the native configuration to a configuration similar to all-trans. This property could indicate that the cis-configuration in the reaction centers does not involve C=C bonds, as C=C-bond cis-isomers of carotenoids usually are reasonably stable against thermal isomerization.

An alternative explanation which possibly could account for the resonance Raman spectra of the reaction center carotenoids in the ground (16-18) and the triplet state (21) as well as the unusual thermal lability (16) would be that the reaction center carotenoids are taking cis-configurations around one or more of the <u>single</u> bonds. A cis-configuration involving only C-C bonds (i.e. s-cis configurations) could well give ground state resonance Raman spectra similar to "normal" cis-isomers, and the barriers for rotation back to the thermal equilibrium called all-trans would certainly be small. Finally, an s-cis configuration could be the reason for the differences between the triplet Raman spectra of reaction center carotenoids (21) and the structural analogue lycopene reported here. The s-cis model for the reaction center carotenoids suggested here does not imply any photoisomerization in the formation of the s-cis configuration. This special configuration could probably be formed easily by the interaction between the protein and the all-trans carotenoid in agreement with experimental observations (17,22).

REFERENCES

1. Pagsberg, P., Wilbrandt, R., Hansen K.B., Weisberg, K.V., Chem.Phys.Letters 39, 538 (1976)

2. Wilbrandt, R., Jensen, N.-H., Pagsberg, P., Sillesen, A.H., and Hansen, K.B., Nature (london) 276, 167 (1978)

3. Jensen. N.-H., Wilbrandt, R., Pagsberg, P., Sillesen, A.H., and Hansen, K.B., J.Am.Chem.Soc. 102, 7441 (1980); Wilbrandt, R., and Jensen, N.-H., J.Am.Chem.Soc. 103, 1036 (1981); Wilbrandt, R., and Jensen, N.-H., Ber.Bunsenges.Phys.Chem. 85, 508, (1981)

4. Cartling, B., and Wilbrandt, R., Biochim.Biophys.Acta 637, 61 (1981)

5. Wilbrandt, R., Jensen, N.-H., Pagsberg, P., Sillesen, A.H., Hansen, K.B., and Hester, R.E., Chem.Phys.Letters 60, 315 (1979); Jensen, N.-H., Wilbrandt, R., Pagsberg, P., Hester, R.E., and Ernstbrunner, E., J.Chem.Phys. 71, 3326 (1979); Wilbrandt, R., Jensen, N.-H., Pagsberg, P., Sillesen, A.H., Hansen, K.B., and Hester, R.E., J.Raman Spectrosc. 11, 24 (1981)

6. Dallinger, R.F., Guanci, J.J., Woodruff, W.H., and Rodgers, M.A.J., J.Am.Chem.Soc. 101, 1355 (1979); Dallinger, R.F., Farquharson, S., Woodruff, W.H., and Rodgers, M.A.J., J.Am. Chem.Soc. 103, 7433 (1981)

7. Tripathi, G.N.R., J.Chem.Phys. 74, 6044 (1981); Tripathi, G.N.R., and Schuler, R.H., J.Chem.Phys. 76, 2139 (1982)

8. Tripathi, G.N.R., Chem.Phys.Lett. 81, 375 (1981)

9. Tripathi, G.N.R., and Schuler, R.H., J.Chem.Phys. 76, 4289 (1982)

10. Lee, P.C., Schmidt, K., Gordon, S., and Meisel, D., Chem. Phys. Lett. 80, 242 (1981)

11. Goedheer, J.C., Biochim.Biophys.Acta 35, 1 (1959); Biochim. Biophys.Acta 172, 252 (1969)

12. Bensasson, R.V., Land, E.J., Moore, A.L., Crouch, R.L., Dirks, G., Moore, T.A., and Gust, D., Nature (london) 290, 329 (1981)

13. Mathis, P., and Cleo, J., Photochem.Photobiol. 18, 343(1973)

14. Farmilo, A., and Wilkinson, F., Photochem.Photobiol. 18, 447 (1973); Rodgers, M.A.J., and Bates, A.L., Photochem.Photobiol. 31, 533 (1980)

15. Krinsky, N.I., Pure and Appl.Chem. 51, 649 (1979)

16. Lutz. M., Agalidis, I., Hervo, G., Cogdell, R.J., and Reiss-Husson, F., Biochim.Biophys.Acta 503, 287 (1978)

17. Agalidis, I., Lutz, M., and Reiss-Husson, F., Biochim.Biophys.Acta 589, 264 (1980)

18. Koyama, J., Kito, M., Takii, T., Saiki, K., Tsukida, K., and Yamashita, J., Biochim.Biophys.Acta 680, 109 (1982)

19. Claes, H., Biochem.Biophys.Res.Comm. 3, 585 (1960); Claes,H., Z.Naturforsch. 16B, 445 (1961); Foote, C.S., Chang, Y.C., and Denny, R.W., J.Am.Chem.Soc. 92, 5218 (1970); Krinsky, N.I., in "The Survival of Vegetative Microbes", Gray, T.G.R., and Postgate, J.R., Eds., Cambridge University Press, Cambridge 1976, pp. 209-239

20. Jensen, N.-H., Nielsen, A.B., and Wilbrandt, R., J.Am.Chem. Soc. (1982) (in press)

21. Lutz, M., Chinsky, L., and Turpin, P.-Y., Photochem.Photobiol. (1982) (in press)

22. Boucher, F., Van Der Rest, M., and Gingras, G., Biochim.Biophys.Acta 461, 339 (1977)

Influence of Excitation Wavelength on the Time-Resolved Resonance Raman Spectroscopy of Deoxy Heme Proteins

M. J. Irwin and G. H. Atkinson

Department of Chemistry
Syracuse University
Syracuse, New York 13210

INTRODUCTION

The structure and dynamics of the deoxy heme proteins which undergo biochemically important ligation reactions have been modeled extensively by the transient deoxy heme proteins formed through the photolysis of their respective ligated heme complexes. The photolytic formation of deoxy hemoglobin (Hb) from the oxyhemoglobin (HbO_2) and carbonoxyhemoglobin (HbCO) complexes are two of the most commonly reported examples. Spectroscopic experiments based on flash photolysis/transient absorption (1) and time-resolved resonance Raman (TR^3) scattering (2-5) make the tacit assumption that the properties of the photolytically-formed deoxy Hb are readily correlated to those of the thermalized deoxy Hb found in biochemical reactions. The validity of this assumption derives from a variety of conditions, only one of which is addressed here. Does the transient deoxy Hb intermediate remain the same when specific changes (e.g., excitation wavelength) are made in the photolytic conditions?

ISBN 0-12-066280-9

Although it is widely accepted that changes in the excitation conditions can have dramatic effects on the photo-products, a wide range of excitation wavelenghts, powers, and pulse widths have been used previously without directly addressing this question. Leaving this issue unresolved introduces great uncertainty into subsequent structural interpretations of spectroscopic data since the different photolytic conditions may have formed different heme intermediates.

The principal uncertainties relate to the excited electronic states of the parent molecule populated by optical excitation and the protoproduct(s) formed as a consequence. The high peak powers typically used in pulsed laser excitation can initiate both multiphoton events leading to the population of highly energetic excited states and multiple excitations (i.e., optical cycling) causing the same molecule to undergo repeated excitation and decay (e.g., photodissociation) during one laser pulse. Although the molecular complexity of hemoglobins precludes obtaining detailed excited-state spectra readily, the same type of information can be obtained empirically by monitoring the photoproducts as the excitation conditions are changed.

Previous studies have had varying degrees of success. Consider for example recent laser photolysis/transient absorption experiments examining HbCO photodissociation in the picosecond time regime. Shank et al. (6) reported that CO was found in less than 0.5 ps using 519 nm photolysis. Noe et al. (7) estimated this photodissociation time to be 11 ps using 530 nm excitation while Greene et al. (8) placed an upper limit of 16 ps on the process with excitation at both 530 nm and 352 nm. Without determining the contributions of the different excited electronic states of HbCO to the photodissociation, it is difficult to evaluate whether the apparent differences in appearance rates has any significance with respect to molecular structure. It is simply not established that the same Hb intermediate is formed by the different excitation wavelengths used in each experiment.

Since TR^3 spectroscopy reflects changes in the vibrational degrees of freedom, it offers a more sensitive method than transient absorption for short-lived heme intermediates. TR^3 spectroscopy has been used widely to characterize the transient deoxy Hb formed by photolysis (9). The great sensitivity of TR^3 spectroscopy to the structure and conformation of deoxy Hb transients provides an excellent opportunity to study the effects of different excitation conditions. In this work, we ulitilize TR^3 spectroscopy to determine whether the same deoxy Hb transient is formed by 532 nm excitation into the α, β absorption bands and by 442 nm excitation into the Soret absorption band.

EXPERIMENTAL METHODS AND TECHNIQUES

The preparation of samples has been described elsewhere (10). Resonance Raman spectra were recorded either by (1) pulsed excitation using a single laser source or (2) TR^3 spectroscopy using two, separately tunable laser sources. In the first case, optical excitation (together with any subsequent photochemistry) and resonance Raman scattering were generated by the same laser wavelength (i.e., 532 nm or 442 nm). In the second case, separate laser wavelengths could be used for each process while the time between laser pulses was varied. In both cases, the pulse widths (FWHM) of the lasers operating at 532 nm and 442 nm were 17 ns and 8 ns, respectively. Further details of the instrumentation are given elsewhere (11,12).

RESULTS

Three regions of the resonance Raman spectrum of deoxy Hb were examined since previous work has shown them to be sensitive to the bonding, electronic, and conformational properties of the heme (2-5). The region studied, together with the excitation wavelengths used to induce resonance Raman scattering, were: (1) 100-450 cm^{-1} (442 nm excitation), (2) 1200-1350 cm^{-1} (442 nm excitation), and (3) 1500-1650 cm^{-1} (532 nm excitation). In order to correlate the spectral features in these three regions, it must be determined whether the excited electronic states populated by pulsed excitation at 442 nm and 532 nm produce the same deoxy Hb transient.

Figures 1A, 1D, and 1G show spectra for these three regions obtained by single, pulsed excitation (at 532 nm or 442 nm as indicated). These spectra are all assigned to a transient, conformational form of deoxy Hb since they contain bands which are significantly shifted with respect to those observed in the resonance Raman spectra of either HbCO or chemically-stable deocy Hb (2-5).

To evaluate whether the transient deoxy Hb was the same, TR^3 spectra were recorded using combinations of pulsed laser excitation at 532 nm or 442 nm. The TR^3 spectra of the 100-450 cm^{-1} and 1200-1350 cm^{-1} regions presented in Figures 1B and 1E were obtained using 532 nm radiation to photolysis HbCO and 442 nm radiation to generate resonance Raman scattering from the transient formed 20 ns later. These spectra (1B and 1E) are identical to those shown in Figures 1A and 1D, respectively. The same results were obtained when the time delay was increased to 400 ns (cf. Figures 1C:1B:1A and 1F:1E:1D).

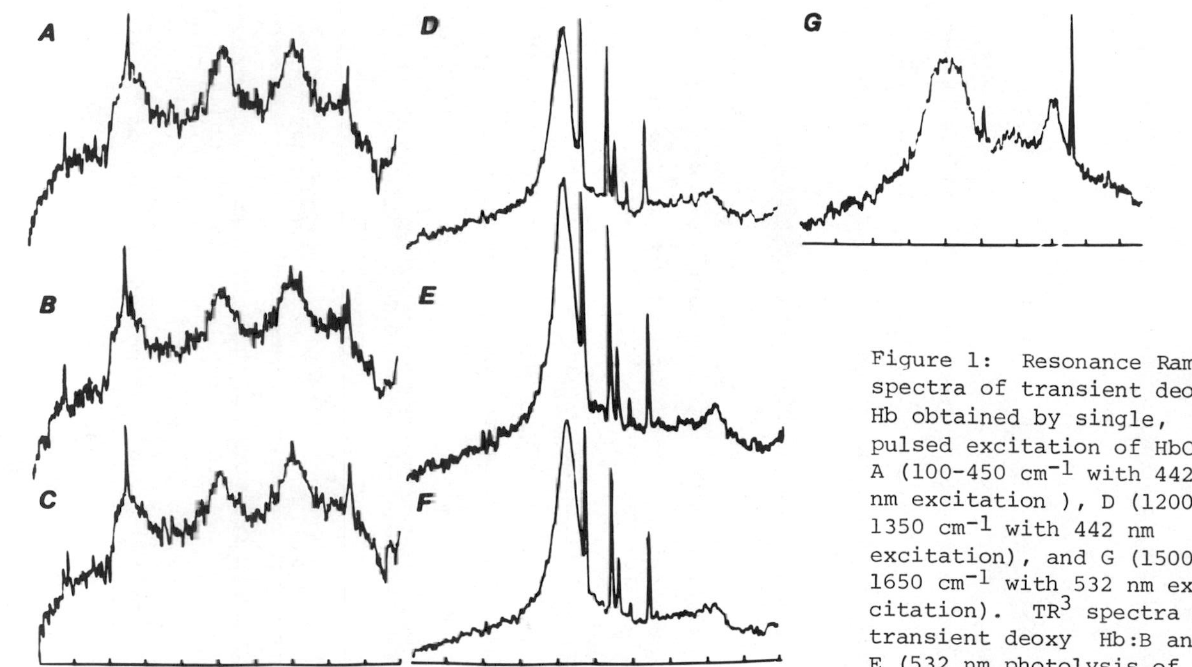

Figure 1: Resonance Raman spectra of transient deoxy Hb obtained by single, pulsed excitation of HbCO: A (100-450 cm^{-1} with 442 nm excitation), D (1200-1350 cm^{-1} with 442 nm excitation), and G (1500-1650 cm^{-1} with 532 nm excitation). TR^3 spectra of transient deoxy Hb:B and E (532 nm photolysis of HbCO, 442 nm probe, 20ns delay; for 100-450 cm^{-1} and 1200-1350 cm^{-1} regions, respectively) and C and F (same as B and E except 400 ns delay).

The analogous experiments were performed in which 442 nm pulsed excitation was used to photolyze HbCO while the 532 nm pulse was delayed in time (20 ns and 400 ns) in order to generate resonance Raman scattering in the 1500-1650 cm^{-1} region from the transient deoxy Hb. In this case, the spectra were again found to be the same as that observed for single pulsed excitation at 532 nm. The conditions relating to buffer, ionic strength, and pH were chosen in these experiments to insure that the conformational relaxation of the deoxy Hb transient was longer than the 400 ns delay chosen (13).

The same type of experiments were also performed on the formation of deoxy Lb from LbCO, a monomeric protein. Figure 2 presents data for chemically-stabilized deoxy Lb (Figure 2A, 2C, and 2E) and for the photoproduct formed by single, pulsed laser excitation of LbCO at 532 nm and 442 nm (Figures 2B, 2D and 2F). No changes in the wavenumber positions of bands in any of the three, structure-sensitive regions was observed. The same results were obtained for another monomeric protein, myoglobin (Mb) and carbonoxymyoglobin (MbCO).

DISCUSSION

The spectra presented in Figure 1 demonstrate that the resonance Raman spectrum generated by 532 nm excitation is the same regardless of whether HbCO was photolyzed at 532 nm or 442 nm. The analogous conclusion can be drawn for resonance Raman spectrum generated by 442 nm excitation. All of these resonance Raman spectra, of course, are significantly different than those observed for HbCO itself and the chemically-stablized conformation of deoxy Hb (2-5). Hence, the frequency shifts in band positions observed in all three spectral regions can be identified with the same transient form of deoxy Hb. The molecular parameters associated with each spectral region can

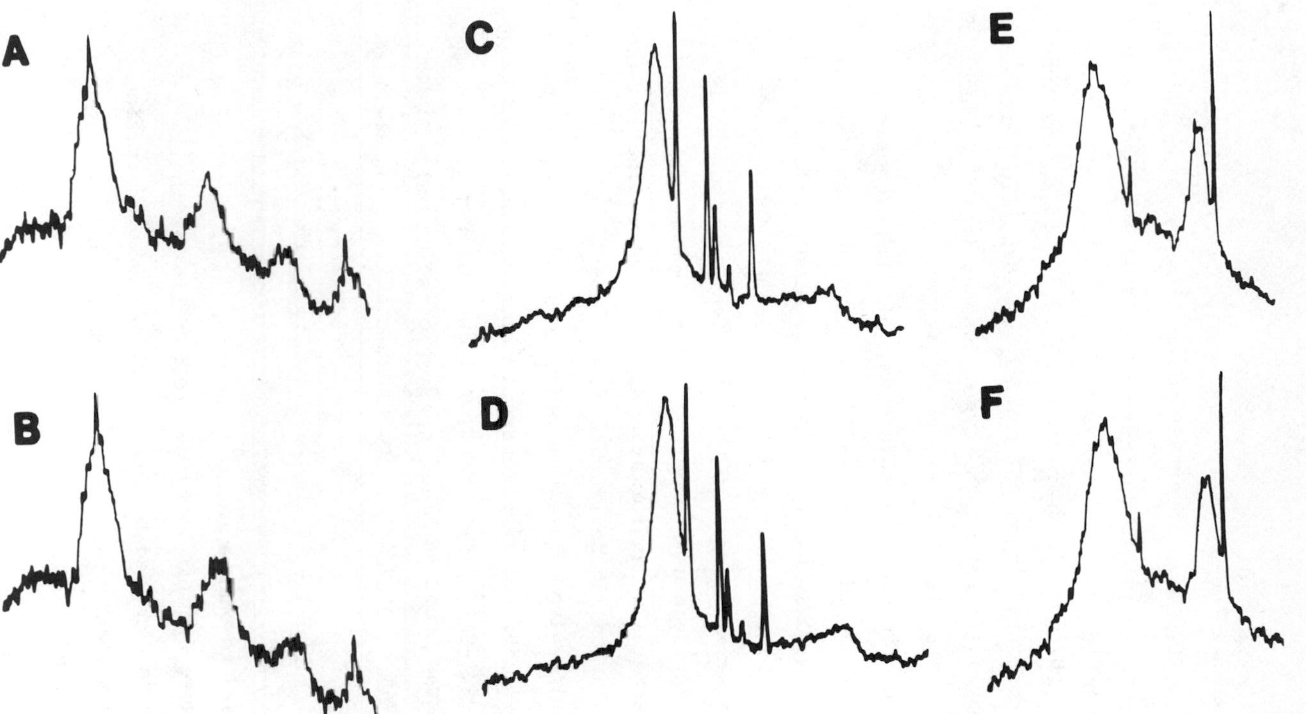

Figure 2: Resonance Raman spectra of deoxy Lb (chemically-stable: A, C and E and photolytically formed from LbCO: B, D, and F) obtained by single, pulsed excitation. A and B (100-450 cm^{-1} with 442 nm excitation), C and D (1200-1350 cm^{-1} with 532 nm excitation), and E and F (1500-1650 cm^{-1} with 532 nm excitation).

be combined then to characterize the same deoxy Hb conformation. This point is especially important since other studies (5,13) have shown this transient conformation relaxes into the chemically-stable conformation of deoxy Hb when the buffer, ionic strength, and pH conditions are controlled. In fact, the transient deoxy Hb formed in photolysis may be identified with an R-like conformation (5,13) of the type proposed in the Perutz model for cooperativity (14,15).

The related resonance Raman spectra of the monomeric protein, Lb, presented in Figure 2 also demonstrates that the conformation of deoxy Lb is the same for both single pulsed and TR^3 excitation. Unlike the tetrameric deoxy Hb, however, the resonance Raman spectra of the transient and chemically-stabilized Lb are the same showing that the monomeric protein does not undergo major conformational change upon ligation (16). The same conclusion can be reached for monomeric myoglobin.

These results also comment on whether these resonance Raman spectra contain any contributions from excited electronic states or whether they reflect conformational and bonding changes only. It is well established that TR^3 spectra of electronically-excited state molecules can be recorded readily using the high peak power excitation conditions used here (9,11). Conversely, minimal excited-state contribtuions are found for the low peak power conditions of cw laser excitation. Thus, the observation that the pulsed resonance Raman Spectra of deoxy Hb (Figure 1), deoxy Lb (Figure 2), and deoxy Mb are the same as their respective spectra obtained by low power, cw laser excitation strongly suggests that the pulsed spectra contain no contritions from excited electronic state species.

A related conclusion can be reached concerning the resonance Raman spectra of the CO complexes of Lb and Mb. The pulsed resonance Raman spectra of deoxy Lb (Figure 2) and photolyzed and LbCO (Figure 2) both are found to be identical to the resonance Raman spectra of deoxy Lb obtained with low power, cw laser excitation (10). Since there is no observed change in the structure or conformation of the monomeric protein upon ligation (in contrast to Hb), one concludes again that either there are no contributions to the TR^3 spectrum from excited electronic states or that the resonance Raman spectra of excited electronic state species are essentially the same as their ground-state counterparts.

CONCLUDING REMARKS

The results of this work show that the unligated conformation of Hb produced by photolysis is the same regardless of whether it is formed by 532 nm or 442 nm excitation. In no case is there any evidence for contributions to resonance Raman scattering from excited electronic states of either the deoxy protein or the related CO complex. The spectral differences observed for the transient deoxy Hb, therefore, are attributed to changes in the protein structure and/or conformation arising from ligation.

Reference

1. Sawicki, C. A. and Gibson, Q. H., *J. Biol. Chem. 251*, 1533-42 (1976)
2. Lyons, K. B., Friedman, J. M. and Fleury, P. A., *Nature 275*, 565-566 (1978)
3. Dallinger, R. F., Nestor, J. R. and Spiro, T. G., *J. Amer. Chem. Soc. 100*,6251-6252 (1978)
4. Terner, J., Spiro, T. G., Nagumo, M., Nicol, M. F. and El Sayed, M. A., *J. Amer. Chem. Soc. 102*,3238-3239 (1980)
5. Irwin, M. J. and Atkinson, G. H., *Nature 293*,317-318 (1981)
6. Shank, C. V., Ippen, E. P. and Bersohn, R., *Science 193*, 50-51 (1976)

7. Noe, L. J., Eisert, W. E. and Rentzepis, P. M., *Proc. Nat'l Acad. Sci. USA 75*, 573-577 (1978)

8. Greene, B. I., Hochstrasser, R. M., Weisman, R. B. and Eaton, W. A., *Proc. Acad, Sci. USA 75*,5255-5259 (1978)

9. Atkinson, G. H. "Time-Resolved Raman Spectroscopy" in Advances in Infrared and Raman Spectroscopy (Clark, J. J. H. and Hester, R. E., eds.) Heyden and Sons. Inc. *Vol 9* pp 1-62 (1982)

10. Irwin, M. J., Ph.D. Thesis, University of Sydney 1982

11. Atkinson, G. H., "Time-Resolved Raman Spectroscopy" in Advances in Laser Spectroscopy (Garetz, B. A. and Lombardi, J. R., eds.) Heyden and Sons, Inc. *Vol 1* p 155-175 (1982)

12. Atkinson, G. H., Gilmore, D. A., Dosser, L. R. and Pallix, J. B., *J. Phys. Chem. 86*, 2305 (1982)

13. Irwin, M. J. and Atkinson, G. H., *Proc. Nat'l. Acad. Sci. USA* (submitted for publication)

14. Perutz, M. F., *Nature 228*, 726-734 (1970)

15. Perutz, M. F., *Annu. Rev. Biochem. 48*,327-386 (1979)

STRUCTURE AND DYNAMICS IN THE PHOTOLYSIS OF CO- AND O_2-HEMOGLOBIN, MONITORED BY TIME-RESOLVED RESONANCE RAMAN SPECTROSCOPY[1]

Thomas G. Spiro
James Terner[2]

Department of Chemistry
Princeton University
Princeton, New Jersey

INTRODUCTION

Hemoglobin, one of the most thoroughly investigated molecules of biology, continues to be a focus of intensive study in many laboratories. Although a great deal is known about hemoglobin, the critical connections between structure and function remain to be elucidated. A question of overriding interest is the molecular mechanism of cooperativity. The four heme groups in the hemoglobin tetramer, although not in direct contact, bind oxygen in a cooperative manner, the affinity for oxygen increasing as the number of O_2 molecules bound increases. It is well established that cooperativity is associated with the existence of two different quaternary structures (two different arrangements of the four subunits), one of which, the T state, is stable in the absence of O_2 or other ligands, the other of which, the R state, is stable when ligands are bound (1,2). The aim of much current research on hemoglobin is to understand how the

[1] This work was supported by NIH Grant HL 12526, and a NIH postdoctoral fellowship to James Terner.

[2] Present address: Department of Chemistry, Virginia Commonwealth University, Richmond, Virginia

ISBN 0-12-066280-9

litgation of the heme group controls the structure of the protein, particularly the switch from the R to the T state. One approach is to monitor spectral signatures of the heme following ligand binding or dissociation, with the aim of detecting and characterizing intermediates that may be linked to changes in the protein structure. Dissociation is conveniently initiated by flash photolysis, especially of the CO complex, which has a high quantum yield for dissociation. It has been determined by picosecond techniques (3,4,5) that laser excitation of HbCO produces a very rapid (<10 ps) absorption change, producing a spectrum that resembles, but is not identical with that of deoxyHb.

Because Raman spectroscopy provides structural information via the vibrational frequencies of the molecular ground state, and because the resonance Raman effect provides selective enhancement of the heme vibrational modes of heme proteins (6-11), time-resolved resonance Raman (RR) spectroscopy has begun to provide a probe of the heme structure after ligand photodissociation from hemoglobin (12-21).

HbCO PHOTOINTERMEDIATE

In an effort to study the rapidly produced COHb photoproduct, which had been detected by picosecond absorption spectroscopy, we employed (15) a synchronously-pumped mode-locked dye laser, operating at 0.8MHz to photolyze a flowing stream of HbCO, and simultaneously excite its RR spectrum. The flow rate was sufficient to ensure that each laser pulse illuminated fresh sample. The 30 ps pulses were quite weak (5 nJ), and, even when tightly focussed produced only partial photolysis (20%). Good quality photoproduct spectra were nevertheless obtained by computer subtraction of the HbCO spectrum. The laser was tuned to 576 nm, near the first π-π* electronic transition (Q) of the porphyrin chromophore. Under these conditions, non-totally symmetric porphyrin vibrational modes (A2g, B1g and B2g under the idealized D4h symmetry of heme) are maximally enhanced, via Q-B vibronic mixing (22). In the high-frequency region, three strong bands are observed, which have been identified as porphyrin skeletal modes ν_{10} (B_{1g}), ν_{19} (A_{2g}), and ν_{11} (B_{1g}) in systematic studies of Ni octaethyl-(23) and proto-porphyrin (24), and of a variety of heme complexes (25). These modes are found at 1605 and 1556 and 1545 cm^{-1} for deoxyHb (15).

The appearance of the HbCO photoproduct RR spectrum is very similar to that of deoxyHb (12-15). There are, however, slight but reproducible downshifts of the band positions, to 1003, 1552 and 1545 cm^{-1}, respectively (15). These shifts persist when the laser pulse is lenghthened to 20 ns (15). Similar shifts had earlier been reported in COHb photoproduct spectra obtained with 10 ns pulses from a YAG laser (503.2 nm) (14). The agreement is encouraging since it indicates that multiphoton effects, which might have been induced by the much higher YAG pulse energies (20 mJ) are not a problem in these experiments. The YAG-excited RR spectrum of MbCO (Mb = myoglobin) photoproduct does <u>not</u> show any detectable shifts relative to deoxyMb (26). At longer times the Q-band RR spectra of deoxyHb and COHb photoproduct are identical, as shown by a flow experiment with a 514.5 nm cw Ar^+ laser (18). From the laser spot size and the flow rate of the HbCO stream, the interaction time of the experiment was estimated to be <300 ns. These results establish that the photolysis of HbCO produces a heme structure which is almost, but not quite the same as that of deoxyHb, whereas the MbCO photoproduct RR spectrum is the same as that of deoxyMb within 10 ns of photolysis. Thus HbCO but not MbCO shows a persistent photointermediate; its lifetime is currently bracketed between 20 and 300 ns. Since the R → T quaternary structure change is known (27) to occur on a longer time scale (tens of μs), the photointermediate is kinetically competent to be on the pathway for the R → T switch. It is plausible that the globin structure of Hb constrains the heme, in order to provide a mechanical coupling between ligation and quaternary structure, whereas Mb, which is monomeric, does not do so.

The RR modes in question are known to respond to the heme ligation and spin-state (28) via their sensitivity to the porphyrin core size (25,29-31). In the absence of specific π-back-donation or doming effects, all of the porphyrin skeletal frequencies above 1450 cm^{-1} show a negative linear dependence on the size of the porphyrin cavity, as enforced by the central metal ion and its axial ligands (25). The dependence is roughly in proportion to the involvement of methine bridge bond stretching in the mode (23b,25) presumably reflecting the weakening of these bonds as the porphyrin is forced to expand (8). High-spin Fe is larger than low-spin Fe because antibonding d_σ orbitals are occupied in the former. For five-coordinate high-spin hemes the core expansion is reduced (32), relative to six-coordinate adducts, because the Fe atom moves out of the heme plane, reducing the antibonding effect. For high-spin Fe^{II}, however, this difference is not fully reflected in the RR frequencies, probably because the pyrrole rings tilt toward the out-of-plane Fe atom in the five-coordinate structure (33), providing an additional frequency lowering (25,31).

The fact that the HbCO photointermediate shows skeletal frequencies lower than those of deoxyHb implies that the porphyrin core is more expanded in the former. Since deoxyHb is high-spin, the photointermediate must be high-spin also. Characteristic times for thermal spin conversion Fe^{II} complexes are 10-100 ns (34); consequently the results of the 30 ps experiment imply that photodissociation occurs via intersystem crossing from the intially excited π-π^* state, to a high-spin ligand field state of the Fe (15). This mechanism had been analyzed by Greene et. al. (5) and is strongly supported by the recent theoretical calculations Waleh and Loew (35) (Strictly speaking the RR frequencies are unconnected with spin-state <u>per se</u>, but only with the structural consequences of occupying antibonding orbitals. This could happen without a change in the electronic spin, and the calcualtions of Waleh and Lowe (35) indicate that the singlet-triplet energy differences of the photodissociating configurations are not large).

The expanded core of the HbCO photointermediate implied by the lowered skeletal frequencies could be produced by constraining the Fe atom in the heme plane following photoexcitation and spin (or configuration) conversion. Indeed two of the three frequencies were satisfactorily reproduced (the third could not be observed) with a six-coordinate high-spin Fe^{II} model compound (18) in which the Fe is held in the heme plane, and the core is expanded (36). However the RR frequencies are not sensitive monitors of the Fe out-of-plane distance because the frequency differences for five- and six-coordinate high-spin Fe^{II} are small, as discussed above. The RR frequencies indicate that the out-of-plane displacement is not as large in the photointermediate as it is in deoxyHb (0.5Å(1)), but they do not rule out an intermediate displacement. In any case, it is not plausible that the Fe atom could be rigidly fixed in the heme plane, since the photodissociation must generate large uncompensated non-bonded repulsions between the proximal imidazole and the pyrrole N atoms (1,8,37). Presumably the Fe atom follows the proximal imidazole as it moves away from heme plane until the non-bonded forces are reduced to the level of the constraining protein forces. The RR frequencies indicate that the position of this temporary equilbrium lies somewhere short of the fully out-of-plane structure of deoxyHb. X-ray crystallography reveals a number of protein structural differences between ligated Hb and deoxyHb (38). Not only does the Fe atom and the proximal imidazole move away from the heme plane, but the entire heme group is displaced by ~1Å relative to the helical polypeptide segment (F helix) anchoring the proximal imidazole, and the latter tilts 11° away from the heme normal. This large scale conformation change might well be associated with the decay of the photointermediate.

<u>HbO_2 Photolysis</u>

While O_2 and CO bind similarly to Hb, the electronic structure of the ligands, and therefore of the adducts are different. The photolysis quantum efficiency is much lower for HbO_2 than for HbCO, presumably reflecting the availability of nondissociative deexcitation pathways via low-lying electronic states (5,39). Nevertheless, Nagumo et al. (19) reported RR spectra of the HbO_2 photoproduct, using ~50 ps pulses from a synchronously pumped mode-locked laser, that appeared to be the same as those previously obtained for COHb (15), albeit with low signal/noise due to the low extent of photolysis. A different result was obtained (20) with a mode-locked YAG laser, producing ~30 ps pulses with 0.2 mj energy (532 nm). The photoprodut spectrum showed substantially shifted and broadened peaks. (Coppey et al. (21) also reported 30 ps YAG-excited HbO_2 spectra, but although some photolysis was noted, the resolution and signal/noise were insufficient to determine the photoproduct spectrum). Polarization measurements allowed assignment of ν_{10}, ν_{19}, and ν_{11} at 1590, 1550 and 1538 cm^{-1}, values which are 13, 2 and 5 cm^{-1} lower than in the HbCO photoproduct spectrum. The HbO_2 photoproduct spectrum was attributed (20) to an electronically excited state of deoxyHb, with π-π^* character, since ν10 and ν11 are known to decrease in frequency when the porphyrin π^* orbital is occupied, via back-bonding from low-spin Fe^{II} (25), or via reduction of ZnEP (EP = etioporphyrin) to the mono- and di-anions (40). This interpretation is in line with the suggestion by Cornelius et al. (41) that MbO_2 photolysis provides an efficient pathway for the production of electronically excited deoxyMb, since a common prompt spectral intermediate, presumably deoxyHb*, was observed upon YAG excitation of MbO_2, MbCO and deoxyMb, with the extent of formation decreasing in the order MbO_2> deoxyMb> MbCO. The excited state RR spectrum can be associated with the HbO_2 optical spectral intermediate observed by Chernoff et al. (42) which relaxes within 90 ps. This would explain the observation by Nagumo et al. (19) of a HbCO-like photoproduct RR spectrum with weak pulses of ~50 ps duration; observation of the electronically excited intermediate requires shorter and more intense pulses. It is possible that this intermediate results from double excitation of HbO_2, via its low-lying excited states (40). The broadened RR bands suggest that some relaxation has taken place even within the 30 ps YAG pulses.

SUMMARY

The picture that emerges from these studies is that photodissociation of HbCO leads to the immediate (<30 ps) production of a high-spin Fe^{II} heme, which is relaxed electronically, but structurally unrelaxed due to constraints of the surrounding protein. These constraints limit the motion of the Fe atom out of the heme plane and thereby produce a slightly expanded porphyrin core, relative to deoxyHb. The relaxation of these constraints occurs with a time constant between 20 and 300 ns, before the R → T quaternary switch. In the case of HbO_2, photolysis with sufficiently short and intense pulses produces electronically excited deoxyHb, possibly via the reexcitation of HbO_2 from its low-lying excited states. Relaxation occurs promptly to the same intermediate observed for HbCO.

REFERENCES

1a. Perutz, M. F. Br Med. Bull 32, 195-207 (1976)

b. Perutz, M. F. Proc. Roy. Soc. London, B208, 135 (1980).

2. Shulman, R. G., Hopfield, J. J., and Ogawa, S., Q. Rev. Biophys., 8, 325 (1975).

3. Shank, C. B., Ippen, E. P., and Bersohn, R., Science, 193, 50-51 (1976).

4. Noe, L. J., Eisert, W. G., and Rentzpis, P. N., Proc. Natl. Acad. Sci., 75, 573-577 (1978).

5. Greene, B. I., Hochstrasser, R. M., Weisman, R. W., and Eaton, W. A., Proc. Natl. Acad. Sci., 75, 5255-5259 (1978).

6. Spiro, T. G. in The Chemical Physics of Biologically Important Inorganic Chromophores, A. B. P. Lever and H. B. Gray, eds., Addison-Wesley, Reading, Mass. (1982) (in press).

7. Kitagawa, T., Ozaki, Y. and Kyogoku, Y., Adv. Biophys., 11, 153 (1978).

8. Warshel, A., Ann. Rev. Biophys. Bioeng., 6, 273 (1977).

9. Felton, R. H. and Yu, N-T. in The Porphyrins Vol. II, Part A., Dolphin D., ed., Academic Press, New York, pp. 347-388 (1978).

10. Asher, S. A. in Methods in Enzymology, Hemoglobin Part A. (Eraldo, L. B., Bernardi, L. B., Cheancone, E. eds.) Academic Press (1981).

11. Rousseau, D. L., Friedman, J. M. and Williams, P. F., Top. Current. Phys., 11, 203 (1979).

12. Dallinger, R. F., Nestor, J. R., and Spiro, T. G., J. Am. Chem. So., 100, 6251, (1978).

13. Woodruff, W. H. and S. Farquharson, Science, 201, 831 (1978).

14. Lyons, K. B., Friedman, J. M., and Fleury, P. A., Nature, 275, 565 (1978).

15a. Terner, J., Spiro, T. G., Nagumo, M., Nicol, M. F., and El-Sayed, M. A., J. Am. Chem. Soc., 102, 3238 (1980).

b. Terner, J., Stong, J. D., Spiro, T. G., Nagumo, M., Nicol, M. F., and El-Sayed, M. A., Proc. Natl. Acad. Sci., 78,1313-1317 (1981).

16. Lyons, K. B. and Friedman, J. M. in Symposium on Interactions Between Iron and Proteins in an Electron Transport, Ho, C. and Eaton, W. A. eds., El-Sevier (1981).

17. Stein, P., Terner, J., and Spiro, T. G., J. Phys. Chem., 86, 168 (1982).

18. Irwin, M. J., Atkinson, G. H., Nature, 293, 317 (1981).

19. Nagumo, M., Nicol, M. F., and El-Sayed, M. A., J. Phys. Chem., 85,2431-2438 (1981).

20. Terner, J., Voss, D. F., Paddock, C., Miles, R. B., and Spiro, T. G., J. Phys. Chem., 86, 859 (1982).

21. Coppey, M., Tourbez, Valat, P., Alpert, B., Nature, 284, 568 (1980).

22. Spiro, T. G., and Strekas, T. C., Proc. Natl. Acad. Sci., 69, 2622 (1972).

23a. Kitagawa, T., Abe, M., and Ogoshi, H., J. Chem. Phys., 69, 4516 (1978).

b. Abe, M., Kitagawa, T., and Kyogoku, Y., J. Chem. Phys., 69, 4526 (1978).

24. Choi, S., Spiro, T. G., Langry, K. C., and Smith, K. M., J. Am. Chem. Soc., 104, 4337 (1982).

25. Choi, S., Spiro, T. G., Langry, K. C., Smith, K. M., Budd, D. L., and La Mar, G. N. J. Am. Chem. Soc., 104, 4345-4351 (1982).

26. Dasgupta, S. and Spiro, T. G. (in preparation).

27a. Sawicki, C. A., Q. H. Gibson, J. Biol. Chem., 251, 1533 (1975).

b. Cho, K. C. and Hopfield, J. J., Biochemistry, 18, 5826 (1979).

28. Spiro, T. G. and Strekas, T. C., J. Am. Chem. Soc., 96, 338 (1974).

29. Spaulding, L. D., Chang, C. C., Yu, N-T., and Felton,

R. H., J. Am. Chem. Soc., 97, 2517 (1975).

30. Huong, P. V. and Pommier, J-C., CR Acad. Sci., Ser. C285, 519 (1977).

31. Spiro, T. G., Stong, J. D., and Stein, P., J. Am. Chem. Soc. ,101, 2648 (1979).

32. Scheidt, W. R. and Reed, C. A., Chem. Rev., 81, 543-555 (1981).

33. Hoard, J. L. and Scheidt, W. R., Proc. Natl. Acad. Sci., (USA) 70, 3919; 70, 1578, (1974).

34a. Beattie, J. K., Sutin, N., Turner, D. H., and Flynn, T. W., J. Am. Chem. Soc., 95, 2052-2054 (1973).

b. Beattie, J. K., Binstead, R. A., and West, R. J., J. Am. Chem. Soc., 100, 3044-3050 (1978).

35. Waleh, A. and Loew, G. H., J. Am. Chem. Soc., 104, 2346 (1982).

36. Reed, C. A., Mashiko, T., Scheidt, W. R., Spartalian, K., and Long, G., J. Am. Chem. Soc., 102, 2302 (1980).

37a. Friedman, J. M., Rousseau, D. L., Ondrias, M. R., Stepnoski, R. A., Science (in press).

b. Friedman, J. M., Rousseau, D. L., Ondrias, M. R., Ann. Rev. Phys. Chem. (1982) (in press).

38. Baldwin, J. and Chothia, C., J. Mol. Biol., 129, 175 (1979).

39. Waleh, A. and Loew, G. H., J. Am. Chem. Soc., 104, 2352 (1982).

40. Ksenofontova, N. M., Maslov, V. G., Sidorov, A. N., and Bobovich, Ya. S., Opt. Spectrosc., 40, 462 (1976).

41. Cornelius, P. A., Steele, W. A., Chernoff, D. A., and Hochstrasser, R. M., Proc. Natl. Acad. Sci. (1982) (in press).

42. Chernoff, D. A., Hochstrasser, R. M., Steele, A. W., Proc. Natl. Acad. Sci., 77, 5606 (1980).

Structure-Function Relationships in Hemoglobin: Transient Raman Studies

J. M. Friedman

Bell Laboratories
Murray Hill, New Jersey 07974

In order to determine the structural basis for the protein control of oxygen binding in hemoglobin (Hb) it is important to probe the changes in the heme environment that are induced by ligation. By comparing the Raman spectra of the deoxy heme from both the equilibrium forms and the transient species occurring within ns or longer subsequent to photolysis, we have systematically examined ligation induced changes in the heme pocket as a function of quaternary structure and solution conditions.

Our initial time resolved resonance Raman (TR^3) studies (1) of the frequency responses of the oxidation state marker band at ~1355 cm^{-1} (also known as either ν_4 or band I) revealed that there is a response in the heme pocket to both ligation and change in quaternary structure which changes the π distribution in the porphyrin. These changes are also seen in the isolated chains but not in myoglobin. A separation of ligand induced tertiary structure changes from those due to quaternary structure changes was detected in a pulse probe TR^3 study (2) of the time evolution of the frequency of ν_4 subsequent to the ns photolysis of HbCO. These results implicate the structural changes associated with the early (<1 μs) "pure" tertiary relaxation with a trigger mechanism for the subsequent R-T switch.

ISBN 0-12-066280-9

To better understand the nature of these structural changes we have examined the low frequency region of the Raman spectra. This spectral region contains modes sensitive to such specific degrees of freedom as the iron proximal histidine (Fe-His) stretching mode. From such studies, which reveal that for both deoxy (3) and transient species (4) the frequency of ν_4 has an inverse linear correlation with that of the Fe-His mode, it is likely that the orientation of the proximal histidine modulates the π density of the heme. In Fig. 1 is shown several representative low frequency Raman spectra of various R and T state deoxy and transient species. In addition to the R-T and ligation induced changes in the frequency of the Fe-His stretching mode, one can also clearly observe the presence or absence of the 345 cm^{-1} band as an indicator of the state of ligation prior to the 10 ns pulse. As can be seen from the figure and as previouosly noted in cryogenic studies (see Rousseau et al., this volume), this peak occurs at 345 cm^{-1} for both R and T deoxy species but occurs as a shoulder at ~355 cm^{-1} for the R and T state transients at 10 ns. Based on recent studies (5) of the time evolution of this band, it appears that the heme or more likely the heme pocket retains the "memory" of the state of ligation for ~1-10 μsec subsequent to photolysis in many Hb's.

In constrast to the 345 cm^{-1} band the Fe-His band displays a range of frequencies that indicates a sensitivity to both ligation state and quaternary structure.(4,6) These are summarized in Fig. 2. As has been shown previously (3,7) we see that the frequency of R state deoxy species are higher than those of T state deoxy species. It can also be seen that additional increases in frequency occur upon going to the corresponding transient species, i.e., Hb* (10 ns). The frequency of the Fe-His stretching mode appears to scale with both the quaternary structure stability and the affinity. As might be expected the frequencies of the transients correlate

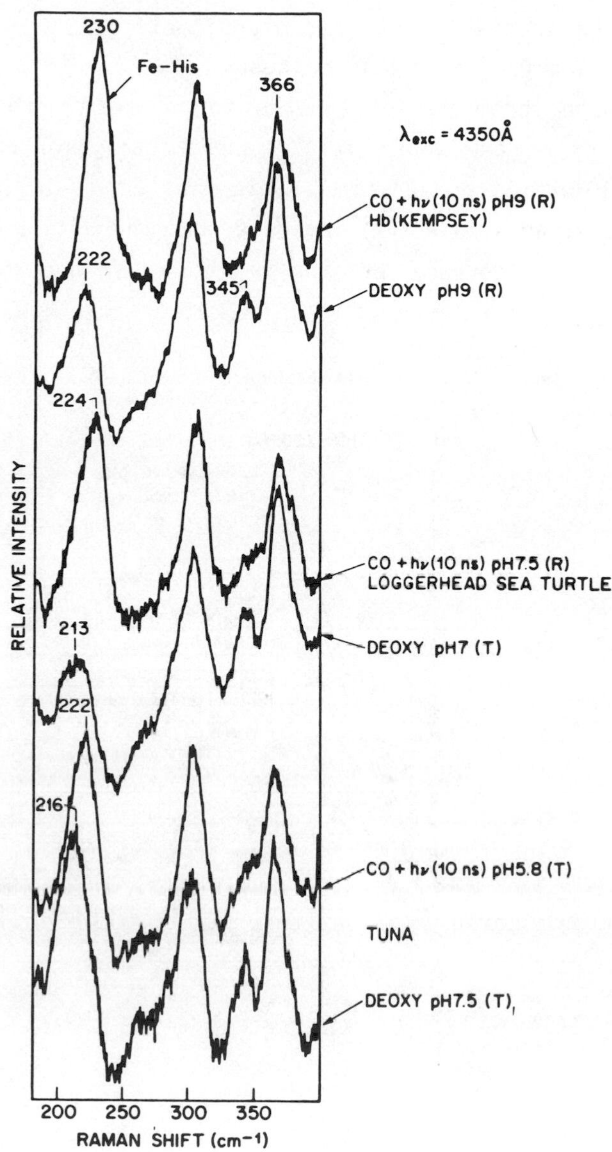

Fig. 1. A deoxy vs. photolyzed (CO, 10 ns) comparison of low frequency Raman spectra of Hb.

best with those ligand related processes that are anticipated to be responsive to the "instantaneous" configuration of the dissociated species. Both macroscopic off rates and geminate recombination fall into this category.

Geminate recombination appears to occur on both the subns and 10's of ns time scale for O_2 and probably for NO but only on the 10's of the ns time scale for CO (8-12). Initial transient Raman studies (1) indicate that there is a connection between the frequency of the ν_4 mode (hence the Fe-His

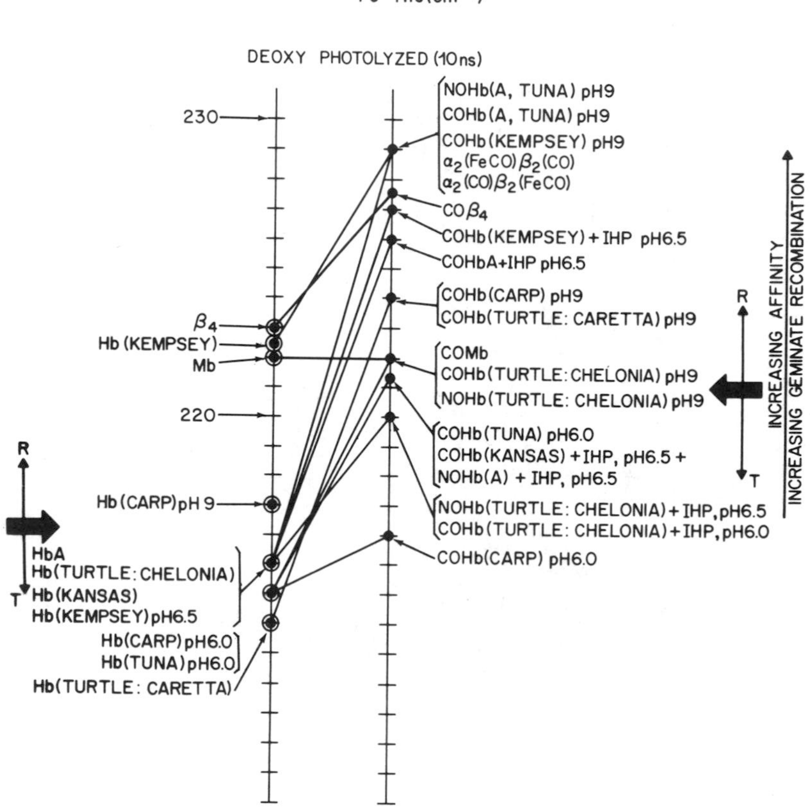

Fig. 2 The frequency of the Fe-His stretching mode in stable deoxy (left) and photolyzed (right) Hb's

stretching mode as well) and both the yield of geminate recombination occurring on the 10's of ns time scale and the very much related time averaged quantum yield of photolysis. More recently (10,13) we have found that for a given ligand the subns process as reflected in the yield of geminate recombination for O_2 and NO averaged over the initial 5 to 10 ns subsequent to photolysis also correlates with the frequency of the Fe-His stretching mode. In all comparisons, the species with the higher frequencies (more R like) are associated with greater yields of recombination. These findings implicate this same structural parameter responsible for the protein induced variation in the frequency of the Fe-His stretching in the mechanism for protein regulation of the potential energy barrier height associated with rebinding. It has been pointed out (10) that this barrier height is likely to be a determining factor in the R-T dependence of the macroscopic off rates. Thus it is of interest to establish the nature of this all important structural parameter.

The large change in the displacement of the Fe in response to the state of ligation is a possible factor in the variation in frequency of the Fe-His stretching mode. However, from a recent analysis (6) it can be concluded that whereas at 10 ns subsequent to photolysis the Fe may not as yet be totally displaced from the plane to the extent seen in deoxy Hb, it must however be sufficiently out of plane to have reduced the large repulsive forces acting against the planar configuration. Instead it was suggested (6) based partly on X-ray data (14) that this frequency variation arises primarily from protein induced tilting (in its own plane) of the proximal histidine with respect to the plane of the porphyrin. Within the small tilt approximation, the greater the tilt the lower is the frequency due to a weakening of the bond because of increased repulsive forces between the imidazole carbon and the nitrogen of the heme.

We now consider whether this proposed histidine tilt can contribute to cooperativity and to species specific variations in oxygen affinity. In Fig. 3 we isolate the contribution of this tilt to the energetics of ligand binding. Following the Goddard and Olafson (15) scheme for separating out the contributions to the energetics of binding, we have three contributions to G, the free energy of binding.

(i) starting with either R or T state deoxy heme which is represented by the displaced (with respect to the Fe coordinate) q state potential surface we move the Fe into the porphyrin plane (energy increases by ($\Delta E_q^{R,T}$);

(ii) we then excite the planar Fe from the q to the t electronic configuration (energy increases by $\Delta E_{qt}^{R,T}$);

(iii) The oxygen is then bound to Fe and forms the ground state of oxyHb (energy decreases by $D_t^{R,T}$).

The tilt of the histidine affects ΔE_q in that the greater the tilt the more energy is required to generate the planar q state Fe. Thus we have $\Delta E_q^T > \Delta E_q^R$. This inequality can become significant if, as the X-ray analysis suggests, (14) the orientation of the histidine is rigidly coupled to the $\alpha_1\beta_2$ interface by the F helix of the protein. This is shown schematically in Fig. 3. When the Fe is out of plane there is only a small energy difference at the heme between the tilted T and less tilted R deoxy Hb's. Upon moving the Fe into the plane the strong repulsive forces set in which act to straighten the tilted configuration thus pulling on the F helix. Because of the tilt, more energy is required to make a planar T state species. If within R and T, the structures were totally rigid the deoxy tilt would persist in the ligand bound configurations, with the free energy of cooperativity (ΔG) localized at the heme. However, as we have shown there are ligand related tertiary changes within R and T; futhermore,

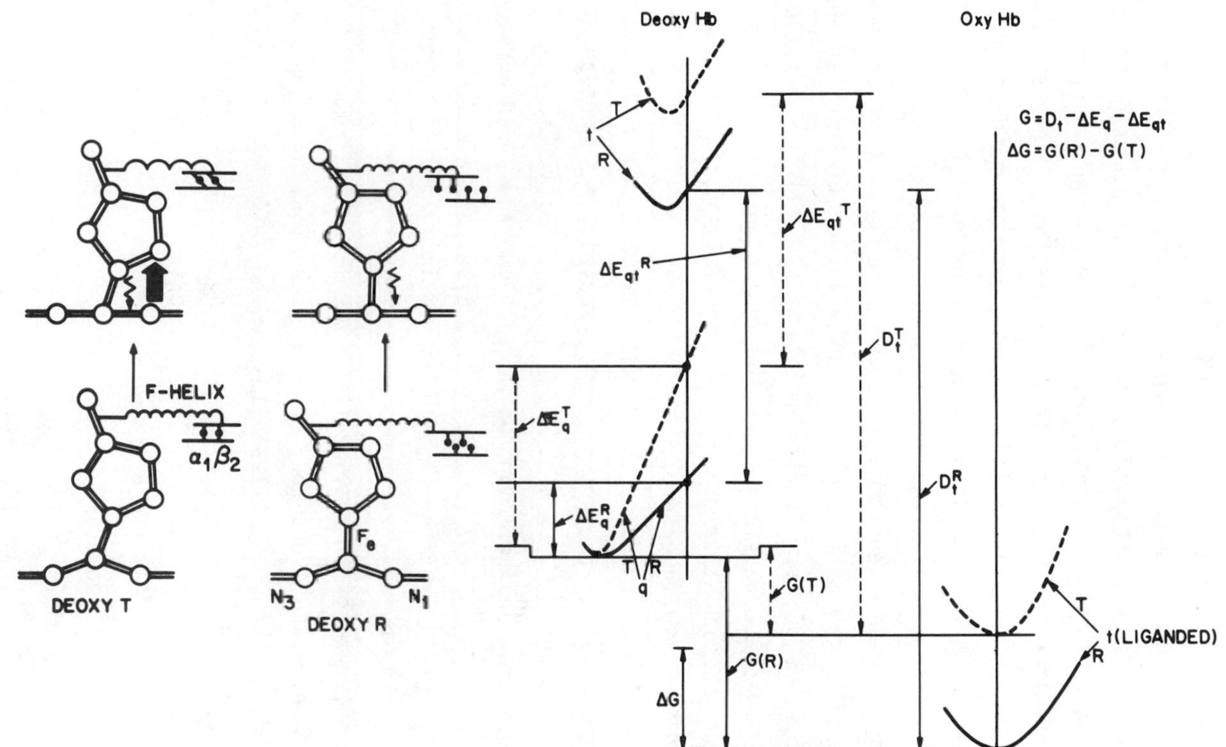

Fig. 3 The contribution of the histidine tilt to the energetics of ligand binding for T(dashed) and R(solid) state Hb.

studies (16) on ligand bound systems have failed to reveal substantial R-T differences on the distal side of the heme. These findings suggest that for both R and T the oxy heme may have an untilted histidine with the strain energy displaced away from the binding site - perhaps to the $\alpha_1\beta_2$ interface. Upon photolysis, the Fe rapidly moves out of plane relieving the strong upright inducing forces which allow the protein induced strain to shuttle back to the histidine. The histidine can now assume (<10 ns) a tilted configuration which reflects the strain associated with the ligand bound protein. It also follows that the more inflexible the F helix and the more fixed the R and T associated configurations of $\alpha_1\beta_2$ interface then the greater is the ability to generate a large tilt induced ΔG. This mechanism suggests that regulation of affinity could be modulated by quaternary structure, by evolution and by solution conditions through changes in the properties of both the F helix and the $\alpha_1\beta_2$ interface.

References

1. Friedman, J. M., Stepnoski, R. A., Stavola, M., Ondrias, M. R. and Cone, R. L. Biochemistry *21,* 2022 (1982).
2. Lyons, K. B. and Friedman, J. M. In "Hemoglobin and Oxygen Binding" (C. Ho, ed.), pp. 333-338, Elsevier, New York (1982).
3. Ondrias, M. R., Rousseau, D. L., Shelnutt, J. A. and Simon, S. R. Biochemistry *21,* 3428 (1982).
4. Friedman, J. M., Rousseau, D. L. and Ondrias, M. R. Ann. Rev. Phys. Chem. *33,* 471-91 (1982).
5. Scott, T. and Friedman, J. M. to be published.
6. Friedman, J. M., Rousseau, D. L., Ondrias, M. R. and Stepnoski, R. A. Science, in press.
7. Nagai, K., Kitagawa, T. and Morimoto, H. J. Mol. Biol. *136,* 271 (1980).
8. Chernoff, D. A., Hochstrasser, R. M. and Steele, A. W. Proc. Nat. Acad. Sci. USA *77,* 5606 (1980).

9. Martin, J. L., Migus, A., Poyart, C., Lecarpentier, Y., Antonetti, A., and Orszag, A. Biochem. Biophys. Res. Comm. *107*, 803 (1982).
10. Friedman, J. M., Stepnoski, R. A. and Noble, R. FEBS Letts. in press.
11. Friedman, J. M. and Lyons, K. B. Nature *284*, 570 (1980).
12. Duddell, N. A., Morris, J. R. Muttucararu, N. J. and Richards, D. A. Photochem. Photobiol. *31*, 479 (1980).
13. Friedman, J. M. to be published.
14. Baldwin, J. M. and Chothea, C. J. Mol. Biol. *129*, 175 (1979).
15. Goddard, W. A. III and Olafson, B. D. in "Biochemical and Clinical Aspects of Oxygen" (W. S. Caughey, ed.) Academic Press, pp. 87-123 (1979).
16. See for example Rousseau, D. L. and Ondrias, M. R. in Ann. Rev. Biophys. Bioeng., in press.

RESONANCE RAMAN SPECTRA OF PHOTODISSOCIATED CARBONMONOXY HEMOGLOBIN (Hb*): ROOM TEMPERATURE TRANSIENTS AND CRYOGENICALLY TRAPPED INTERMEDIATES

D. L. Rousseau
M. R. Ondrias[1]
J. M. Friedman

Bell Laboratories
Murray Hill, New Jersey

I. INTRODUCTION

The molecular dynamics of the binding and release of oxygen by hemoglobin (Hb) involves a series of complex events affecting both the heme binding site and the surrounding protein. Under physiological conditions electronic and conformational changes in the heme evolve on the subpicosecond to the several microsecond time scale and are thereby very difficult to probe. However by lowering the temperature to ~10°K before photodissociating oxygen or carbon monoxide it is possible to "freeze in" metastable conformations (Hb*) of hemoglobin for extended periods of time. In the past these cryogenically stabilized forms of hemoglobin have been studied by infrared (1) and optical absorption spectroscopies (2) as well as Mössbauer (3,4) and EPR (5,6) spectroscopies. Although it was recognized that the metastable species were related to transient forms of hemoglobin, structural interpretations were not possible. We are now able to examine metastable forms of hemoglobin at cryogenic temperatures (7) and compare

[1] *Present address: Universtiy of New Mexico, Albuqerque, N.M.*

ISBN 0-12-066280-9

the results to transient resonance Raman a studies of hemoglobin obtained at physiological temperatures (8). We focus here on the low frequency spectrum which contains the line assigned as the iron-histidine (Fe-His) stretching mode in deoxyhemoglobin as well as several modes of the porphyrin macrocycle.

II. RAMAN SPECTRA AT CRYOGENIC TEMPERATURES

Deoxyhemoglobin

The spectrum of deoxyhemoglobin at room temperature obtained with 441.6 nm excitation is shown in Fig. 1E. The line at 217 cm^{-1} has been assigned as the Fe-His stretching mode (9) and has been observed to shift to ~222 cm^{-1} on conversion to the R structure (10-14). Moreover, there is a large subunit heterogeneity in the frequency of this mode (11,14). Thus it appears to be very sensitive to protein conformation. The remaining lines in the low frequency region have been assigned as porphyrin ring modes (15) and from isotopic substitution studies also have large contributions from the peripheral substituents (16,17).

Upon freezing deoxy Hb (spectrum D in Fig. 1) the Fe-His stretching mode becomes more symmetrical and shifts to ~219 cm^{-1} (18). Small shifts in some of the other lines also take place. Further reduction in temperature to 15°K results in a shift of the Fe-His stretching mode to 235 cm^{-1} (Fig. 1A) and a substantial reduction in relative intensity. The intensity of the Fe-His stretching mode varies sharply with the exciation wavelength. Indeed, with 457.9 nm excitation the line is not detected in the spectrum (7). (The sharp line at 230 cm^{-1} in the frozen spectrum reported by Ondrias, et al. (12) does not come from the Hb but instead is a lattice mode of the ice.) The intensity dependence of the Fe-His stretching mode has been interpreted as evidence that this mode is not enhanced by the porphyrin π-π* transitions

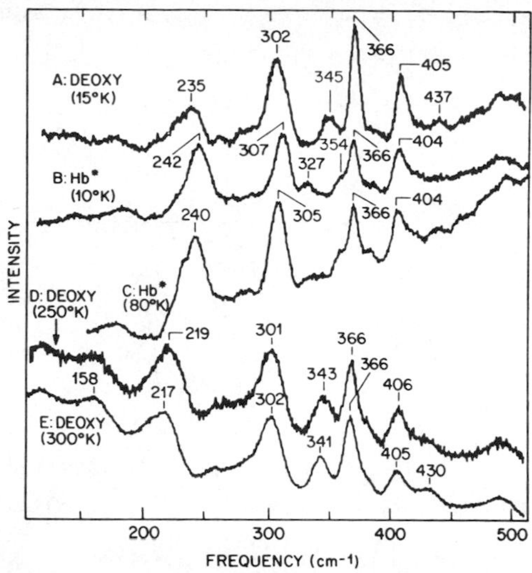

Figure 1. Resonance Raman spectra of hemoglobin (Hb) obtained with 441.6 nm excitation. A: Deoxy Hb at 15°K. B: Photodissociated carbonmonoxy Hb (Hb) at 10°K. C: Photodissociated carbonmonoxy Hb at 80°K. The shoulder at 230 cm^{-1} on the 240 cm^{-1} line is the H_2O translational mode in ice. D: Deoxy Hb at 250°K. E: Solution deoxy Hb at 300°K.*

but instead by d-d or charge transfer transitions. The change in intensity is a consequence of changing the orientation of the histidine with respect to the porphyrin. This change may result from protein contraction at low temperature or may result from a reduction in the non-bonded interactions as the excited state vibrational populations are reduced with the lowered temperature. Only small shifts are detected in the porphyrin modes as the temperature is lowered from 250°K to 15°.

Photodissociated Hemoglobin (Hb*)

Spectrum B of Fig. 1 results from HbCO frozen at 10°K then

photolyzed with the 441.6 nm laser. It is evident that the spectrum is quite similar to that obtained from deoxy Hb although there are several quantitative differences. First the Fe-His

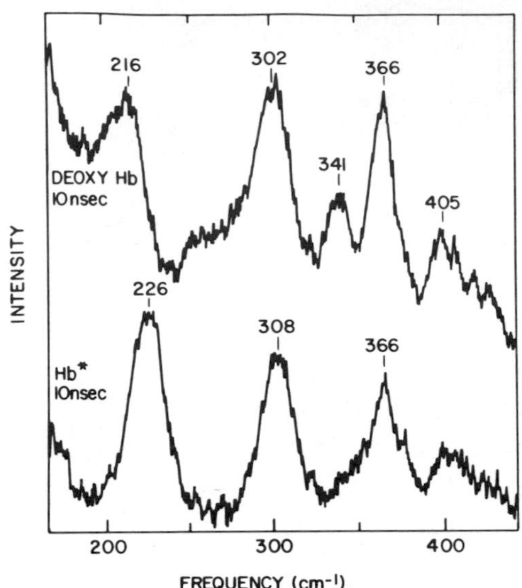

Figure 2. Resonance Raman spectra of deoxy Hb and photo-dissociated Hb (Hb) obtained at 35°C with 10 nsec pulses at 435.0 nm. Top: deoxy Hb. Bottom: Hb* generated by photolysis of HbCO.*

stretching mode has shifted to 242 cm^{-1} and is significantly more intense than it is in the deoxy Hb low temperature spectrum. Several of the porphyrin modes also change frequency. Most notable is the line at 345 cm^{-1} which shifts to 354 cm^{-1} in the photolyzed spectrum. On warming the sample to 80°K the Fe-His stretching mode shifts slightly to lower frequency as does the mode at 307^{-1}. In addition the mode at 327 cm^{-1} which is pronounced at 10°K is absent in the 80°K spectrum. The shoulder at ~230 cm^{-1} in the 80°K spectrum results from ice. Until more refined normal mode analyses are completed it cannot be determined what the structural basis is for the changes that are detected in the porphyrin modes. Possible origins of the changes in the Fe-His stretching mode will be discussed below.

RAMAN SPECTRA WITH 10 NANOSECOND PULSES

In Fig. 2 time resolved spectra of Hb in the low frequency region are shown. Within the signal-to-noise ratio the spectrum of deoxy Hb obtained with 10 nsec pulses agrees with the room temperature continuous wave spectrum in Fig. 1. The lower spectrum in Fig. 2 is of photodissociated HbCO. Just as for the low temperature samples, the line at 341 cm^{-1} becomes a shoulder on the line at 366 cm^{-1}. In addition the line in the 300 cm^{-1} region shifts to higher frequency.

The Fe-His stretching mode in the 10 nsec spectrum of photodissociated HbCO is at 226 cm^{-1}, 10 cm^{-1} higher than it is in the spectrum of the deoxy preparation, and it has increased intensity. Thus, the change in frequency detected between the deoxy and the photodissociated preparations is the same in the room temperature transient experiments as in the experiments on the low temperature in intermediates. The absolute frequency is higher in the low temperature preparations due, presumably, to the above mentioned temperature dependent effects. However, from the Raman spectral changes which we report here, it appears that the conformational properties of the heme detected 10 nsec subsequent to photolysis are the same as those of the low temperature intermediate stabilized at 80°K. At 10°K an addition line at 327 cm^{-1} is present in the spectrum. It is already possible that the 10°K spectrum may represent a form of Hb* that, under physiological conditions, can only be probed at times shorted than 10 nsec.

CONCLUSIONS

There are significant differences between the spectra of photodissociated HbCO and deoxy Hb in both the low temperature preparations and the room temperature samples probed with 10 nsec pulses. The differences in the porphyrin modes probably result from changes in orientation of peripheral substituents, although changes in heme doming or ruffling can not be ruled out at present until a better analyses of the normal modes becomes available. The changes in frequency and intensity of the Fe-His stretching modes have been interperted as resulting from a change in conformation between the histidine and the porphyrin (7,8,19). Thus, preparations in which the Fe-His stretching mode is at high frequency, correspond to Hb conformations in which the histidine is more upright and the non-bonded repulsions are thereby reduced. This interpertation, which can account for the frequency changes in the Raman data, is also consistent with x-ray crystallographic results (19).

There have been many studies, in the past, on metastable forms of hemoglobin stabilized at low temperatures (1-6). However, the physiological applicability of these studies has been uncertain owing to the possible constraints imposed on the protein and heme by the frozen solvent. The analogous behavior of the Hb* species generated at room temperature and 80°K confirms the physiological relevance of the low temperature spectra. Thus, these preliminary results indicate that the parallel examination of time-resolved and temperature resolved spectra could lead to a synergy that allows for a determination of the dynamics of ligand binding and its dependence on heme structural parameters.

REFERENCES

1. Alben, J. O., Beece, D., Bowne, S. F., Doster, W., Eisenstein,

L., Frauenfelder, H., Good, D., McDonald, J. C., Marden, M. C., Moh, P. P., Reinisch, L., Reynolds, A. H., Shyamsunder, E. and Yue, K. T., Proc. Nat. Acad. Sci. (USA) *79*, 3744 (1982).

2. Iizuka, T., Yamamoto, H. and Yonetani, T., Biochem. Biophys. Acta. *371*, 126 (1974)
3. Spartalian, K., Lang, G., and Yonetani, T., Biochem. Biophys. Acta. *428*, 281 (1976).
4. Marcolin, H.-E., Reschke, R. and Trautwein, A., Eur. J. Biochem. *96*, 119 (1979).
5. Iizuka, T., Yamamoto, H., Kotani, M. and Yonetani, T., Biochem. Biophys. Acta. *351*, 182 (1974).
6. Yonetani, T., Yamamoto, H., Iizuka, T., J. Biol. Chem. *249*, 2168 (1974).
7. Ondrias, M. R., Rousseau, D. L. and Simon, S. R., submitted for publication.
8. Ondrias, M. R., Friedman, J. M. and Rousseau, D. L., submitted for publication.
9. Hori, H. and Kitagawa, T., J. Am. Chem. Soc. *103*, 3608 (1980).
10. Nagai, K., Kitagawa, T., and Morimoto, H. J., J. Mol. Biol. *2136*, 273 (1980).
11. Nagai, K. and Kitagawa, T., Proc. Nat. Acad. Sci. (USA) *77*, 2033 (1980).
12. Ondrias, M. R., Rousseau, D. L., and Simon, S. R., Proc. Natl. Acad. Sci. (USA) *79*, 1511 (1982).
13. Ondrias, M. R., Rousseau, D. L., and Simon, S. R., Biochemistry *21*, 3428 (1982).
14. Ondrias, M. R., Rousseau, D. L., Kitagawa, T., Ikeda-Saito, M., Inubushi, T., and Yonetani, T., (1982) J. Biol. Chem. *257*, 8766 (1982).
15. Abe, M., Kitagawa, T. and Kyogoku, Y., J. Chem. Phys. *69*, 4526 (1978).
16. Choi, S. and Spiro, T. G., submitted for publication.
17. Rousseau, D. L., Ondrias, M. L., LaMar, G. N. and Smith, K. M., submitted for publication.

18. Ondrias, M. R., Rousseau, D. L., and Simon, S. R. Science, *213*, 657 (1981).
19. Friedman, J. M., Rousseau, D. L., Ondrias, M. R. and Stepnoski, R. A. (1982) Science, in press.

TIME RESOLUTION BY TRANSIENT CHROMOPHORE CREATION: DEFINING THE SCISSILE BONDS IN AN ENZYME-SUBSTRATE COMPLEX TO 0.03 Å[1]

P.R. Carey, C.P. Huber, H. Lee, Y. Ozaki and A.C. Storer

Division of Biological Sciences
National Research Council of Canada
Ottawa, Canada

STRATEGY

In chemical or biochemical reactions there is great interest in detecting and characterising reaction intermediates. However, major difficulties in the spectroscopic characterisation of intermediates are that there are usually many species present, the species of interest are short-lived and in biochemical work often involve complex macromolecules. An elegant means of overcoming these problems is to create a resonance Raman (RR) probe solely at a key position in the intermediate of interest. Then the RR spectrum automatically provides the required temporal and spatial definition. This approach can be used to study an area of major biochemical importance - the mechanisms of enzyme reactions. The strategy of <u>transient</u> chromophore generation is seen in the reaction sequence involving the breakdown of thiono-ester substrates by the enzyme papain:

[1]*NRCC publication no. 20665.*

ISBN 0-12-066280-9

$$\mathrm{RC(=O)NHCH_2C(=S)OCH_3} + \text{HS-papain} \rightarrow \mathrm{RC(=O)NHCH_2C(=S)\text{-}S\text{-}papain} + \mathrm{CH_3OH}$$

dithioacyl papain λ_{max} 315 nm

$$\xrightarrow{H_2O} \mathrm{RC(=O)NHCH_2C(=S)OH} + \text{HS-papain}$$

In this reaction mixture a transient dithioester chromophore is formed from the C=S group of the substrate and the HS- group from the cysteine residue in papain's active site. The dithioester chromophore has a λ_{max} near 315 nm and is the only species in solution with an absorbance to the red of 300 nm. Thus it is possible to excite specifically the RR spectrum of the dithioester group. The spectrum contains features which have large contributions from the stretching motions of the C=S bond, which is undergoing nucleophilic attack in the active site, and features which have contributions from C-S stretching motions, i.e. from the bond which is being cleaved during enzymolysis (1). Hence, we have an accessible 'handle' on the catalytic process at the time and place of greatest importance.

COMPARISON OF SCANNING MODE AND MULTIPLEX DETECTION

RR studies of enzyme-substrate intermediates provide a good test case for comparing the advantages of multiplex detection over recording spectra in the conventional scanning mode. We have completed the construction of a 'rapid Raman' diode-array based system with high sensitivity in the UV and can make a preliminary comparison of this with our existing scanning instrument for the characterisation of enzyme-substrate transients. The comparison is made using the following set of conditions:

Sample. 1 ml solution containing $\sim 10^{-4}$ M enzyme (papain), 10^{-2} M substrate (the thionoester $\mathrm{RC(=O)NHCH_2C(=S)OCH_3}$) and 20% CH_3CN (to keep the substrate in solution). The enzyme-substrate intermediate has a half life of about 1 second and by using a hundred fold excess of substrate a quasi steady-state population of intermediates is formed for a few minutes. Rotating or stirred cell.

Excitation Source. 324 nm Kr^+ line ∿50 mW, CW.

Scanning Spectrometer. Spex 1/2 m double, 1800 g/mm holographic gratings blazed at 300 nm. Spectral slit width ∿8 cm^{-1}. EM1 9789QB photomultiplier, DC detection.

Multiplex Spectrometer. Spex Triplemate, gratings blazed at 300 nm; filter stage 600 g/mm, spectrograph stage 3,600 g/mm holographic, spectral slit width ∿8 cm^{-1}. Detection, Tracor Northern TN-6132 diode array (1024 detectors) and TN-1710 multichannel analyser.

POINT-BY-POINT COMPARISON

	Scanning	*Multiplex*
Spectral range of interest 400-1300 cm^{-1}	Requires 5 fresh samples to scan range in 5 x 200 cm^{-1} increments.	Complete spectrum from 1 sample. 900 cm^{-1} range.
Acquisition time	20 mins scanning plus one hour for washing and filling cell.	Equivalent S/N in 20 secs (20 accumulations of 1 sec exposures).
Spectral overview	Hard to 'sew' spectral fragments together due to sample aging.	Uniform spectral trace.

	Scanning	*Multiplex*
Reproducibility	Difficult to achieve due to sample aging and the inherent errors of a scanning instrument.	Highly reproducible.
Background fluctuations (often encountered in reaction mixtures)	Cause spurious 'peaks'.	Adds small increment to total background, inconsequential.
Fluorescence	Long exposure to laser beam causes fluorescence background due to build-up of photoproducts.	Short exposure, little or no fluorescence observed.
Rapid-mixing, rapid-flow to study intermediates on the 10^{-2} sec time scale	Requires unobtainably large amounts of enzyme.	Requires modest amount of material.

SUMMARY OF THE PRESENT KNOWLEDGE ON DITHIOACYL PAPAINS

The RR spectra of the intermediates have provided precise structural information on the critical region in the enzyme's active site. However, to reach the present level of understanding it has been necessary to undertake wide-ranging spectroscopic investigation into dithioesters. These have included:

1. Developing a force field for simple dithioesters of the type $CH_3C(=S)SCH_3$ and $CH_3C(=S)SC_2H_5$ (2). The normal coordinate analysis is being extended (in collaboration with Dr. J. Teixeira-Dias of the University of Coimbra, Portugal) to dithioesters which serve as models for the enzyme-bound substrate, e.g. $C_6H_5C(=O)NHCH_2C(=S)SC_2H_5$ and isotopically substituted analogs.
2. Physical chemical, Raman, RR and FTIR studies on N-acyl glycine and other dithioesters to elicit the conformational preferences of these molecules (3-5). Again, these serve as models for the enzyme-substrate complexes.
3. Combined X-ray crystallographic Raman analysis of single crystals of N-acyl glycine dithioesters to provide very precise structure-spectra correlations (6).
4. Detailed investigations of the enzyme kinetics to relate structural to mechanistic information.
5. Further investigation of a novel nitrogen-sulfur interaction found in the above work concerning N-acyl glycine dithioesters (7).

As a result of these studies (1-8) we have been able to define changes in length of the bond being broken in the active site to within 0.03 Å. At this level of precision we can begin to understand how very subtle changes in biological structures determine chemical reactivity.

REFERENCES

1. Storer, A. C., Murphy, W. F., and Carey, P. R., *J. Biol. Chem.* *254*, 3163 (1979).
2. Teixeira-Dias, J. J. C., Jardim-Barreto, V. M., Ozaki, Y., Storer, A. C., and Carey, P. R., *Can. J. Chem.* *60*, 174 (1982).
3. Verma, A. L., Ozaki, Y., Storer, A. C., and Carey, P. R., *J. Raman Spectrosc.* *11*, 390 (1981).

4. Ozaki, Y., Storer, A. C., and Carey, P. R., *Can. J. Chem. 60*, 190 (1982).
5. Storer, A. C., Ozaki, Y., and Carey, P. R., *Can. J. Chem. 60*, 199 (1982).
6. Huber, C. P., Ozaki, Y., Pliura, D. H., Storer, A. C., and Carey, P. R., *Biochemistry 21*, 3109 (1982).
7. Ozaki, Y., Storer, A. C., and Carey, P. R., in preparation.
8. Ozaki, Y., Pliura, D. H., Carey, P. R., and Storer, A. C., *Biochemistry 21*, 3102 (1982).

TIME RESOLVED FLUORESCENCE SPECTRA OF CHLOROPHYLL A DIMERS AND AGGREGATES USING SELECTIVE FLUORESCENCE QUENCHING[1]

A C de Wilton, J A Koningstein and L V Haley

Department of Chemistry
Carleton University
Ottawa, Canada

In dry benzene solutions Chlorophyll a (Chl a) exists predominantly as the dimeric species (1), but interpretation of absorption or emission spectra of such solutions is complicated by the presence of monomeric and more highly aggregated species of Chl a in equilibrium with the dimer. Indeed, the fluorescence spectrum of a dry solution of 10^{-5}M Chl a in benzene excited with 3ns laser pulses at 444nm shows the well known strong fluorescence band at ∿670nm accompanied by a vibronic band at ∿740nm characteristic of monomeric Chl a in solution. However, other fluorescent species also contribute to the spectrum in this region. There is a weak band at ∿625nm, together with stronger emission to the red of the main band (670-800nm) (Fig 1C). These bands increase in intensity if the concentration of Chl a is increased, but diminish in intensity, relative to the emission at 670nm, if water is added to the dry solution. As the intensities at ∿625nm and ∿740nm increase the main band broadens as a result of the presence of another emission band at ∿690nm. Neither the emission bands at ∿625nm and ∿690nm, nor the intensity to the red of the main band which is additional to the monomeric emission were observed for Chl a in more dilute wet benzene solutions or in pyridine. These bands are therefore assigned to dimeric or aggregated species of Chl a.

1 Research supported by the Natural Sciences and Engineering Council of Canada.

ISBN 0-12-066280-9

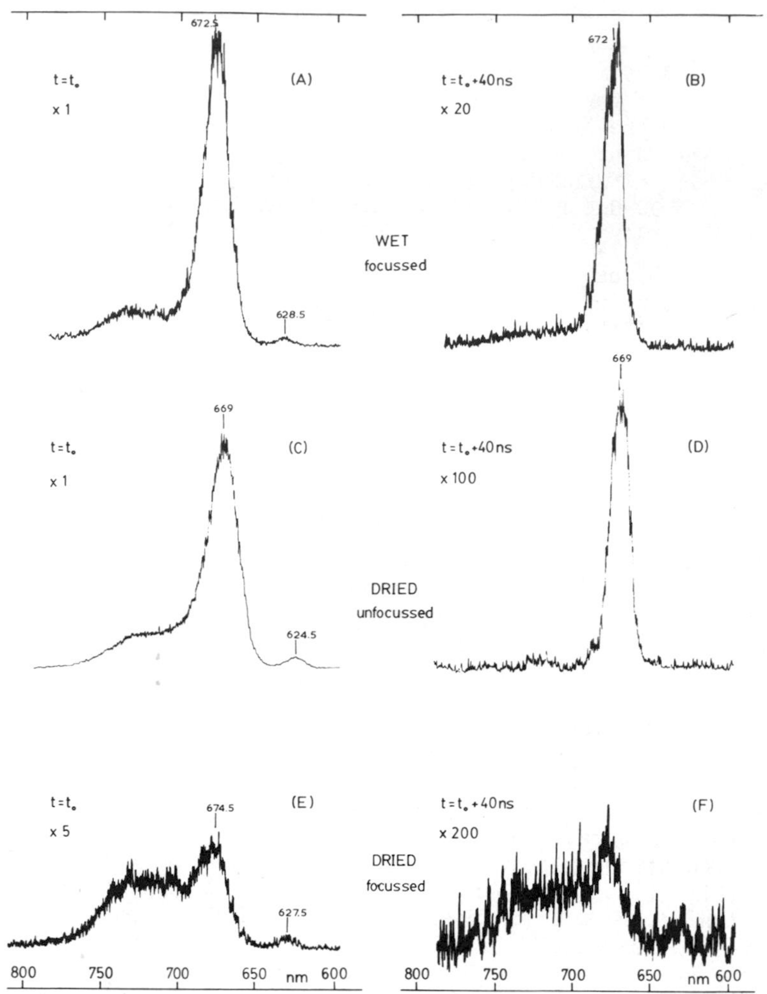

FIGURE 1. Time resolved fluorescence spectra of 3×10^{-5}M Chl a in benzene solutions excited by 3ns laser pulses at 444nm. Fluorescence spectra are shown for $t = t_o$ (the time at which the fluorescence pulse reaches maximum intensity) and for $t = t_o +$ 40ns: (A), (B) wet solution, focussed conditions; (C), (D) and (E), (F) dry solution, under unfocussed and focussed conditions respectively.

Comparison of the intensities of the time resolved fluorescence spectra showed that the fluorescence decay times of the emission at ~625nm and at ~740nm are of the same order of magnitude as that of the monomeric emission at 670nm, but could not be resolved with the 2ns gate of our detection system. The lifetime of the emission at ~690nm is significantly shorter.

Fluorescence from a longer lived species is apparent in the spectrum recorded at t_o + 40ns, in both wet and dry benzene solutions (Fig. 1B and 1D). This fluorescence is characterised by a band at 670nm, the same wavelength as the peak of the monomer emission, but without vibronic fine structure and having a narrower bandwidth. A similar long lived fluorescence band was also observed in wet pyridine solutions (τ = 145ns) (2). In benzene solutions this band increases in intensity as the water content of the solution is increased. Since the concentration of large hydrated aggregates of Chl a at a concentration of 10^{-5}M and at room temperature should be small (3) we believe that this band belongs to a dimeric species of hydrated Chl a.

In order to separate the spectra of the short lived species use was made of the following observation. When the laser radiation was focussed into the sample cell the fluorescence appeared to be quenched. Figure 1(E) shows that under irradiation at 444nm and at light levels of higher ~10^{14} photons per pulse emission from all bands decreases in intensity. However, emission at ~670nm from the monomeric species is quenched more effectively. Under focussed conditions all bands shift to the red, which may be attributed to a change in environment due to creation of excited state species in solution. Transmission studies on Chl a in pyridine showed that under these conditions the photon density was sufficient to cause population of the ~100ps lived excited state, S_4, of monomeric Chl a at 22,650cm^{-1} (4). In benzene, from the disappearance of the monomeric fluorescence together with an increase in the transmission of the solution, we conclude that population of the S_4 excited state of the monomeric species takes place, followed by absorption from the populated state. We do not at present have enough information about higher lying excited states to ascertain whether de-excitation occurs from a higher lying excited singlet state or whether photoionization or other processes are involved in the observed quenching of monomeric fluorescence.

These results demonstrate clearly that selective fluorescence quenching may be used to separate fluorescence spectra of different molecular species in equilibrium in solution having radiative lifetimes which are not sufficiently different to be resolved in time by the detection system. At a particular wavelength the absorption strengths of the different species may be different enough that by judicious choice of laser wavelength and power it is possible to selectively quench fluorescence from one or more species and thus further separate the spectra shown in Figure 1(E). In fact, further experiments using with laser pulses at 607nm resulted in quenching of the emission bands at ∿625nm and ∿690nm with respect to the monomer emission. However, excitation over a range of wavelengths either in the red or the blue, at different power levels, showed that the quenching of the emission at ∿740nm was independent of quenching of these bands.

Since calculations show that the larger the oligomer the more red shifted the excited states, the broad spectral feature centered around ∿740nm is believed to contain contributions from higher aggregates. From consideration of theoretical calculations for the lifetimes of the Chl a dimer (5) it is thought that the bands at ∿625nm and ∿690nm originate from the Chl a dimer in dry benzene. To first order, such an assignment leads to an exciton splitting of 1500cm^{-1} for the Chl a dimer in dry benzene in which the interaction occurs via C=O...Mg linkages.

REFERENCES

1. K. Ballschmitter, K. Truesdell and J.J. Katz, Biochim. Biophys. Acta 184, 604 (1969).
2. A. de Wilton, L.V. Haley and J.A. Koningstein, Can. J. Chem. 60, 2198 (1982).
3. K. Sauer, J.R. Lindsay-Smith, A.J. Schultz, J. Am. Chem. Soc. 88, 2681 (1966).
4. M. Asano and J.A. Koningstein, Chem. Phys. 57, 1 (1981).
5. L.V. Haley and J.A. Koningstein, Can. J. Chem., to be published.

EXCITED STATE POPULATION EFFECTS ON COHERENT RAMAN SPECTROSCOPY[1]

R. Bozio
P. L. Decola
R. M. Hochstrasser

Department of Chemistry
and
Laboratory for Research on the Structure of Matter
University of Pennsylvania
Philadelphia, Pa. 19104

1. INTRODUCTION

The efficiency of four wave mixing (FWM) processes whereby three coherent electromagnetic waves at frequencies ω_1, ω_2, and ω_3 interact through the third order susceptibility $(\chi)^{(3)}$ of a material system to produce a fourth coherent wave at $\omega_4 = \omega_1 + \omega_2 + \omega_3$ is resonantly enhanced whenever some transition energy of the material is matched by the incoming frequencies or by their binary sums or by the outcoming frequency ω_4. Based on this principle, a number of coherent non-linear spectroscopies has been developed in the last few decades[1].

Coherent anti-Stokes Raman spectroscopy (CARS) and its Stokes analogue (CSRS) correspond to the case in which only two incoming frequencies are used and their difference $|\omega_1-\omega_2|$ is tuned around the vibrational transition frequencies of the system. Spectra are then obtained by measuring the intensity of a generated wave at $\omega_4=2\omega_1-\omega_2$. In CARS $\omega_1>\omega_2$ and the generated frequency is higher than both incoming frequencies - the opposite holds for CSRS. The advantages and drawbacks of CARS and CSRS compared with ordinary spontaneous Raman scattering have been extensively discussed in the literature[2]. Multiresonant CARS and CSRS in which a single-photon electronic resonance occurs simultaneously with the Raman resonance has received less attention so far although the possibility of obtaining a strong signal enhancement was realized for some time.

[1]This research was supported by a grant from the National Institutes of Health (GM12592) and in part by the NSF/MRL DMR7923647.

ISBN 0-12-066280-9

Strictly full (triply) resonant conditions for CARS and CSRS can only be achieved by the use of three different incoming frequencies. However, if the difference between ground and excited state vibrational frequencies is notmuch greater than the width of the electronic and vibrational transitions, the CARS or CSRS signal is further enhanced by the resonance of the outcoming frequency with a vibronic transition. Whenever the conditions for fully resonant excitation are met the four levels involved in the resonances can be sorted out of the complex manifold of excited states and the material system can be represented as a statistical ensemble of four-level systems[3].

Signal enhancement and selectivity are not however the only important features of multiresonant FWM processes. The ability to give Doppler-free lineshapes has been theoretically predicted for doubly-and triply-resonant CSRS measurements of gas phase samples[4]. In the condensed phase the possibility of narrowing inhomogeneous lineshapes depends on the existence of special correlations between the inhomogeneous distributions of the levels[5]. Detailed lineshape analysis can thus provide insights into such correlations[6].

Another important possibility of multiresonant CSRS which has been recently demonstrated for gas and condensed phase samples is that of displaying additional narrow resonances at the excited state vibrational frequency whenever adiabatic interactions induce fluctuations of the transition energies involved in the resonance enhancement (pure dephasing)[7,8]. Two major implications are: (i) lineshape parameters of the excited state vibrations not affected by the dynamics of the purely electronic transition can be obtained and (ii) the rate of pure dephasing can be evaluated directly from the intensity of the dephasing induced coherent emission (DICE).

One problem with resonant FWM measurements is the possible complexity of the spectra. Perturbative calculations of the nonlinear susceptibility $\chi^{(3)}$ ($\omega_3=2\omega_1-\omega_2$) according to the density matrix or diagrammatic techniques show that in a scan of $|\omega_1-\omega_2|$ two resonances should be expected in resonant CARS and up to four in resonant CSRS for each vibrational mode of the molecular system. Furthermore, the very occurrence of resonances between laser frequencies and molecular transitions might cause the build up of real population in the excited states. This in turn would act as the starting point for additional FWM generation processes at later times in the pulse envelope.

The present paper deals with a theoretical and experimental investigation of the effects of an initial excited state population on the multiresonant CARS spectrum. We briefly outline a diagrammatic technique by which the multiresonant terms of the susceptibility can be calculated in a straightforward way. We then present and discuss the results for multiresonant CARS and CSRS of a four-level system whose initial population is shared between ground and excited electronic states. The comparison of spectra obtained at various detunings of the ω_1 field from exact resonance with the electronic transition is proposed as a method for the unequivocal assignment of the observed resonances.

Experimental results are presented for the multiresonant CARS spectra of pentacene in benzoic acid single crystals at low temperature. They allow us to demonstrate the presence of excited state population and to verify the predicted effects.

2. TIME-ORDERED DIAGRAMS FOR CARS AND CSRS

It is convenient to represent nonlinear response functions by means of time ordered diagrams. These diagrams arise because of a particular mathematical technique - iteration - used commonly to solve the Liouville equation of motion for the density operator of a system. The diagrams thus represent only one of many pictures for the interaction with radiation. In this article our concern is mainly with resonant phenomena, such that each successive light field is used to couple a pair of states separated by the energy $\hbar\omega$ of a combination of light quanta. The diagrams then take a particularly simple form. In the following V_i and $V_i^\dagger$ with i = 1,2 are the molecule field interactions corresponding to the anihilation and creation of a photon at ω_i: In a semiclassical picture V_i and $V_i^\dagger$ involve $e^{i\omega_i t}$ and $e^{-i\omega_i t}$. Starting with the initial state a the first field (ω_i) to interact may cause coherence to be introduced into the level pair ac by the two mechanisms shown in Figure 1a:

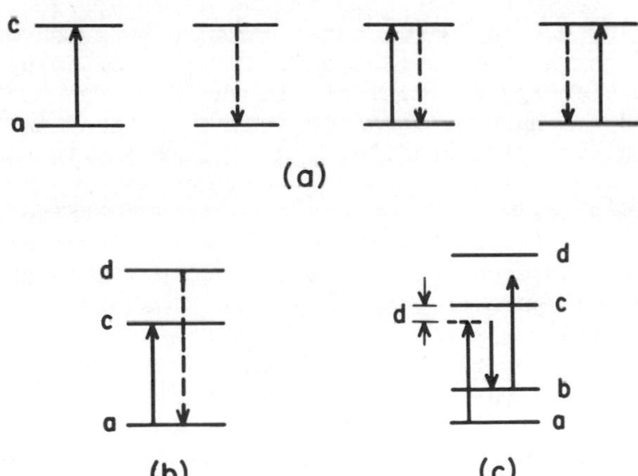

Figure 1. Diagrammatic representation of resonant molecule-radiation interactions (see text) - (a) Light absorption; (b) Difference frequency generation; (c) Coherent anti-Stokes generation.

In the first diagram a photon at ω_1 is anihilated in the transition amplitude $\langle a|V_1|c\rangle$. The second involves creation of an ω_1 photon in the step $\langle c|V_1^\dagger|a\rangle$. The combination of these two steps viz. $\langle c|V_1^\dagger|a\rangle \langle a|V_1|c\rangle$. represents the process of light absorption $a \rightarrow c$ and is represented by the last two diagrams labelled 1(a). The second step could have involved $-\omega_2$, meaning that a photon of frequency ω_2 was created, in for example the amplitude $\langle d|V_2^\dagger|a\rangle\langle a|V_1|c\rangle$. This process may give rise to an induced dipole, oscillating at the frequency $\omega_1-\omega_2$, resonance enhanced by the coherence in the level pair cd. In the diagrams the creation or anihilation of a photon is signified by a down or up arrow. A full or dashed line signifies whether the bra or the ket evolution is involved. This makes a fundamental difference for two reasons: The bra sequence $\langle a|V_1|c\rangle\langle c|V_2^\dagger|d\rangle$ yields ad-coherence resonance enhanced by $\omega_1-\omega_2$, whereas in the previous example the cd level pair was required to match $\omega_1-\omega_2$ in order to have resonance enhancement. Secondly, the signs of the real parts of the energy denominators are different when solid and dashed line processes are involved, and this influences the result of an ensemble average of the polarization waves (see below).

Third order processes require three driving fields which in the present case of CARS and CSRS involve two fields at ω_1 and one at ω_2, with the generated field at $\omega_4=2\omega_1-\omega_2$. For CARS to be fully resonant beginning with state 'a' it is necessary to evolve to the ad coherence exclusively with solid lines as in $\langle a|V_1|c\rangle$ $\langle c|V_2^\dagger|b\rangle$ $\langle b|V_1|d\rangle$. This diagram is shown in Figure 1b With ω_1 detuned by d. Reading from left to right this diagram says: ac coherence resonantly introduced by ω_1, ab coherence resonant with $\omega_1-\omega_2$, and ad coherence resonant at $2\omega_1-\omega_2$. Stokes emission at $2\omega_1-\omega_2$ (with $\omega_2>\omega_1$) is resonant with the cb separation, cb coherence can be introduced in the three resonant ways shown in Figure 2. Each of these steps provides a recipe for obtaining the relevant resonance energy denominator: In third order each term contains three denominators, one from each step taken in time order starting with the left handside of a diagram in Figures 1 and 2. If the first step is $\langle\alpha|V_i|\beta\rangle$ the corresponding denominator is $[\omega_{\alpha\beta}+\omega_i+i\Gamma_{\alpha\beta}]$ where $\omega_{\alpha\beta}=\omega_\alpha-\omega_\beta$, and ω_i is a positive number. In the case that the first step is $\langle\beta|V_i^\dagger|\alpha\rangle$ the denominator is $[\omega_{\beta\alpha}-\omega_i+i\Gamma_{\alpha\beta}]$. For subsequent steps in the evolution of the density operator the real part of the energy denominator is $\omega_{\mu\nu}-\Sigma\omega_i$, where μ and ν are the states reached on the ket and bra sides respectively, and $\Sigma\omega_i$ is the sum of the field frequencies involved each taken with the appropriate sign. The difference in the sign of the real parts in these factors is critical in evaluating the linenarrowing capability of each resonant term. It turns out that in fully correlated systems only those terms can be linenarrowed that have both solid and dashed arrows. A fully correlated four level system is one in which the environmental shifts for each of the levels are in direct proportion.

Diagrams containing only full or dashed lines describe processes that are linenarrowed when anticorrelation occurs between pairs in the four level system.

In this paper we study the coherent Stokes and Anti-Stokes emission that originates from populations in both the ground state a, and an excited state, c. A total of 8 quasi-resonant diagrams, shown on Figure 2, contribute to the signal[9]. The resonant CARS susceptibility of the four level system can be rearranged to the form:

$$\chi^{(3)}(\omega_{as}) = \left\langle \frac{-\mu_{ac}^{(1)}\mu_{cb}^{(2)}\mu_{bd}^{(1)}\mu_{da}^{(as)}}{6\hbar^3(d - i\,\Gamma_{ac})} \left\{ \frac{\rho_{aa}^{o}}{(\omega_{ba}-\Delta+i\,\Gamma_{ab})(\omega_{dc}-d-\Delta + i\,\Gamma_{ad})} - \frac{\rho_{cc}^{o}}{(\omega_{ba}+d-\Delta+i\,\Gamma_{cb})(\omega_{dc}-\Delta + i\,\Gamma_{dc})}\left[1 + \frac{i\,\Gamma_1}{(\omega_{dc}- d - \Delta + i\,\Gamma_{ad})} + \frac{i\Gamma_2(\omega_{dc} - \Delta + i\Gamma_{dc})}{(\omega_{dc}-d-\Delta + i\,\Gamma_{ad})(\omega_{ba}-\Delta+i\,\Gamma_{ab})}\right]\right\}\right\rangle \quad (1)$$

We have assumed, as in Figure 1c, that ω_1 is detuned from the ω_{ca} resonance by $d = \omega_1-\omega_{ca}$. The experimental variables are d, $\Delta \cong \omega_1-\omega_2$, in addition to ρ_{aa}^{o} and ρ_{cc}^{o} - the initial populations in states a and c. The three diagrams that appear in CARS for $\rho_{cc}^{o} = 1$, and in CSRS for $\rho_{aa}^{o}=1$, have DICE resonances. For

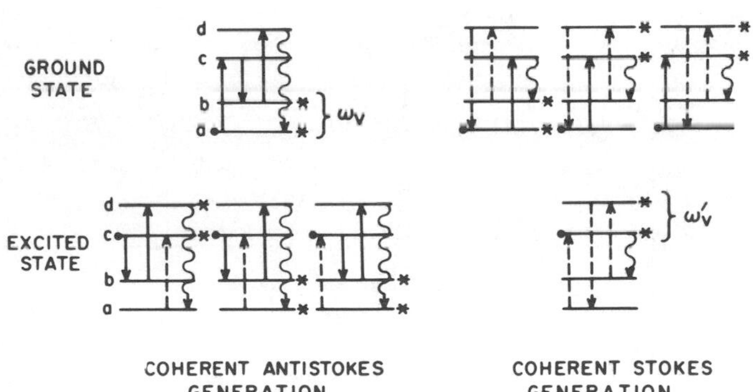

Figure 2. Diagrams representing the time ordered interaction sequences for Stokes and anti-Stokes generation starting from the ground and excited electronic states.

CARS one depends on the existence of $\Gamma_1 = \Gamma_{dc} + \Gamma_{ca} - \Gamma_{da}$ and shows a resonance at ω_{dc}-d, the other depends on $\Gamma_2 = \Gamma_{cb} + \Gamma_{ca} - \Gamma_{ba}$ and the resonance occurs at ω_{ba}. At very low temperature when there is no dephasing $\Gamma_1 = \Gamma_2 \cong \gamma_{cc}$, so in principle these resonances are not entirely dephasing induced but have a remnant contribution due to the finite lifetime of state c. It should be noted that the steady state solutions to the Liouville equation leading to equation (1) are valid only if $1/\gamma_{cc}$ is large compared with the laser pulsewidth.

3. EXPERIMENTAL

Crystals of benzoic acid containing 5 x 10^{-5} mole/mole pentacene were grown by the Bridgman technique and cleaved parallel to the ($\underset{\sim}{a}$ $\underset{\sim}{b}$) plane to yield platelets ca. 1 mm thick. Measurements were performed at 4.2 and 1.6K.

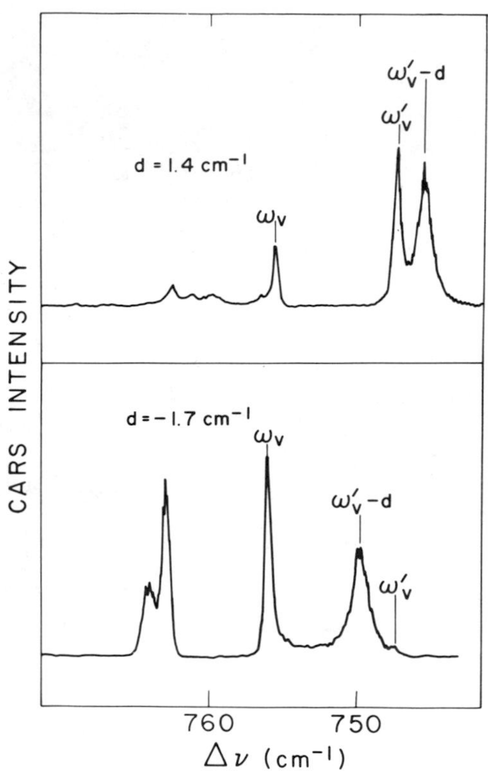

Figure 3. CARS spectrum of Pc/BZA at 1.6K for two different detunings d. Ground state frequency, ω_v = 755 cm^{-1}; excited state frequency, ω_v' = 747 cm^{-1},

The details of the experimental set up have been previously described[10]. Two dye lasers (Molectron DL300 and DL II) were pumped by a Molectron UV1000 nitrogen laser. Rhodamine 6G and Rhodamine B were used as dyes for the excitation at ω_1 and ω_2 respectively, Coumarin 540A was used for the excitation at ω_2 in CSRS experiments. Pulse energies of both dye lasers were typically 3-10 μ Joule. The ω_1 laser bandwidth was narrowed to 0.03 cm^{-1} by introducing an etalon in the cavity. The bandwidth of the ω_2 laser was kept at ca. 0.3 cm^{-1} for wide range scans. High resolutions scans were taken on some resonances by using the pressure tuning technique with an ω_2 laser bandwidth of 0.03 cm^{-1}. Spectra were taken with both lasers having polarization either parallel or perpendicular to the $\underset{\sim}{b}$ crystal axis.

4. RESULTS AND DISCUSSION

Typical multiresonant CARS spectra of pentacene in benzoic acid (P/BZA) at 1.6K are shown in Fig. 3. During each scan the ω_1 frequency was kept fixed at the value $\omega_{ac} + d$ where ω_{ac} = 17,003.6 cm^{-1} is the frequency of the strongest $S \rightarrow S_1$ vibronic transition of pentacene and d is the detuning from exact resonance. The ω_2 frequency was scanned so that the difference $\Delta=\omega_1-\omega_2$ ranged between 800 and 740 cm^{-1}. One Raman active frequency of benzoic acid (at 797 cm^{-1}) and three ground state frequencies of pentacene (at 787, 763 and 753 cm^{-1}) occur in this region. When Δ is nearly resonant with one of the pentacene frequencies the CARS signal generation can be described considering only four levels: the electronic ground state $|0\rangle \equiv |a\rangle$, the ground state vibrational level $|0v\rangle \equiv |b\rangle$, the electronic excited state $|0'\rangle \equiv |c\rangle$ and the excited state vibrational level $|0'v'\rangle \equiv |d\rangle$. The comparison of absorption and fluorescence spectra indicates the following pairing of ground and excited state vibrational frequencies (cm^{-1}): $(\omega_v{}'\omega_v{}')$=(787, 791); (763, 761); (756, 747).

In Fig. 3 the spectra obtained with two different values of the detuning d are compared. As expected, the peak located at the ground state vibrational frequencies (756 and 763 cm^{-1}) do not shift with the detuning. With d = 1.4 cm^{-1} (upper spectrum) one strong and one weak peak are observed at the excited state frequencies 747 and 761 cm^{-1} respectively. With negative detuning (d = -1.7 cm^{-1}, lower spectrum) the intensity of these peaks becomes vanishingly small. Additional broader bands are observed in both spectra at the frequencies ($\omega_v{}'$ - d). In agreement with the theoretical predictions (eqn. 1), these peaks are assigned to resonances of the outcoming field ω_3 with $|o\rangle \rightarrow |o'v'\rangle$ transitions.

The results of a thorough analysis of the effect of detuning on the CARS spectrum are presented in Fig. 4 where the observed peak frequencies are plotted against the ω_1 frequency[11]. Horizontal lines have been drawn in correspondence to the frequencies of the ground and excited state vibrations. Straight lines of unit slope cross the horizontal lines for excited state frequencies at the frequency ω_{ac} of the $|o\rangle \rightarrow |o'\rangle$ transition. It is evident that for each of the three vibrational modes of pentacene the CARS spectrum displays three resonances at ω_v, ω_v' and, $(\omega_v' - d)$. Eqn. (1) shows that the presence of peaks at the excited state vibrational frequencies which do not shift with detuning implies a finite value of the zeroth order diagonal density matrix element $\rho_{cc}^{(o)}$ and thus demonstrates the build up of excited state population during the laser pulses. This has been

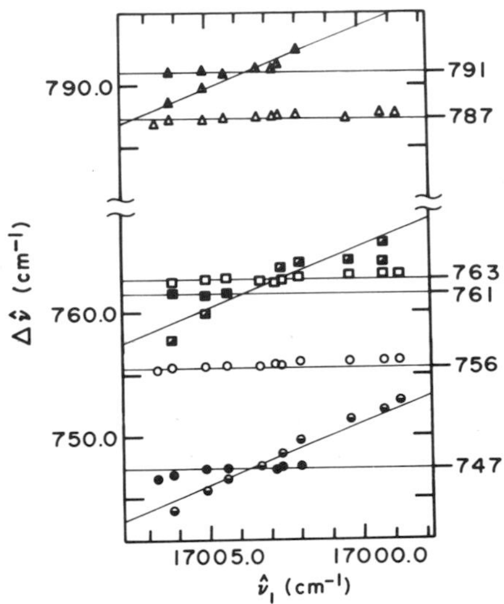

Figure 4. Plot of the observed CARS frequencies versus the ω_1 frequency.

further checked by studying the ω_1 power dependence of the peak at 747 cm^{-1}. Its intensity decreases faster than the expected ω_1 squared power dependence. In fact, in addition to the signal generation process, also the rate of population transfer is dependent on the ω_1 power. Figs. 3 and 4 show that the intensity of the resonances attributed to excited state population decreases much faster for negative detuning than for positive detunings. This is likely related to an asymmetry of the $|0\rangle \rightarrow |0'\rangle$ absorption lineshape due to the presence of phonon sidebands. That is, phonon sideband absorption leads to excited state population.

We have also carried out a study of the detuning dependence of the CSRS spectra with quite similar results as to the presence of excited state population and its dependence on detuning. The signal due to excited state population in the CSRS case was already at the limit of detectability with a detuning of $-5\ cm^{-1}$. Considering the large detuning ($d = -16.8\ cm^{-1}$) used in previous experiments aiming at investigating the DICE process[8], the present results allow us to exclude that excited state population may have contributed to that signal. However, temperature dependent dephasing processes would make the transfer of population possible at larger detunings and studies at elevated temperatures are needed to fully understand the DICE effect.

Closer examination of eqn. 1 shows that an additional resonance proportional to the excited state population should be observed at $(\omega_v + d)$. This resonance is missing or extremely weak in the spectra of Fig. 3. Its intensity relative to that of the resonance at $(\omega_v' - d)$ should be given approximately by $(\rho_{cc}^{(o)}/\rho_{aa}^{(o)})(\Gamma_{ad}/\Gamma_{cb})$. A better understanding of this feature will be provided by the detailed theoretical fitting of the spectra which is currently in progress.

5. CONCLUSIONS

This work has shown that under commonly employed excitation conditions, and near resonance, four wave mixing (CARS and (CSRS) signals are obtained from both the ground and zeropoint level of the excited state of pentacene in benzoic acid. There is every reason to suppose this result is generally applicable. The situation can be particularly confusing on-resonance, at which condition the outgoing wave and Raman signal frequencies are equal, so that lineshape analysis is intractable. However the lineshape analysis is possible when d exceeds the widths.

It may not be a simple matter to obtain CARS and CSRS resonant signals under excitation conditions for which the steady state, 3rd order polarization results are really valid. To circumvent this problem we have recently developed a non-perturbative approach to the resonant nonlinear response[12]. The approach also allows us to understand the effects of laser power on the observed spectral widths.

One of the most interesting aspects of these resonant processes is their potential for exploring inhomogeneous distributions and linenarrowing effects. On the basis of the results presented here for CARS we were able to learn that the assumption of fully correlated states made previously for PC/BZA is clearly not correct. Although not apparent from Figure 3, we have found that the ground state Raman transition has a full width at half maximum of $0.23\ cm^{-1}$, which is about one-half the value obtained by Andrews and Hochstrasser using CSRS[8]. These linewidth results will be discussed in more detail elsewhere in relation to our recent theoretical models of correlations in the inhomogeneous distribution[6].

While the inevitable occurrence of excited state population introduces complications into the interpretation of the resonant nonlinear response, there are some benefits. One of these is that the techniques easily allow the recording of excited state Raman spectra. In the present case the Raman spectra of a state having a lifetime of 15ns are obtained with high signal to noise even in the presence of strong fluorescence of the state in question.

ACKNOWLEDGEMENTS

One of us (R.B.) is grateful to the Conciglio Nazionale delle Richerche of Italy for financial support during the period of this research.

REFERENCES

1. M. D. Levenson, "Introduction to Nonlinear Laser Spectroscopy", (Academic Press, New York, 1982).
2. S. Druet and J. P. Taran in "Chemical and Biochemical Applications of Lasers", ed. by C. B. Moore (Academic Press, New York, 1979) Vol. 4; H. C. Andersen and B. S. Hudson in "Molecular Spectroscopy - A Specialists Periodical Report" (The Chemical Society, London 1978) Vol. 5.
3. J. R. Andrews, R. M. Hochstrasser and H. P. Trommsdorff, Chem. Phys. 62, 87 (1981).
4. S. A. J. Druet, J.-P. E. Taran, and Ch. J. Bordè, J. Phys. (Paris) 40, 819 (1979); J.-L. Oudar and Y. R. Shen, Phys. Rev. A22, 1141 (1980).
5. B. Dick and R. M. Hochstrasser, "Spectroscopic and Linenarrowing Properties of Sum and Difference Frequency Generation", J. Chem. Phys., in press.
6. R. Bozio, P. L. DeCola, and R. M. Hochstrasser, to be published.
7. N. Bloembergen, H. Lotem, and L. T. Lynch Jr., Ind. J. Pure Appl. Chem. 16, 151 (1978).
8. J. R. Andrews and R. M. Hochstrasser, Chem. Phys. Lett. 83, 427 (1981); ibid 82, 381 (1981).
9. There are other partially resonant contributions involving even higher excited states but these are expected to be small for the case pentacene/benzoic acid.
10. R. M. Hochstrasser, G. R. Meredith, and H. P. Trommsdorff, J. Chem. Phys. 73, 1009 (1980).
11. Given the limited dispersive character of the observed lineshapes, the peak frequency represents a good approximation to the true resonance frequency.
12. B. Dick and R. M. Hochstrasser, to be published.

NEW RESULTS ON VIBRATIONAL DEPHASING AND POPULATION RELAXATION FROM PICO-SECOND TECHNIQUES

H.R. Telle, H. Graener and A. Laubereau

Physikalisches Institut, Universität Bayreuth, Bayreuth, West Germany

Vibrational relaxation processes have received increasing interest in recent years. With ultrashort light pulses and novel experimental techniques we are in the position to study vibrational dynamics in the electronic ground state of simple liquids on the subpicosecond and picosecond time scale. In this brief summary two topics will be discussed.

NONEXPONENTIAL DEPHASING IN THE SUBPICOSECOND TIME DOMAIN

In the past several investigations of the vibrational dephasing time T_2 in liquids have been performed (1). In these studies exponential decay of the coherent vibrational excitation has been observed on the time scale of several ps or longer. No direct information has been available on the short time behaviour, $t \lesssim 1$ ps. Similarly, for theoretical work on stimulated Raman scattering purely exponential time behaviour was assumed. It is, however, known from theory of relaxation processes that this approximation does not hold for very short times which compare with the time scale of the dephasing mechanism (2). We have generalized the theory of transient stimulated Raman scattering to include non-exponential dephasing (3). Our approach avoids the two-level approximation used previously and introduces the vibrational dynamics via the vibrational autocorrelation function (4). Using stochastic theory the following expression for this function may be derived (5):

$$\phi = \exp\left\{-\frac{|t|}{T_2} - \frac{\tau_c}{T_2}\left[\exp\left(-\frac{|t|}{\tau_c}\right)-1\right]\right\} \qquad (1)$$

The correlation time τ_c denotes the time scale of the interaction. For $t \lesssim \tau_c$ the time dependence of ϕ is non-exponential with horizontal slope at t=0. For large t, Eq. 1 yields the expected

ISBN 0-12-066280-9

exponential decay with time constant T_2. The limiting cases of homogeneous and inhomogeneous broadening are included for $\tau_c \ll \delta\nu^{-1}$ and $\tau_c \gg \delta\nu^{-1}$, respectively, where $\delta\nu$ denotes the spectral linewidth. Simple exponential dephasing considered previously is contained in the present treatment for $\tau_c \to 0$. Experimental information on τ_c is highly desirable for the understanding of the dephasing process. This can be achieved measuring the vibrational excitation by coherent Raman scattering of delayed probing pulses (3).

We have carried out numerical calculations of the molecular excitation by stimulated Raman scattering and of the coherent Raman probe scattering using the more general description of vibrational dynamics of Eq. 1. Our results are briefly summarized as follows: (i) Keeping the spectroscopic linewidth $\delta\nu$ constant (FWHM of the Fourier transform of Eq. 1) we note only little effect of τ_c on the stimulated amplification of the Stokes pulse of the pumping process. (ii) The time evolution of the vibrational excitation and of the coherent probe scattering signal, however, are notably effected by τ_c ($\delta\nu$ = const). (iii) For experimental investigations special conditions appear to be advantageous: excitation by stimulated Raman scattering with moderate amplification factor $G \simeq 2$ and distinct frequency shift of the stimulated excitation process $\Delta\nu = \nu_L - \nu_S - \nu_o \simeq 3\ \delta\nu$. The small amplification is highly favourable because of the reduced nonlinearity of the process; intensity fluctuations of the pump consequently have only

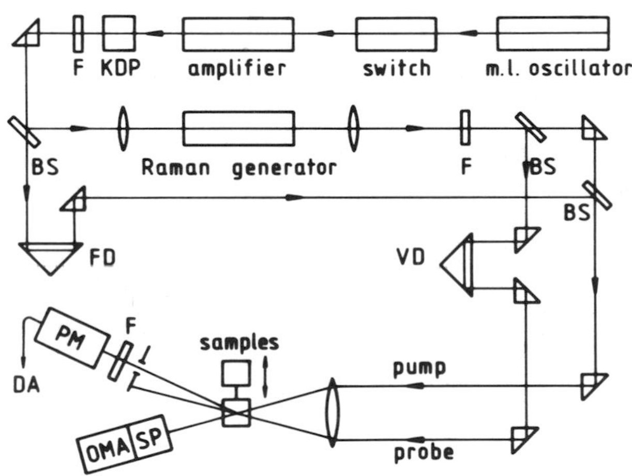

FIGURE 1. Schematic of double-channel setup for the measurement of τ_c; beam splitter BS, filter F, fixed delay FD, variable delay VD, data acquisition DA.

little influence. Frequency detuning $\Delta\nu \neq 0$ causes the excitation process to terminate more quickly and improves experimental time resolution.

We have devised a double-channel picosecond system for the direct measurement of τ_c, which is depicted schematically in Fig. 1. Single bandwidth limited pulses are generated by a mode-locked Nd:glass system and converted to the second harmonic by a KDP crystal. The pulse is subsequently divided into two parts by a beam splitter BS. One part serves as pump pulse with frequency $\tilde{\nu}_L$= 18990 cm^{-1}. The second fraction produces a Stokes-shifted pulse by efficient stimulated scattering in 2,2'-dichlorodiethylether ($\tilde{\nu}_S$= 16026 cm^{-1}). Nonlinear autocorrelation measurements and spectral observations yield the pulse parameters t_L=7.7±.5 ps, t_S=4.8±.5 ps, $\delta\tilde{\nu}_L$=3.1±.8 cm^{-1} and $\delta\tilde{\nu}_S$=6.4±.8 cm^{-1}. Part of the Stokes pulse is combined with the pulse at $\tilde{\nu}_L$ and excites the vibrational transition $\tilde{\nu}_o$ in the sample cell (length 5 mm) by stimulated Raman amplification. The second part of the Stokes pulse is properly delayed and serves as probe pulse via coherent Raman Stokes scattering with off-axis geometry. The probe scattering emission $S^{coh}(t_D)$ at $\tilde{\nu}_S-\tilde{\nu}_o$ is observed in phase-matching direction by the help of a photomultiplier and filters.

For proper control of experimental time resolution, the instrumental response curve $R(t_D)$ is measured under identical geometrical conditions. To this end the sample is automatically replaced every second laser shot by a reference cell generating a reference scattering signal R. As reference substance we use ethanol which shows three broad Raman bands around 2900 cm^{-1} leading to almost instantaneous response of these molecules.

We have investigated several CH_2- and CH_3-stretching modes in the neat liquid at room temperature (3). Data on the symmetric ν_1-mode at 2987 cm^{-1} of CH_2BrCl are presented in Fig. 2. A small amplification of the incident Stokes pump pulse by a factor of G = 1.6±.4 has been adjusted for the excitation process. The normalized probe scattering signal, $N^{coh}(t_D)$, is plotted versus delay time t_D, which represents the ratio of the coherent probe scattering signal $S^{coh}(t_D)$ to the reference signal $R(t_D)$ measured at the same delay setting. Values of $N^{coh} \neq 1$ reflect the delayed response of the vibrational transition under investigation.

The solid curve in Fig. 2 has been calculated for τ_c=0.4 ps and T_2=1.5 ps. The latter value is consistent with the observed asymptotic time dependence of the scattering signal $S^{coh}(t_D)$. The good agreement with the experimental points should be noted. The results for τ_c and T_2 are in excellent accordance with data on the spontaneous Raman linewidth. For comparison, two broken lines are also shown in the Fig. which are calculated for constant $\delta\nu$ and values of τ_c=0.2 ps and 1.0 ps, respectively; these curves notably deviate from the experimental points.

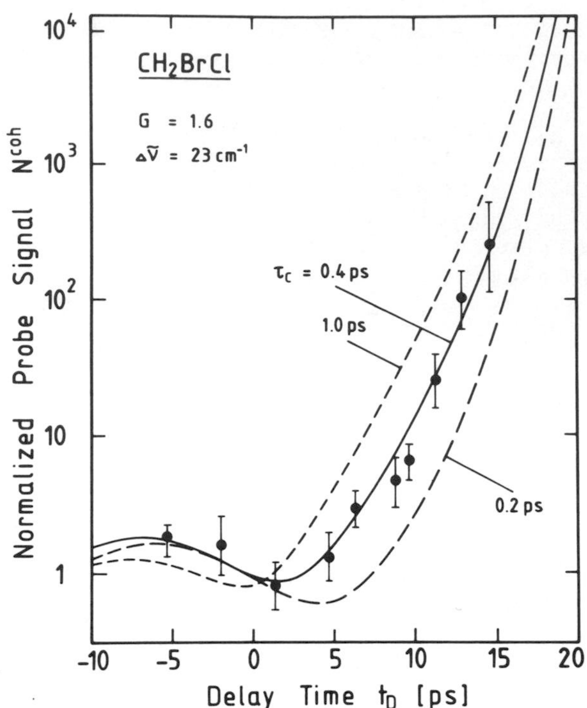

FIGURE 2. Normalized probe signal $N^{coh}(t_D)$ versus delay time t_D for the ν_1-CH_2-mode of neat CH_2BrCl; experimental points; theoretical curves

Similar data were obtained for the ν_1-stretching modes of CH_3I and CH_3CN yielding values of τ_c=0.5 and 0.7 ps, respectively. The time constants in the subpicosecond range indicate that the relaxation mechanism occurs close to the limit of homogeneous broadening. Our results represent the first direct observation of non-exponential dephasing on the subpicosecond time scale. Assuming dephasing via the repulsive part of the intermolecular potential to be predominant the constant τ_c represents the time scale of translational motion; i.e. the elastic collision time. Simple models of the liquid state suggest this time to be 10^{-12} to 10^{-13}s in satisfactory agreement with our experimental data.

VIBRATIONAL POPULATION LIFETIMES OF METHYLENE HALIDES

Population decay and transfer processes were directly observed recently on the picosecond time scale in a number of systems (6-8). Systematic studies for a larger variety of molecular systems are highly desirable for an improved understanding of the physical mechanisms. We have devised a picosecond Raman spectrometer for the measurement of population lifetimes in liquids, which is based on a mode-locked YAG-laser with single picosecond pulses of 23 ps and a repetition rate of several Hz. In comparison to previous investigations applying Nd:glass lasers the pulse repetition rate is increased by 1-2 orders of magnitude yielding a significant reduction of measuring time and/or improvement of experimental accuracy. The experimental setup is schematically shown in Fig. 3 (9).

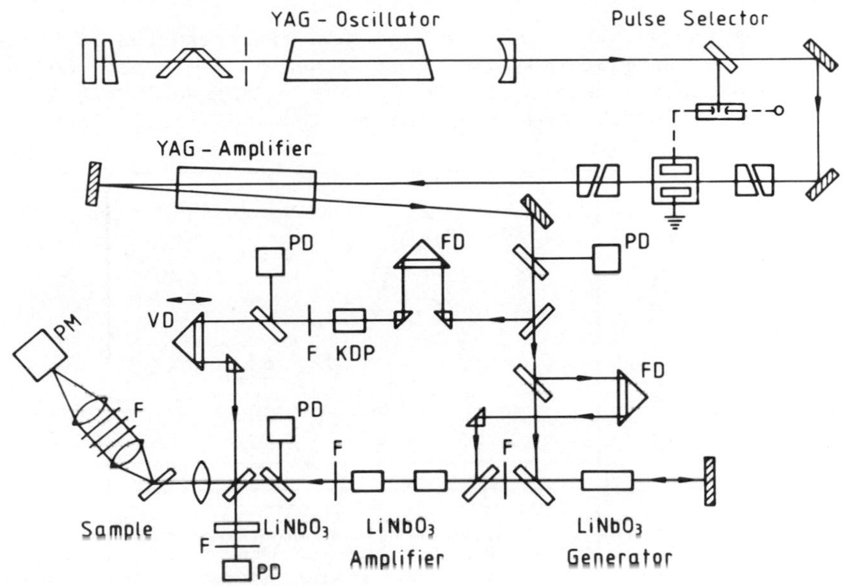

FIGURE 3. Schematic of the experimental system to measure vibrational population lifetimes; fixed delay FD, variable delay VD, filter F, photo-detector PD, photomultiplier PM

The infrared excitation pulse is generated by a multiple step parametric generator - amplifier setup yielding tunable pulses of 10 ps and 200 µJ around 3 µm. The excited molecules are interrogated by a green probing pulse via spontaneous anti-Stokes Raman scattering (1). We routinely control the time resolution of the setup measuring the cross correlation of excitation and probing pulses. The system provides values of the population lifetime $T_1 \gtrsim 4$ ps with large dynamic range and high measuring sensitivity.

We have investigated CH_2-modes of several methylene halides and of $(CH_2Cl)_2$ in the neat liquid and dissolved in CCl_4 (9). CH_2Cl_2 and $(CH_2Cl)_2$ and the solvent CCl_4 were spectral grade "Uvasol". The other samples were standard grade with a purity of 99 %, the remaining 1 % predominantly made up by other CH_2-halides.

The first excited state of the symmetric stretching mode ν_1 was directly excited by the resonantly tuned IR pulse. Anti-Stokes probe scattering of the first excited state of the ν_1-mode was detected which has the largest Raman cross-section. Possible scattering contributions from the asymmetric CH_2-mode and of adjacent overtones and combination modes are negligible because of the small Raman cross-sections and the spectral discrimination of the detection system.

Examples for the time-resolved data are presented in Fig. 4. The measured anti-Stokes scattering signal $S(t_D)$ is plotted on a semi-logarithmic scale versus delay time t_D between probe and excitation pulse (full and open points). The individual experimental

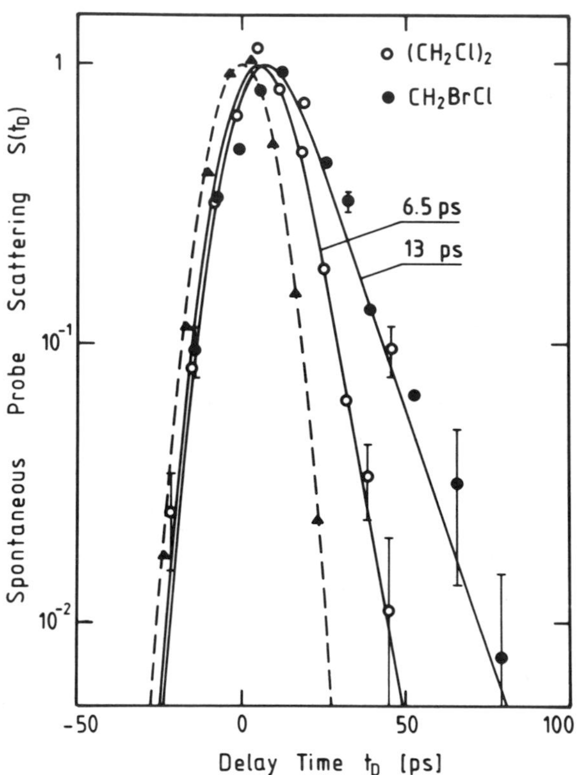

FIGURE 4. Spontaneous anti-Stokes scattering signal $S(t_D)$ of the probe pulse versus delay time t_D for the symmetric CH_2-stretching modes of $(CH_2Cl)_2$ (o) and CH_2BrCl (● ; same vibration as Fig. 2); broken line and full triangles represent the instrumental response of the $(CH_2Cl)_2$ measurement

points indicate the accumulated information of 200 to 600 laser shots and extend over a factor of ∿ 100. The data have been corrected for small background signals detected when one beam (probe or pump) was blocked. The error bars represent the statistical photon counting fluctuations of the corrected scattering signal. The full triangles in the figure show some results of the simultaneous cross correlation measurement.

The curves in the figure are calculated assuming Gaussian shape of the light pulses and using the population lifetime T_1 as fitting parameter. It is interesting to see that the delay of the signal maxima and the asymptotic decay are fully accounted for by the theoretical curves. Results on T_1 are listed in Table I, which indicates values of 6.5 to 50 ps for the investigated symmetric CH_2-stretching modes. Data on $(CH_2ClCH_2)_2O$ are included to demonstrate the time resolution of the measuring system.

The population lifetimes are explained by vibrational energy transfer via anharmonic coupling to neighbouring combination bands and overtones. Numerical estimates have been performed considering Fermi resonance with the adjacent overtone of the CH_2-bending mode (8) yielding only fair agreement. It is concluded that higher order combination bands in the immediate neighbourhood of the symmetric stretching mode give a notable contribution to the observed population decay.

TABLE I. Measured population lifetimes T_1

	concentration vol %	$\tilde{\nu}\,|cm^{-1}|$	$T_1\,|ps|$
CH_2Br_2	100	2987	7 ± 1
	30	2987	7 ± 1
CH_2Cl_2	100	2989	12 ± 2
	10	2989	12 ± 2
CH_2I_2	50	2967	45 ± 5
CH_2ClBr	30	2987	13 ± 2
CH_2ClI	30	2979	14 ± 2
$(CH_2Cl)_2$	10	2950	6.5 ± 1
$(CH_2ClCH_2)_2O$	5	∿ 2964	2 ± 2

In conclusion we point out that we have directly observed the time scale τ_c of vibrational dephasing in liquids. Our time constants in the subpicosecond range are consistent with spectroscopic data (10). Theoretical dephasing models suggest that the measured τ_c characterizes the translational motion of the molecules.

We have also carried out a systematic study of the population decay of the ν_1-modes of CH_2-halides. Comparison of the measured decay times with theoretical estimates gives evidence for rapid V-V transfer. Our data suggest that anharmonic coupling to neighbouring third order combination-tones notably contributes to the investigated depopulation processes.

REFERENCES

1. For a review, see A. Laubereau, W. Kaiser, Rev. Mod. Phys. 50, 607 (1978).
2. See, for example, P.C. Martin, "Measurements and Correlation Functions" (Gordon and Breach, New York, 1968).
3. H.R. Telle and A. Laubereau, to be published.
4. R. Kubo, "Fluctuations, Relaxation and Resonance in Magnetic Systems", editor D. Ter Haar (Penum, New York, 1962).
5. W.G. Rothschild, J. Chem. Phys. 65, 455 (1976).
6. A. Laubereau, D. von der Linde, W. Kaiser, Phys. Rev. Lett. 28, 1162 (1972);
   R.R. Alfano, S.L. Shapiro, Phys. Rev. Lett. 29, 1655 (1972).
7. K. Spanner, A. Laubereau, W. Kaiser, Chem. Phys. Lett., 44, 88 (1976);
   A. Laubereau, S.F. Fischer, K. Spanner, W. Kaiser, Chem. Phys. 31, 335 (1978).
8. A. Fendt, S.F. Fischer, W. Kaiser, Chem. Phys. 57, 55 (1981); Chem. Phys. Lett. 82, 350 (1981).
9. H. Graener and A. Laubereau, Appl. Phys. B 29, in press.
10. G. Döge, R. Arndt, A. Khuen, Chem. Phys. 21, 53 (1977);
    J. Yarwood, R. Arndt, G. Döge, Chem. Phys. 25, 387 (1977).

RAMAN SPECTRA OF SHORT-LIVING MOLECULAR STATES BY CARS

A. Lau, H-J. Weigmann, W. Werncke, K. Lenz, M. Pfeiffer

Central Institutes of Optics and Spectroscopy
Academy of Sciences of GDR
1199 Berlin-Adlershof
Rudower Chaussee 5
German Democratic Republic

The investigation of vibrational spectra of short-living molecular states demands a method with high time response or time resolution and with high sensitivity because the species under investigation occur at low concentration only.

The first demand can be easily realized using pulse lasers to generate coherent Raman effects. The time response corresponds to the pulse duration. A high sensitivity needs the use of resonance enhancement.

There exist different coherent Raman methods: Inverse Raman Scattering (IRS), Coherent Antistokes Raman Scattering (CARS), Raman induced Kerr Effect (RiKE) and modifications of these three types to which they can be related.

A detailed consideration (1) shows, that in contrast to the case far off resonance under strong resonance conditions CARS has highest sensitivity. This is due to the fact that at one photon resonance an enhancement of both the Raman term χ^{Ra} of the susceptibility and other non-Raman terms χ^{e} of the electronic background occurs.

All χ^{Ra} and χ^{e} terms are dependent on the frequency difference between the frequencies of the molecular absorption and laser or signal frequencies. This frequency dependence and the limiting values for strong resonance are very different for IRS, RiKE and CARS, yielding to very large χ^{e} terms for IRS and RiKE in contrast to CARS.

ISBN 0-12-066280-9

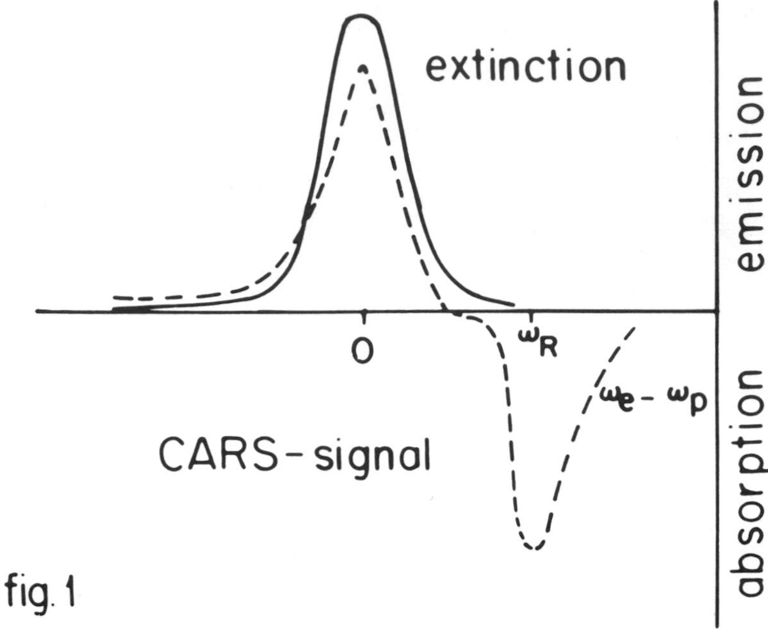
extinction
emission
absorption
0
ω_R
$\omega_e - \omega_p$
CARS-signal

fig. 1

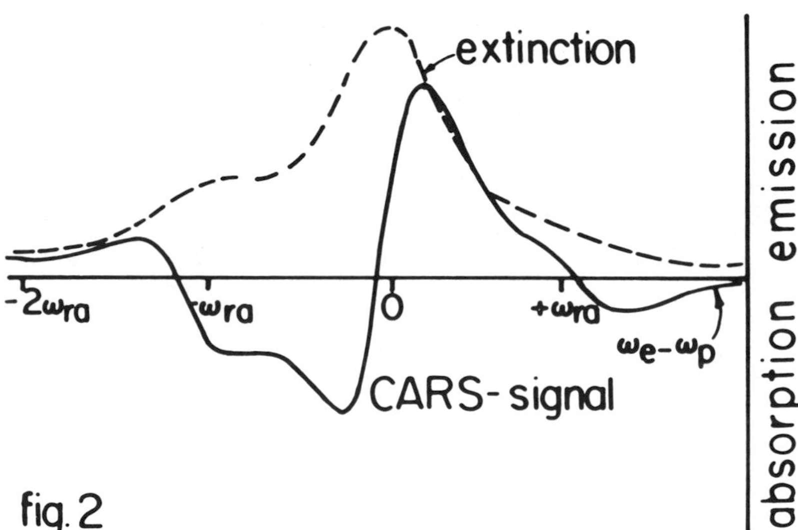
extinction
emission
absorption
$-2\omega_{ra}$
$-\omega_{ra}$
0
$+\omega_{ra}$
$\omega_e-\omega_p$
CARS-signal

fig. 2

Corresponding to the equations for the intensity I

$$I \propto \text{im}\ \frac{\chi^{Ra}}{\chi^{n} + \chi^{e}} \qquad \text{for IRS}$$

and

$$I \propto \left|\frac{\chi^{Ra}}{\chi^{n} + \chi^{e}}\right|^2 \qquad \text{for RiKE and CARS}$$

large χ^{e} values can yield to a masking of Raman lines. These considerations are valued in the ns-region and in qualitative agreement with experimental results, favoring by this CARS over IRS and RiKE.

To get highest sensitivity for CARS and to use it for investigations of species under low concentrations, consideration of optimal resonance conditions for low concentrations ($\chi^{Ra} \simeq \chi^{n} + \chi^{e}$) has to be made.

In tuning the two exciting laser frequencies and by this automatically the third beam (CARS-signal) through the extinction curve of the molecule under investigation, different amplitudes and lineshapes for the CARS lines are detected. To calculate these CARS lines, different models had been used:

(1) a 3 or 4 level model, (2) a vibronic model for Franck-Condon or A-type resonance scattering (neglecting the influence of the interaction between the vibration under consideration and other vibrations of the model), and (3) a transformation technique yielding the expected values for the amplitude and lineshape of the Raman lines by the absorption spectrum.

In the 3 and 4 level model, CARS lines can occur as absorptions in the electronic background signal if the laser frequencies are at the low frequency side of the extinction curve (Fig. 1). At higher laser frequencies only emission signals as CARS signals are to be expected.

In the vibronic and transformation model as well, additional absorptions for the CARS lines can occur if the CARS pump radiations are positioned at the high frequency side of the extinction curve which have higher amplitudes than at the low frequency side (Fig. 2).

A comparison between these theories with experimental results are difficult, because in many cases during the tuning of the laser frequencies through the extinction curve a bleaching of the molecular one photon absorption takes place. This is due to a depopulation of the ground state and population of excited states. The population of excited states is dependent on the laser intensity and lifetime of the excited states. In many cases, the usually used laser powers are sufficient to depopulate most of the molecules from the ground state so that the main part of the spectrum is caused by molecules in excited states (2). Taking this possibility into mind, a comparison between calculated and measured excitation profiles gives assertions whether the CARS spectra are due to molecules in the ground or excited state.

Independent of details of theories, maximum CARS intensity should be expected for the frequency of the pump laser at the maximum of the extinction curve or at slightly lower frequencies. The theories have to be improved by including the possibility of the "non-Franck-Condon" resonance (B-type scattering) non-constant linewideth in vibronic states, influence of χ^e-terms, and inhomogeneous absorption profiles for molecules in order to get quantitative agreement with experimental results.

For our investigations of short living molecular states, we are using two different CARS arrangements schematically shown in Fig. 3 and 4. The excited states are populated either by 337 nm or 347 nm laser radiation or by the CARS pump beams themselves.

In comparing these arrangements with regard to their qualification to investigate short-living excited molecules, two main points must be considered: the time resolution and the possibility of adaptation to the resonance condition of the molecule under investigation. In the multiplex set-up, the pulse width of the ruby laser is 20 ns, with a possible time delay of 0-40 ns. This means that this arrangement is best fitted for the investigation of excited singlet states and short-living triplet molecules, but with one important limitation: the absorption of the excited states must be positioned in the range 600-700 nm due to the resonance enhancement condition. The main advantage of the multiplex set-up is given by possibility to record a CARS spectrum spread over 200 cm^{-1} with one laser pulse. This is important in investigating irreversible processes especially. In the narrow band case, the CARS pump radiation can be adapted to molecular absorptions positioned between 400 and 850 nm. The pulse width is usually

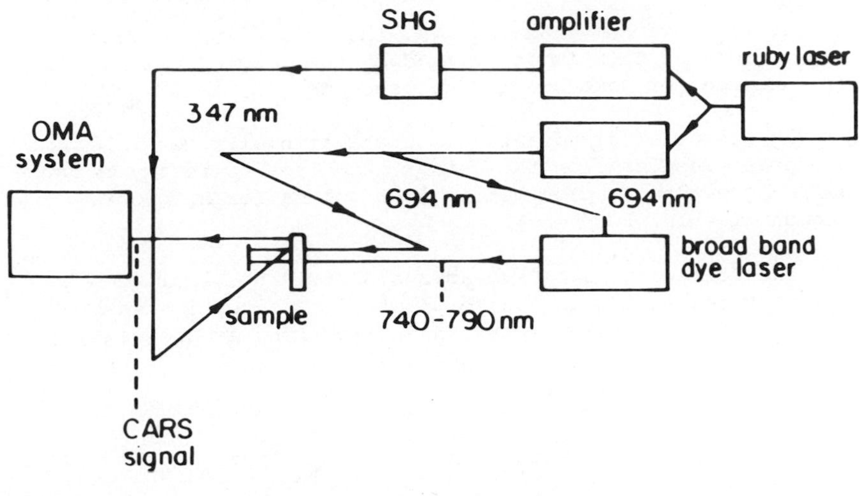
SHG
amplifier
ruby laser
347 nm
OMA
system
694 nm
694 nm
broad band
dye laser
sample
740-790 nm
CARS
signal

fig 3

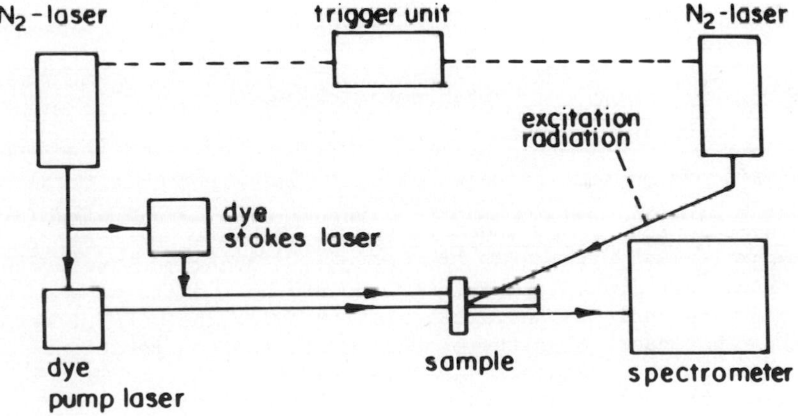
N_2-laser
trigger unit
N_2-laser
excitation
radiation
dye
stokes laser
sample
dye
pump laser
spectrometer

fig. 4

in the range of some nanoseconds, the delay time between excitation and CARS is determined by the trigger unit and is of the order of some hundred nanoseconds.

An essential disadvantage investigating irreversible processes results from the number of pulses necessary to get a CARS spectrum (approximately 1000 pulses for a spectral region of about 200 cm^{-1}).

The following table summarizes shortly the excited molecular states in liquid and solid matrices investigated by CARS spectra using multiplex and narrow band arrangements.

S_1 State	T_1 State	Exciplex
chrysene (3)	chrysene (2)	pyrene/
1,2,5,6-dibenzanthracene	diazachrysene	N,N-
	N-methylacridone	diethylaniline (1)
pyrene (4)	N-(n-propyl)-acridone	
rhodamine 6G (2)		
rhodamine B (2)		
4-dimethylaminobenzaldehyde		

Stable Radical

N-n-tetramethyl-p-phenylendiaminecation-radical

Discussing some of these results, it is possible to demonstrate the well known possibilities to obtain information about changes in bond order and structure by vibrational spectra. In particular in the case of chrysene, a detailed interpretation can be given, taking into account the results of a normal coordinate analysis and a calculation using the QCFF/PI method (5). The observed shifts of the CC-stretching vibrations arising in the resonance CARS spectra of the S_1 and T_1 molecule are in good agreement with the calculated wavenumbers and changes in bond lengths.

Summarizing the results it may be seen that resonance CARS is a well suited method in the field of time-resolved vibrational spectroscopy, giving advantages especially by the investigation of fluorescent short-living states e.g. molecules in the S_1 state.

REFERENCES

1. Pfeiffer, M., Lau, A., Werncke, W., and Holz, L., Optics Commu. 363, 41 (1982).
2. Lau, A., Konig, R., and Pfeiffer, M., Optics Commun. 32, 175 (1980).
3. Weigmann, H. J., Lenz, K., Lau, A., Pfeiffer, M., and Werncke, W., Chem. Phys. Lett. 73, 175 (1980).
4. Kaminski, T., Moriyama, R., Igerashi, R., Adschi, S., and Maeda, S., J. Chem. Phys. 3600, 75 (1980).
5. Jung, Ch., Lau, A., Werncke, W., and Pfeiffer, M., published in Chem. Phys.

RESONANCE CARS OF EXCITED MOLECULES[1]

Shiro Maeda
Haruhiko Kataoka
Toshio Kamisuki
Yukio Adachi

Research Laboratory of Resources Utilization
Tokyo Institute of Technology
Nagatsuta, Yokohama, Japan

INTRODUCTION

CARS is, alongside of CSRS, very suited for time-resolved spectroscopy in the intrinsic characteristics; the nonlinear dependence on the field makes the signal acquisition increasingly more efficient for pulsed lasers of shorter duration and higher peak power, in contrast with the ordinary Raman, and the coherent beam output delivers the required signal from the concurrent fluorescence and other hostile emissions, that may be met more often when studying short-lived excited species.

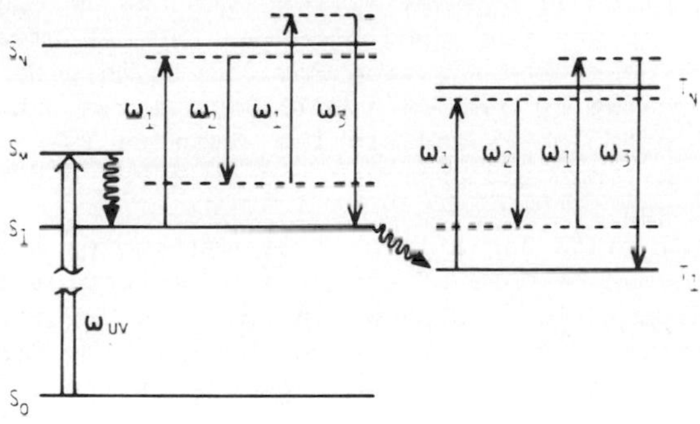

FIGURE 1. Resonance CARS schemes for excited states.

[1] *This work was supported by a Grant-in-Aid for Scientific Research No.56430006 from the Ministry of Education, Science and Culture, Japan.*

ISBN 0-12-066280-9

Complication caused by the non-resonant background is usually deemed to be an essential drawback, but it is not necessarily insurmountable and even can be effectively utilized in some cases.

Incorporation of the electronic resonance effect is very profitable, as in the spontaneous Raman, to relieve the required CARS signal from the rest. It is also of much significance in resonance CARS that the line shape changes **remarkably** depending upon the resonance conditions.

In this paper, those characteristic features of resonance CARS are demonstrated in the Raman spectroscopic study of electronic excited states of pyrene and some other systems in solution state.

METHOD OF OBSERVATION

The method consists of generating an excited state population by the pulsed irradiation of N_2 laser(337 nm, 10 nsec duration) and observing the CARS signal from the instantaneous population by scanning the ω_2 dye laser, with the ω_1 dye laser tuned close to the transient absorption, both pumped by the same N_2 laser as used for the excitation.(cf. Fig.1) Then, the observed CARS process was generated within a few nsec after the maximum of UV pulse. The input laser power was kept as low as possible, typically 30 kW of 337 nm focused to a spot of 1 mm diameter and 20 kW of dye lasers tightly focused, if a reasonably good S/N ratio is obtained.

LOWEST SINGLET(S_1) AND TRIPLET(T_1) STATES OF PYRENE

With the pulsed 337 nm irradiation, pyrene exhibits intense transient absorption in visible region as shown in Fig.2, which was recorded under the same conditions as the CARS measurement. The spectrum of deaerated solution(Fig.2a) is in good correspondence with the result by Post ét al.(1), who showed that the formation of triplet monomers and singlet excimers is neglisible in the present conditions and assigned the observed bands to the $S_N \leftarrow S_1$ transitions.

Since the S_1 state has a long lifetime(fluorescence lifetime $\sim$ 300 nsec), resonance CARS of the excited molecules is obtained in fairly good conditions as shown in Fig.3, where the signal light was collected through two stages of iris for discriminating from the intense fluorescence. The behavior of the resonance is obviously associated with the above $S_N \leftarrow S_1$ transitions, indicating the S_1 origin of CARS signals convincingly.

According to Richards et al.(2), most of the S_1 population produced above is rapidly transfered to the lowest triplet state (T_1) due to the effect of dissolved oxygen within about 20 nsec, yielding high T_1 population because of the much slower T_1 quenching. In accordance with this, the transient spectrum of air-saturated solution(Fig.2b), which was obtained at the same timing as

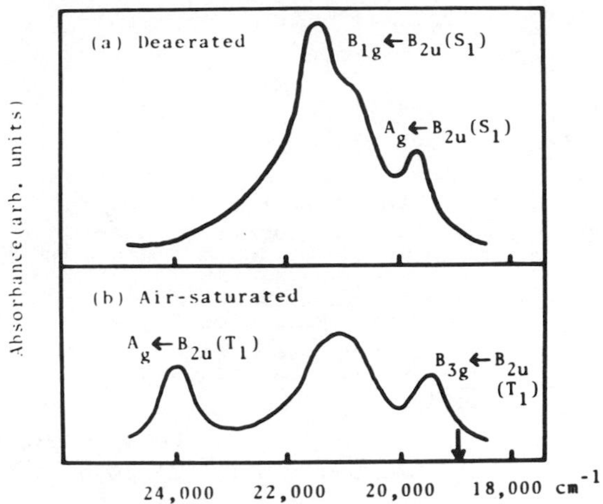

FIGURE 2. *Transient absorption spectra of pyrene in cyclohexane.*

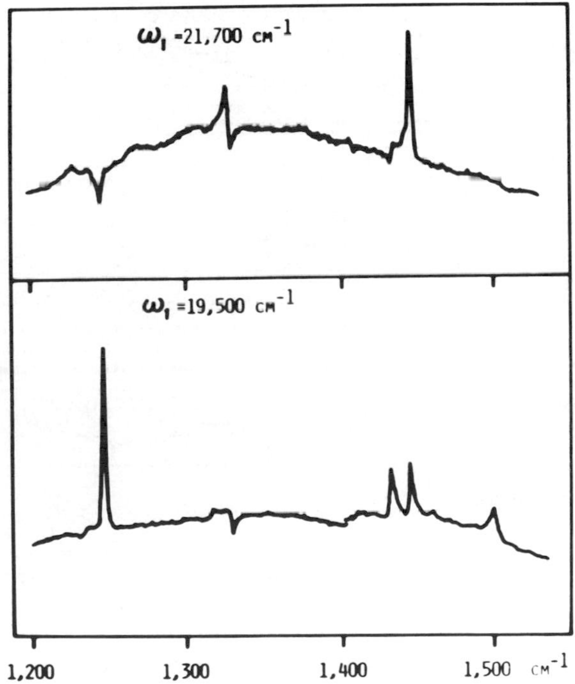

FIGURE 3. *Resonance CARS of pyrene S_1 state. 10^{-3} Mdm^{-3} pyrene in deaerated n-C_5D_{12} at $-70°C$.*

Fig.2a, shows an additional band at 24,000 cm^{-1} as well as a slight change due to another weak $T_N \leftarrow T_1$ absorption at 19,000 cm^{-1}.

The CARS spectra are shown in Fig.4 for ω_1 values near those $T_N \leftarrow T_1$ transitions. The lower frequency band appears to contribute resonance enhancement very little. Essentially equal spectra were obtained with a delay time longer than 100 nsec.

Observed S_1 and T_1 vibrational frequencies are listed in Table 1 along with the ground state and other related values. The S_1 values are in good agreement with absorption result at low temperature by Bree et al.(3). The T_1 frequencies as resonantly observed are considerably different from those of S_1, contrary to a simple consideration from their common parent electron configuration, but rather show close correspondence with pyrene anion. This observation, together with the fact that the π-bond order obtained by an elaborate calculation is very much alike between T_1 and anion, suggests closely similar bonding schemes in both states.

Fig.5 shows the resonant behavior of S_1 signals in two different polarization schemes. It is noticeable that the dispersive shape of 1433 cm^{-1} signal is reversed from the parallel(χ_{1111}) to perpendicular(χ_{1221}) at ω_1=21,200 cm^{-1} in contrast with 1445 cm^{-1} signal, whose behavior is understood by assuming Franck-Condon type mechanism of resonance associated with the two transitions in Fig.2a. It can be shown that the behavior of 1433 cm^{-1} signal

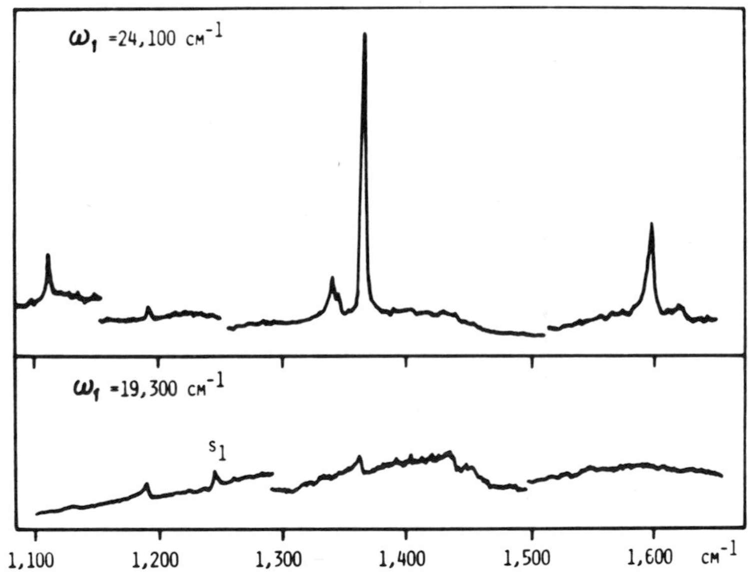

FIGURE 4. Resonance CARS of pyrene T_1 state. $10^{-3}Mdm^{-3}$ pyrene in air-saturated n-C_7H_{16} at R.T.

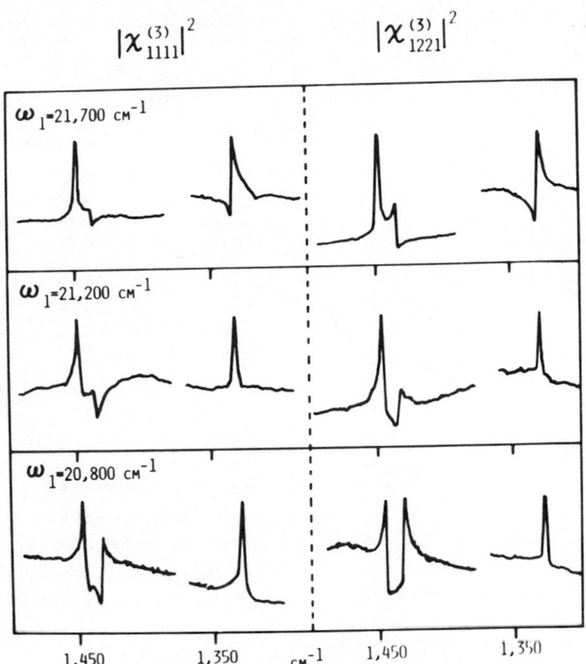

FIGURE 5. Resonance CARS line shapes of S_1 pyrene.

is very likely attributed to a B_{1g} mode, which vibronically couples those closely lying excited levels. Then, there should be a rather complicated 'anomalous' Raman polarization in the resonant region, because the two coupling states are in close proximity.

EXCIPLEX AND CT EXCITED STATE

When an electron donor, e.g., N,N'-diethylaniline(DEA) is added in pyrene solution, the photo-excitation as above gives rise to the formation of pyrene-DEA exciplex(4). The exciplex is intensely fluorescent and shows transient absorption corresponding to a combination of pyrene anion and DEA cation spectra, as shown in Fig.6. It is interesting to investigate the above spectral correspondence from the vibrational point of view.

Resonance CARS of pyrene-DEA system was observed in essentially the same way, as shown in Fig.7. Observed exciplex frequencies were practically equal to those of chemically prepared pyrene anion, as shown in Table 1, confirming that the nature of the exciplex is closely alike the ion pair. From the difference in the spectral features by solvent, it is also confirmed that the exciplex is readily dissociated into component ions in polar acetone solution.

Another example of exciplex spectrum is given in Fig.8 on tetramethyl paraphenylenediamine(TMPD)-biphenyl(BP) system in n-heptane. Here both the donor and acceptor signals were observed

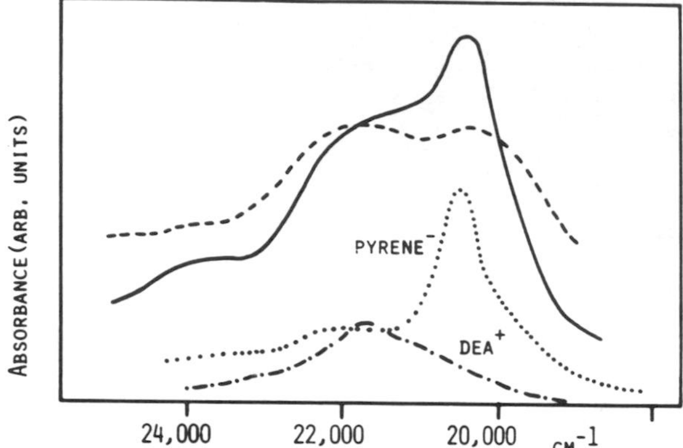

FIGURE 6. Absorption spectra of pyrene-DEA exciplex. Pyrene $5x10^{-4}$ Mdm^{-3}, DEA 0.1 Mdm^{-3}. —— in cyclohexane, --- in acetone.

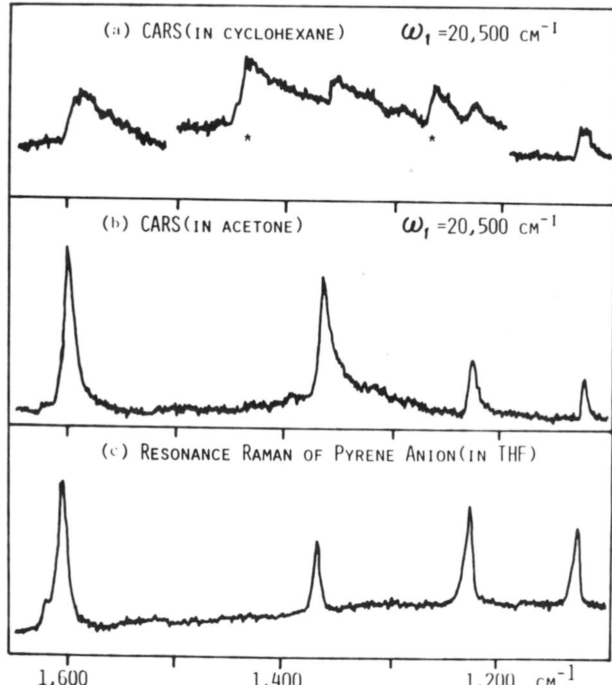

*FIGURE 7. Resonance CARS of pyrene-DEA exciplex, with resonance Raman of pyrene anion. * Solvent signals.*

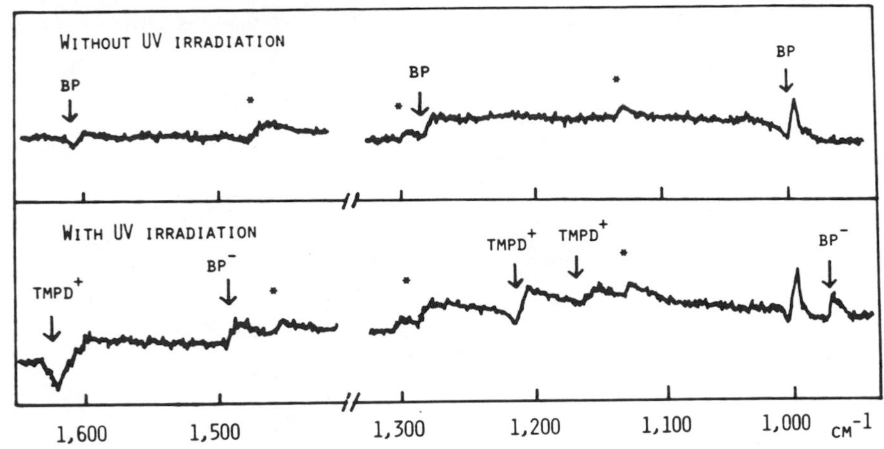

*FIGURE 8. Resonance CSRS of BP-TMPD exciplex in n-heptane. TMPD 2×10^{-3} Mdm^{-3}, BP 0.1 Mdm^{-3}. * Solvent signals.*

TABLE I. Raman Frequencies of Pyrene(cm^{-1})

State	S_1		T_1	*Pyr-DEA*	*Pyr⁻*	S_0
Mode of obs.	*CARS*	*Abs.*[a]	*CARS*	*CARS*	*Raman*	*Raman*[b]
	1565	1564	1595	1597	1604	1630(A_g)
	1500					1596(B_{1g})
	1445	1452				1408(A_g)
	1433					1373(B_{1g})
	1329	1339	1365	1362	1364	1352(A_g)
			1340			1243(A_g)
	1246	1249	1188	1223	1225	1144(A_g)
		1113	1110	1125	1123	1109(B_{1g})
	1032	1030				1067(A_g)

[a]*Ref.3.* [b]Bree, A., et al., *Spectrochim. Acta, 27A*, 2315(1971)

with the frequencies also very close to those of respective ions.

Quite a similar result was obtained on the charge transfer excited state of the weak complex between TCNB and benzene, where two intense CARS signals were observed at 1458 cm^{-1} and 1273 cm^{-1} in close coincidence with $TCNB^-$.

REFERENCES

1. Post, M., Langelaar, J., and Van Voorst, J., *Chem. Phys. Lett. 10*, 468 (1971).
2. Richards, J., West, G., and Thomas, J., *J. Chem. Phys., 74*, 4137 (1970).
3. Bree, A., and Vilkos, V., *Spectrochim. Acta, 27A*, 2333 (1971).
4. Potashnik, R., Goldschmidt, C., and Ottolenghi, M., *J. Chem. Phys., 55*, 5344 (1971).

SURFACE ENHANCEMENT MECHANISMS

Richard K. Chang

Section of Applied Physics
and
Center for Laser Diagnostics
Yale University
New Haven, Connecticut

INTRODUCTION

Since initial observation of surface enhanced Raman scattering (SERS) on roughened Ag surfaces (1,2), time-resolved vibrational spectroscopy (TRVS) of molecular adsorbates on metallic surfaces has been reported by only a few research groups. Picosecond Raman gain spectroscopy has been applied to CN^{-1} and carbon-related adsorbates on roughened Ag surfaces (3,4). Other groups have reported the observation of nanosecond hyper-Raman spectroscopy of SO_3^{2-} adsorbates on Ag powder (5) and picosecond spontaneous Raman spectroscopy of CN^{-1} adsorbates on roughened Ag electrodes beyond laser-induced surface damage and desorption limits (6).

In contrast, publications reporting non-TRVS involving SERS of many different molecular adsorbates on roughened Ag, Au, and Cu surfaces are presently very extensive. The reasons for the limited reports on TRVS of molecular adsorbates on metal surfaces which exhibit SERS may be twofold. First, the intense laser fields associated with TRVS can destroy the surface morphology which gives rise to surface enhancement and/or can induce molecular desorption and even molecular decomposition. Second, existing controversies about SERS mechanisms and uncertainties in the enhancement factor for several of the less controversial mechanisms may have delayed the merging of SERS and TRVS. The purpose of this article is to briefly review the SERS mechanisms in the hope that experimentalists engaged in TRVS will find it easier to adapt various SERS techniques to their research and be stimulated to read some of the recent literature which more completely summarizes the status of SERS (7-9).

ISBN 0-12-066280-9

THEORY

From an experimental point of view, the metal surface is best classified as a surface having deliberately been made smooth or roughened. However, it should be noted that microscopic roughness on a "smooth surface" is often not detectable with conventional scanning electron microscopes which have a spatial resolution of ∿250 Å. Following is a short description of the physical origin of each enhancement mechanism associated with ideally smooth and intentionally roughened surfaces.

A. Smooth Surfaces

1. Image Force Effect. The original dipole plus the image dipole can give rise to a very large electronic polarizability derivative relative to the molecular vibration coordinate (i.e., the Raman scattering cross section) when the molecule-metal separation distance is on the order of a few angstroms (10). Depending on the assumptions used in this model, the enhancement factor for molecules some 1.6 Å away from the Ag "effective boundary" can range anywhere from 1 - 10^6.

2. Reflectivity Modulation. Physisorbed or chemisorbed molecules modulate the electrons within the metal and hence the metallic reflectivity at the discrete molecular vibrational frequencies (11). Quantitative and reliable estimations of the enhancement factor have not yet been made (12).

3. Propagating Surface Plasmons. By appropriately matching the light propagation vector along the surface of the metal, thin film, or grating, propagating surface plasmon waves can be excited. The electromagnetic field on the metal-air interface becomes large when the incident laser beam excites such surface plasmon waves (13,14). Furthermore, the propagating surface plasmons can also increase the Raman radiation at a specific angle which is commensurate with the propagation vector-matching conditions for the Stokes shifted wave. The overall enhancement for the inelastic radiation can be as large as 100 times.

B. Roughened Surfaces

1. Increased Adsorbate Density. Surface roughness increases the effective surface area relative to the geometric area of the metal film or electrode. However, by limiting the electrochemical oxidation current [e.g., to less than (50 mC/cm^2)], the effective surface area is not more than 2 times the geometric area (15). With successive anodization, the effective surface can indeed increase by several orders of magnitude and hence larger numbers of molecular adsorbates can result.

2. Localized Surface Plasmons. In the electrostatic limit, which assumes the size of the metallic particle, a, to be much smaller than the incident wavelength, λ_o, (i.e., $a < 0.01\ \lambda_o$), the surface-averaged electromagnetic intensity can be extremely large when the incident photon energy, $\hbar\omega$, is tuned to the localized surface plasmon resonance mode. For small micro-objects such as spheres, prolate spheroids, and oblate spheroids embedded in a homogeneous medium, the enhancement factor for the incident and Raman shifted photon energies can be readily calculated and has been estimated to be in the $10^6 - 10^{11}$ range for Ag spheroids with large ratios of major and minor axes (2a/2b) (16,17). Depending on 2a/2b and the refractive index of the surrounding medium (n_{sur}), the intensity at the tips of the spheroid, and to a lesser extent the surface-averaged intensity, rapidly increase with 2a/2b and n_{sur}. The Raman scatterer is not required to be in direct contact with the surface but can be located as far away as the outer limit of the enhanced electromagnetic field that can polarize this molecule. The Raman dipole can polarize the metallic micro-object, which can also enhance the Raman radiation reaching the detector.

In the electrostatic limit, the enhancement factors are shape dependent only and not size dependent, with the proviso that the dielectric constant $[\varepsilon(\omega) = \varepsilon_1(\omega) + \varepsilon_2(\omega)]$ of the microstructure is equal to that of bulk. Metals such as Ag, Cu, Au, and Al are quite unique because $\varepsilon_1(\omega)$ is in the -10 to -1 range in the visible part of the electromagnetic spectrum and $\varepsilon_2(\omega)$ is remarkably small. Although other Group VIII metals have $\varepsilon_1(\omega)$ in the -10 to -1 range, their $\varepsilon_2(\omega)$ values are too large for large electromagnetic enhancement on the microstructure surface. Consequently, while the localized dipolar surface plasmon resonances can be excited for most metals, only those with low $\varepsilon_2(\omega)$ will have a sufficiently high Q to enhance the surface-averaged intensity and thus act as an efficient receiving antenna at the incident wavelength and as a transmitting antenna at the Stokes wavelength. When the smallest dimension of the microstructure is less than the electron mean free path in the bulk, $\varepsilon_2(\omega)$ for the microstructure will become larger than that for the bulk and the overall electromagnetic enhancement of the microstructure surface will be less.

When the dimensions of the microparticle exceed $a > 0.01\ \lambda_o$, the electrodynamic solution must be used to calculate the surface-averaged intensity or the intensity at any zone on the spheroid surface (18). The multipole contributions with their appropriate phase retardation add destructively to the strong dipolar term. Consequently, the surface-averaged intensity and the intensity at any zone on the surface rapidly decrease as the dimension increases (18). The overall electromagnetic enhancement factors for the incident and Raman radiation have been solved only for spheres as a function of radius and wavelength (19,20), while the enhancement factor for the incident radiation has been solved for a spheroid (with $2a/2b = 2$) as a function of a and wavelength when the spheroid is immersed in air, water, or cyclohexane (18). The latter calculations have been compared with the experimental results of CN^- adsorbed on a uniform array of Ag spheroids ($2a/2b = 2$, $a = 100$ nm), evaporated on SiO_2 posts lithographically produced, and immersed in these three surrounding media (21).

3. Charge-Transfer Process. The enhancement factor associated with the charge-transfer process involving metallic adatoms or clusters of adatoms with adsorbed molecules has not yet been quantified. Nevertheless, this process is receiving increasing attention (22). The two essential concepts which are invoked in the charge-transfer process are: First, because the adatoms or clusters have such small dimensions, intraband transitions within the sp-band need not be restricted to be vertical in the $E(\vec{k})$ dispersion curve of the metal, and interband transitions need not initiate from the bulk d-level but can initiate from the d-level (located closer to the Fermi level) associated with the adatoms/clusters (22). Second, because the photoemitted electrons from the metal can be trapped by the adsorbed molecule to form a temporary negative molecular ion (23), the charge-transfer process resembles a resonance Raman process in which the photon directly excites the electron from the ground state into the first excited electronic state.

In Fig. 1, the incident photon creates an electron (solid circle) and a hole (open circle) in the metal. The momentum selection rule is relaxed because of the small dimensions of the metal microstructures, and intraband transitions (filled sp-band $\leftrightarrow$ empty sp-band above the Fermi level E_F) are increased. Even though the photoexcited electron has less energy than the work function energy, ϕ, this electron can charge transfer into the adsorbed molecule (with nuclei coordinate, Q) to form a temporary negative molecule. The electron can also charge transfer back into the metal and radiatively recombine with the hole while the adsorbed molecule is left in the excited vibrational state (V=1) of the ground electronic state. The energy alignment between E_F and the molecular states is a function of chemisorption and of the applied potential (relative to V_{SCE}) at the electrode-electrolyte interface. Similar processes can take place for photon-created electron-hole pairs which originate in the d-level (with energy E_d). However, unless the incident photon has considerably larger photon energy (i.e., in the near UV), these photoexcited electrons involving the d-level will have energy much smaller than ϕ and the probability for charge transfer will consequently be smaller than in the case of photoexcited electrons which originate in the empty sp-band.

The main difference between Raman spectroscopy and electron energy loss spectroscopy (EELS) is the origin of the low energy electrons. Photoemission from the metal adatoms/clusters provides the electrons needed for the SERS process; for the EELS process the electrons are produced by the electron gun. Both forms of

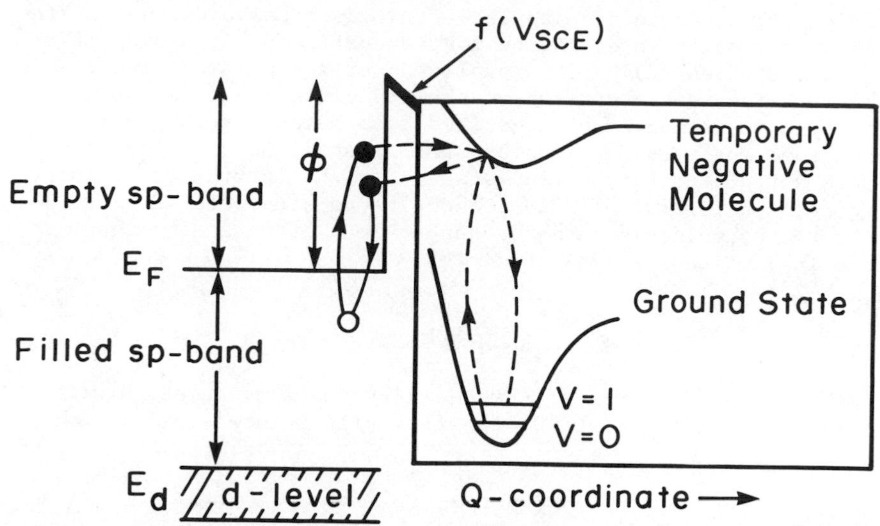

FIGURE 1. Schematic representation of the charge-transfer process.

spectroscopy bear close resemblance otherwise and together can provide complementary information on the enhancement mechanisms (24).

CONCLUSION

TRVS in biological samples exploits the resonance Raman effect to increase the Raman signal and to localize the optical probe within the more interesting portion of the molecule with large molecular weight. Along with resonance Raman effect, interfering fluorescence effects often occur and provide a source of noise. The adaptation of SERS techniques which have mainly been limited to simpler molecules with smaller molecular weight can indeed further increase the resonance Raman signal because of the surface enhanced electromagnetic fields and can also increase the Raman polarizability because of the charge-transfer process. Furthermore, because of the metal-molecular interaction, the nonradiative decay channels, such as those that can excite localized and propagating surface plasmons, can decrease the undesirable fluorescence emerging from resonant Raman adsorbates. Therefore, metallic electrodes in electrolytes, colloids in aqueous solutions, island films in UHV or air, and optionally sized and shaped uniform microstructures on a glass slide can all serve as convenient surface enhancing substrates for TRVS studies. In particular, biological molecules adsorbed on colloids, which are flowing from a syringe, may be most amenable for combining TRVS and SERS studies (25). The localized surface plasmons and charge-transfer processes occurring on the colloid surface can be more effective in producing photoexcited adsorbates. Furthermore, these two processes can increase the Raman signals during the subsequent photointerrogation stage. In the future, we envision frequent application of SERS techniques to study the TRVS of photoexcited molecules and resonance Raman effects of large biological molecules (26).

ACKNOWLEDGMENT

The partial support of this work by the Army Research Office (Grant No. DAAG29-82-K-0040) is gratefully acknowledged.

REFERENCES

1. Fleischmann, M., Hendra, P. J., and McQuillan, A. J., Chem. Phys. Lett. 26, 163 (1974).
2. Jeanmaire, D. L. and Van Duyne, R. P., J. Electroanal. Chem. 84, 1 (1977).
3. Heritage, J. P. and Bergman, J. G., Opt. Commun. 35, 373 (1980).
4. Heritage, J. P., Bergman, J. G., Pinczuk, A., and Worlock, J. M., Chem. Phys. Lett. 67, 229 (1979).
5. Murphy, D. V., Von Raben, K. U., Chang, R. K., and Dorain, P. B., Chem. Phys. Lett. 85, 43 (1982).
6. Voss, D. F., Paddock, C. A., and Miles, R. B., Appl. Phys. Lett. 41, 51 (1982).
7. Van Duyne, R. P., in "Chemical and Biochemical Applications of Lasers," Vol. 4 (C. B. Moore, ed.), p. 101. Academic Press, New York, (1979).
8. Furtak, T. E. and Reyes, J., Surf. Sci. 93, 351 (1980).
9. "Surface Enhanced Raman Scattering" (R. K. Chang and T. E. Furtak, eds.), Plenum Press, New York, (1982).
10. Schatz, G. C., in "Surface Enhanced Raman Scattering" (R. K. Chang and T. E. Furtak, eds.), p. 35. Plenum Press, New York, (1982).
11. Otto, A., Timper, J., Billmann, J., Kovacs, J., and Pockrand, I., Surf. Sci. 92, L55 (1980).
12. McCall, S. L. and Platzman, P. M., Phys. Rev. B 22, 1660 (1980).
13. Chen, Y. J., Chen, W. P., and Burstein, E., Phys. Rev. Lett. 36, 1207 (1976).
14. Jha, S. S., in "Surface Enhanced Raman Scattering" (R. K. Chang and T. E. Furtak, eds.), p. 129. Plenum Press, New York, (1982).

15. Bergman, J. G., Heritage, J. P., Pinczuk, A., Worlock, J. M., and McFee, J. H., Chem. Phys. Lett. 68, 412 (1979).
16. Metiu, H., in "Surface Enhanced Raman Scattering" (R. K. Chang and T. E. Furtak, eds.), p. 1. Plenum Press, New York, (1982).
17. Gersten, J. I. and Nitzan, A., in "Surface Enhanced Raman Scattering" (R. K. Chang and T. E. Furtak, eds.), p. 89. Plenum Press, New York, (1982).
18. Barber, P. W., Chang, R. K., and Massoudi, H., to be published.
19. Messinger, B. J., Von Raben, K. U., Chang, R. K., and Barber, P. W., Phys. Rev. B 24, 649 (1981).
20. Kerker, M., Wang, D. S., and Chew H., Appl. Opt. 19, 4159 (1980).
21. Liao, P. F., Bergman, J. G., Chemla, D. S., Wokaun, A., Chem. Phys. Lett. 82, 355 (1981).
22. Otto, A., in "Light Scattering in Solids III" (M. Cardona and G. Güntherodt, eds.), Springer-Verlag, Heidelberg, in press.
23. Schulz, G. J., in "Principles of Laser Plasmas" (George Bekefi, ed.), p. 33. Wiley, New York, (1976).
24. Schmeisser, D., Demuth, J. E., and Avouris, Ph., Chem. Phys. Lett. 87, 324 (1982).
25. Terner, J., Strong, J. D., Spiro, T. G., Nagumo, M., Nicol, M., and El-Sayed, M. A., Proc. Natl. Acad. Sci. USA 78, 1313 (1981).
26. Cotton, T. M., J. Am. Chem. Soc. 102, 7960 (1980).

TIME RESOLVED STUDIES OF ELECTROCHEMICAL SERS: THE PYRIDINE / Cl^- / Ag MODEL SYSTEM

A.M. Stacy
R.P. Van Duyne

Department of Chemistry
Northwestern University
Evanston, Illinois

ABSTRACT

Both growth and decay transients of the SERS intensity of the 1008 cm^{-1} band of pyridine adsorbed on a silver electrode occur in response to the application of a double potential step waveform. The final potential E_1 of the applied waveform determines the shape of the transient response. If E_1 is chosen negative of -0.7V vs. SCE, a transient decay is observed while for E_1 positive of -0.7V, the signals only grow in time. Contributions to the origin of electrochemical SERS transients include: 1) relaxation of metastable surface roughness; 2) adsorption/desorption kinetics; and 3) restructuring of the adsorbate overlayer with respect to the substrate. All of these processes are probably potential, laser wavelength and intensity, and impurity dependent.

INTRODUCTION

An avalanche of both experimental and theoretical work on surface enhanced Raman scattering (SERS) has been directed toward understanding the 10^6 fold enhancement in the apparent Raman scat-

ISBN 0-12-066280-9

tering cross section for molecules adsorbed on silver surfaces. Since the initial experiment of pyridine adsorbed on a silver electrode [1] many systems have been characterized, including many interfacial situations and a wide variety of adsorbates [2,3]. In spite of this extensive research, SERS has been studied almost exclusively under steady state conditions.

In this paper we will focus exclusively on time-resolved studies of the SERS intensity for the 1008 cm^{-1} band of pyridine adsorbed on a silver electrode in response to the application of a double potential step waveform. We have observed both growth and decay transients of SERS immediately following a standard anodization procedure used to create submicron surface roughness.

Jeanmaire and Van Duyne [1] reported the first application of SERS for the investigation of a time-dependent surface process. They monitored the time response of the intensity due to the 1008 cm^{-1} band of pyridine adsorbed on a silver electrode for a potential waveform consisting of a forward step from -0.1V to -0.6V vs. SCE and subsequent return after 10 ms. In contrast to a bulk diffusional response, the time profile for SERS tracks the excitation waveform with only a small amount of time lag, attributed to the double layer charging time. Recently, Pemberton and Buck [4] have analyzed the time dependence of a surface enhanced resonance Raman scattering (SERRS) signal to obtain quantitative adsorption parameters and adsorption rates for dithizone anion adsorption on silver.

Chen *et al* [5] studied the SERS intensity for pyridine on silver as a function of time immediately following the completion of an anodization cycle. They report a large rise in the SERS intensity followed by a rapid decay in response to a potential step from the anodization potential to -0.8V vs. SCE. The work reported here explores this transient response in more detail and demonstrates that the observed time profile is sensitive to the specific double potential step waveform applied. Such transient behavior suggests that under carefully chosen anodization conditions the enhancements obtained in a transient mode may be even larger than the 10^6 steady-state enhancements [6].

EXPERIMENTAL

Solutions of 0.05 M pyridine and 0.1 M KCl were used for the SERS experiments. Spectrophotometric grade pyridine (99 + %) was obtained from Aldrich Chemical Company and used both directly from the bottle and after a triple distillation with no observable differences. Solutions were prepared with distilled, deionized water (Milli - Q, Millipore Corp.) and were deaerated by bubbling with prepurified nitrogen.

The silver electrode (~99.98% silver from Goldsmith, Co.) was mechanically polished with 0.05 micron alumina from Buehler LTD. The electrode was placed in the spectroelectrochemical cell, described elsewhere [6], with a standard calomel reference electrode and held at an initial potential E_1 between -0.2V and -1.0V. A potential step waveform from the chosen initial potential to E_2 = +0.15V was applied to the silver electrode with the pyridine present and the laser illuminating the electrode. The current passed in the forward step was integrated and the potential was returned to E_1 after 25 mCoul/cm^2 of charge were passed.

The time-evolution of the 1008 cm^{-1} band of pyridine adsorbed on the silver electrode was followed both during and after the anodization pulse. The Raman spectrometer and data collection system are described in detail elsewhere [6]. All Raman experiments were excited with the 647.1 nm line of a krypton ion laser (Coherent Radiation Model CR-500K). Since the ring breathing mode of pyridine shifts 2-3 cm^{-1} with potential changes, a 6 cm^{-1} spectrometer band pass was used. For 10mW 647.1nm illumination with a 6 cm^{-1} band pass, a neutral density filter of 0.5 OD was required directly in front of the monochromator to reduce the number of counts per second (cps) below the present count rate limit of our PMT protection system (viz., I_{max} = 60,000 cps). The counting intervals used to capture transient behavior were 0.01 and 0.1 seconds. However, for convenient comparisons, intensities will be quoted as counts per second.

RESULTS

The SERS time response of the 1008 cm^{-1} band of pyridine adsorbed on a silver electrode, the time response of the background, and the applied potential waveform are shown in Fig. 1. For the

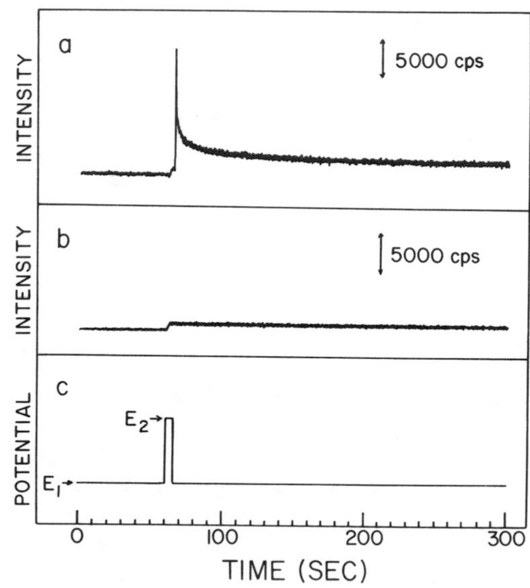

Figure 1

a) *SERS time response of the intensity due to the 1008 cm^{-1} band of pyridine adsorbed on a silver electrode where the initial potential E_1 is -0.85V vs. SCE and the step potential E_2 is to +0.15V. After 25mCoul/cm^2 of charge are passed, the potential is returned to E_1. The solution contains 50 nM pyridine and a 0.1M KCl supporting electrolyte. Laser excitation wavelength = 6471 Å. The counting interval = 0.1 counts per second.*

b) *Time response of the background intensity under the same conditions.*

c) *Double potential step waveform input to the potentiostat.*

first 60 seconds, the intensity of the 1008 cm^{-1} band on the freshly polished electrode at -0.85V was only a few hundred counts per second above the background. After passing 25 mCoul/cm^2 charge at +0.15V and returning to -0.85V, a large transient response of the SERS intensity at 1008 cm^{-1} is observed, over a time spread of a few hundred seconds, similar to that reported by Chen et al [5]. The background intensity did not show any transient behavior within the limits of our detection.

For better time-resolution, the transient response can be observed with a 0.01 second counting interval as shown in Fig. 2.

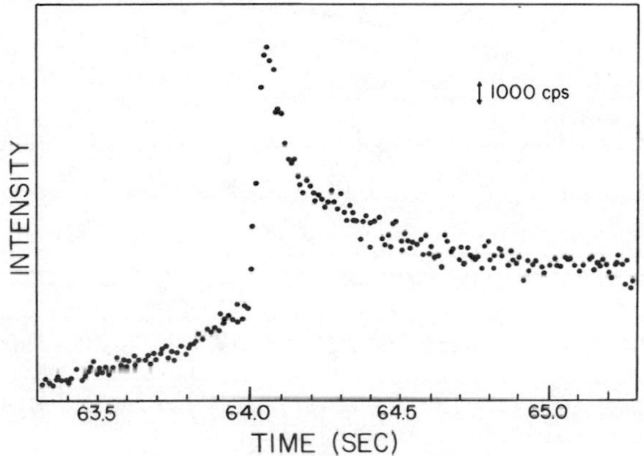

Figure 2

SERS time response of the intensity due to the 1008 cm^{-1} band of pyridine adsorbed on a silver electrode. The conditions are identical to those of Fig. 1 except the transient is shown in more detail on a 0.01 second counting interval.

The initial large intensity at 1008 cm^{-1} decays to one-half of the peak intensity in 0.2 seconds. The data can be fit to a double exponential decay suggesting that at least 2 processes are contributing to the observed behavior [7]. The detailed nature of the decays is currently under further study.

The variation in the shape of the transient response can be correlated to the choice of the final potential E_1 as displayed in Fig. 3. If E_1 is chosen negative of -0.7V, transient decay is

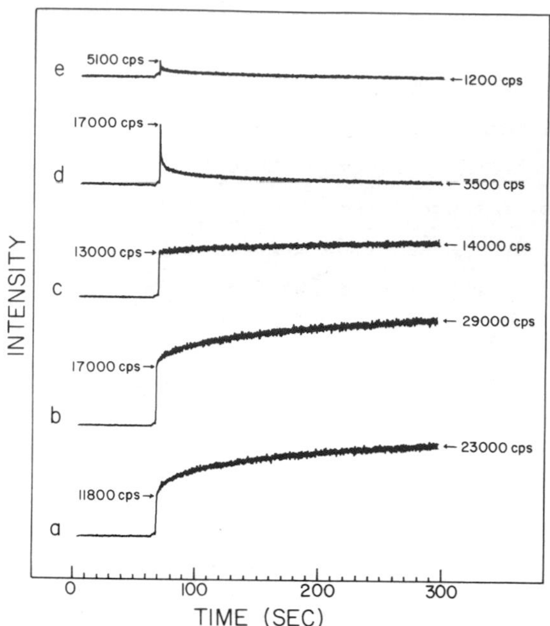

Figure 3

SERS time response of the intensity due to the 1008 cm^{-1} band of pyridine adsorbed on a silver electrode. The conditions are the same as Fig. 1 except the initial and final potential E_1 is varied.
(a) $E_1 = -0.4V$. (b) $E_1 = -0.6V$. (c) $E_1 = -0.7V$.
(d) $E_1 = -0.85V$. (e) $E_1 = -1.0V$.

observed. In contrast, for E_1 positive of -0.7V, the signals only grow with time. The intensity of the 1008 cm^{-1} band for $E_1 = -0.6V$ grows from 17000 cps directly following the anodization to 29000 cps in 300 seconds. $E_1 = -0.7V$ is a special case for which the intensity profile tracks the potential step from +0.15V to -0.7V and there is neither a transient decay nor a significant growth in intensity.

The long-time count rates at t = 300 seconds, noted in Fig. 3, also vary with E_1. The largest intensity of the 1008 cm^{-1} band is observed for $E_1 = -0.6V$ (viz., 29,000 cps), while the counts obtained in the steady state for $E_1 = -.0.85V$ (viz., 3500 cps) are almost an order of magnitude lower. In contrast to these results, Billman and Otto [8] report a maximum intensity for the 1008 cm^{-1} band at -0.85V for 647.1 nm excitation.

In one further experiment, we observed an increase in the intensity of the 1008 cm^{-1} band if we apply a single potential step waveform from -0.6V to -0.85V to an electrode previously anodized with E_1 = -0.6V, thereby confirming Billmann and Otto's results. However, the intensity at -0.85V then shows a rapid transient decay, essentially identical to the transient in Fig. 3d. A subsequent return of the potential to -0.6V does not recover the original intensity. The decay is similar to the quenching of SERS that has been observed after the application of a cathodic potential waveform [9-12]. Although this quenching of SERS has been reported as irreversible, we find that we can recover the original intensity at -0.6V by applying a potential step waveform from -0.6V to -0.2V and back to -0.6V.

DISCUSSION

The observed growth and decay transients (Fig. 3) for SERS intensities suggest that the initial surface microstructure of the electrode generated by double potential step anodization is not stable. We propose that during the time interval following the anodization pulse and before steady state conditions are achieved, changes occur in the surface particle size/shape distribution and/or in the adsorption geometry of the pyridine/Cl^- monolayer. The SERS intensities are very sensitive to these changes.

Studies of the electrodeposition of silver indicate that the surface microstructure is strongly potential dependent [13,14]. The more negative the deposition potential E_1, the larger the nucleation density of silver particles; with a large nucleation density, the growing silver particles coalesce to produce a smooth surface. Perhaps the silver deposited at -0.6V results in submicron surface roughness with silver particles of the proper size and shape to give intense SERS; whereas, for E_1 = -1.0V, a smoother surface is created and therefore less intense SERS. In order to test this hypothesis, it will be necessary to obtain a series of scanning electron micrographs for surfaces subjected to the same anodization conditions used to generate the data in Fig. 3.

One can envision a number of possible ways that adsorbed pyridine/Cl^- could rearrange with time thereby yielding SERS transients. The SERS intensities are expected to change as adsorption/desorption processes vary the surface excess of pyridine or as pyridine on the surface migrates to or from SERS active sites. Two classes of trace impurities can disrupt the adsorption geometry of the pyridine/Cl^- monolayer: 1) other adsorbates can compete with pyridine for surface sites or 2) trace metal cations can be electrodeposited onto the silver surface. Finally, laser illumination may induce photochemical processes that restructure both the surface microstructure and the pyridine/Cl^- overlayer.

Perhaps the most important aspect of this study is the clear demonstration that it is necessary to control several parameters simultaneously in order to obtain a reproducible, high intensity SERS active surface. The most significant anodization parameters to control include: 1) the waveform of the applied potential; 2) the amount of charge passed; 3) the amount of charge recovered; 4) the illumination of the electrode with the laser; 5) the bulk concentrations of the adsorbate and supporting electrolyte; and 6) the presence of trace impurities. A systematic study varying these parameters is in progress in order to more fully elucidate the mechanistic details of time-resolved SERS.

ACKNOWLEDGMENTS

This research was supported by the National Science Foundation (Grant Nos., CHE78-24866 and CHE82-05801) and the Office of Naval Research (Contract No. N00014-79-C-0794).

REFERENCES

1. D.L. Jeanmaire and R.P. Van Duyne, *J. Electroanal. Chem. 84*, 1 (1977).

2. R.K. Chang and T.E. Furtak, Eds., "Surface Enhanced Raman Scattering," Plenum Press, New York, 1982.

3. H. Seki, *J. Electron Spec. Relat. Phenom.*, in press.

4. J.E. Pemberton and R.P. Buck, *J. Electroanal. Chem. 136*, 201 (1982).

5. C.K. Chen, T.F. Heinz, D. Ricard and Y.R. Shen, *Chem. Phys. Letts. 83*, 455 (1981).

6. R.P. Van Duyne, in "Chemical and Biochemical Applications of Lasers," Vol. 4, C.B. Moore, Ed., Academic Press, New York, 1979, pp. 101-185.

7. B.K. Johnson, private communication.

8. J. Billman and A. Otto, *Sol. St. Comm. 44*, 105 (1982).

9. L. Moerl and B. Pettinger, *Sol. St. Comm. 43*, 315 (1982).

10. T. Watanabe, "Proceedings of the Surface Enhanced Raman Scattering Symposium," *Inst. of Sol. St. Phys.*, Univ. of Tokyo, Dec. 17-18, 1981, p. 13.

11. M.J. Weaver, F. Barz, J.G. Gordon II, M.R. Philpott, preprint.

12. H. Wetzel, H. Gerischer, B. Pettinger, *Chem. Phys. Lett. 80*, 159 (1981)

13. M. Fleischmann and H.R. Thirsk, in "Advances in Electrochemistry and Electrochemical Engineering," Vol. 3, P. Delahay and C.W. Tobias, Eds., John Wiley and Sons, New York, 1963. pp. 123-210.

14. G. Gunawardena, G. Hills, I. Montenegro and B. Scharifker, *J. Electroanal. Chem. 138*, 225 (1982).

COMBINED SURFACE ENHANCED AND RESONANCE RAMAN SCATTERING BY DABSYL ASPARTATE ON COLLOIDAL SILVER

O. Siiman
L.A. Bumm
R. Callaghan
M. Kerker

Department of Chemistry
Clarkson College of Technology
Potsdam, New York

Excitation profiles of enhanced Raman bands (SERS) of citrate adsorbed on Carey Lea silver hydrosols varied with sol preparation and with fractions collected either after centrifugation or from chromatography of the sol on a polyacrylamide gel column. The various sol samples differed in particle size distribution and absorption spectroscopy. The initial red sol (1) which was quite aggregated gave a very broad maximum enhancement $6x10^5$ in the 460-680 nm excitation region. Newly prepared yellow sols consisted mainly of unaggregated, nearly spherical particles which had diameters in the 10 to 40 nm range. These preparations and ultracentrifuged samples prepared from them showed somewhat sharper but still quite broad SERS maxima in the 475-525 nm range. A chromatographed sample prepared from the yellow sol

ISBN 0-12-066280-9

showed a sharp peak in the SERS excitation profile at 490 nm. SERS drops off sharply in all cases beyond 650 nm in the red and between 406.7 and 547.9 in the violet. Analysis of adsorbed citrate by a radioactive exchange technique has provided values of absolute SERS enhancement.

Citrate was displaced on the silver surface by the chromophoric dicarboxylate, dabsyl aspartate (2) (dabsyl ≡N-4-dimethylaminobenzene-4'-sulfonyl), in order to observe both resonance Raman (RRS) and SERS as well as the interaction of these two effects. In basic solutions this chromophore absorbs maximally at 472 nm. Thus, the combined RRS-SERS in the blue green, SERS alone in the red and RRS in solution could be observed for comparison. Two silver sols with adsorbed citrate, the chromatographed and the centrifuged yellow sol, were treated in this way with dabsyl aspartate. With the former sol addition of dabsyl aspartate split the original 400 nm absorption band into two components, one at 385, another at 525 nm. A similar effect in silver-pyridine hydrosols was attributed (3) to aggregation of particles into chains. In the region, 450 to 515 nm, the RRS-SERS spectra closely resembled the RRS spectra of dabsyl aspartate and other (4) azobenzene derivatives. As the excitation wavelength was moved progressively further to the red into the region of SERS only spectra, some major changes were observed. The azo group ν(N=N) band at 1417 cm^{-1}, which showed the greatest intensity in the RRS spectrum became steadily weaker. The 1139 cm^{-1} phenyl ring band grew in intensity to be the most intense band. A new substituent sensitive phenyl group band appeared at 746 cm^{-1} in addition to one already present at 820 cm^{-1}. Qualitatively, the centrifuged sol with added dabsyl aspartate showed similar Raman spectra. However, this sol did not exhibit any change in the absorption band at 400 nm upon addition of dabsyl aspartate. The concentration of dabsyl aspartate adsorbed to the silver particle surface was obtained by measuring the difference in absorbance at 472 nm between a blank consisting of an aqueous solution of dabsyl aspartate and the supernatant of the sol after removal of the silver particles containing adsorbate by ultracentrifugation at 40,000 rpm. A quantitative estimate of the SERS enhancement factor at various excitation wavelengths was then obtained by comparing integrated intensities of phenyl and azo group Raman bands near 1400 cm^{-1} for sol and blank, both containing the same concentration of dabsyl aspartate. The 1415 cm^{-1} Raman band of 1.35 M aqueous sodium citrate was also used as an external standard each time. The SERS excitation profiles peaked near 514.5 nm in each of the sol plus dabsyl aspartate systems. SERS enhancement factors of 1400 and 2200 above the RRS enhancement were obtained at 514.5 nm for respective centrifuged and chromatographed sols to which dabsyl aspartate had been added. These values are about two orders of magnitude less than the maximum measured enhancements of 2 to $4x10^5$ for citrate adsorbed on the same particles.

Our observed SERS enhancement with citrate on colloidal silver agrees well with the predicted one from electrodynamic theory calculations (5) for silver spheres of 5 nm radius. Furthermore, the splitting of the absorption band is consistent with theoretical predictions either for deviation of the silver particle from spherical symmetry (6) or for interaction of the absorption band of a silver colloid with that of an absorbing coating (7). Among the remaining problems is the low SERS enhancement of 10^3 superimposed upon the RRS enhancement and the lack of coincidence of the SERS and absorption peaks of the sols. These are subjects of continuing experimental and theoretical investigations.

ACKNOWLEDGMENTS

This work was supported in part by NSF Grant CHE-801144, by Army Research Office Grant DAAG-29-82-K-0062, and by NIH Grant GM-30904.

REFERENCES

1. Kerker, M., Siiman, O., Bumm, L.A., and Wang, D.-S., Appl. Opt. 19, 3253 (1980).

2. Lin, J.-K., and Chang, J.-Y., Anal. Chem. 47, 1934 (1975).

3. Creighton, J.A., Blatchford, C.G., and Albrecht, M.G., J. Chem. Soc. Faraday Trans. 75, 790 (1979).

4. Kumar, K. and Carey, P.R., Can. J. Chem. 55, 1444 (1977).

5. Wang, D.-S., Kerker, M., and Chew, H., Appl. Opt. 19, 2315 (1980); also Kerker, M., Wang, D.-S., and Chew, H., Appl. Opt. 19, 4159 (1980).

6. Wang, D.-S., and Kerker, M., Phys. Rev. B. 24, 1771 (1981).

7. Wang, D.-S., and Kerker, M., Phys. Rev. B. 25, 2433 (1982).

TRANSIENT RAMAN SCATTERING STUDIES OF CHEMICAL KINETICS IN AQUEOUS MICELLAR AND SEMICONDUCTOR COLLOIDAL SOLUTIONS

S. M. Beck
L. E. Brus
Bell Laboratories
Murray Hill, New Jersey

I. INTRODUCTION

A. Transient Raman Spectroscopy

Transient absorption spectroscopy ("flash photolysis") has been the primary tool of fast kinetics in condensed phases for over 30 years. This technique has superb sensitivity and has been pushed to the 10^{-13} sec time scale with modern laser technology. However, the method provides little structural information about short-lived chemical species in the typical situation of diffuse optical spectra.

A practical form of transient vibrational spectroscopy would be a clear advance. Many groups have attempted to develop transient resonance Raman scattering since the early work of Bridoux and Delhaye on multichannel detection,[1] and Wilbrandt and coworkers on chemical applications.[2]

Transient Raman scattering obviously offers direct structural information about short lived species. It also often enables one to separate and identify multiple

ISBN 0-12-066280-9

transient species in cases where the corresponding optical spectra severely overlap. It allows use of isotopic substitution schemes designed to unravel reaction mechanisms. These advantages have been previously demonstrated in homogeneous solution chemistry.[3]

The chief experimental problem is low detection sensitivity. This problem is partially alleviated if one effectively uses resonance Raman enhancement by probing over a wide spectral range. The use of multichannel scattering detection (Figure 1) helps to keep the optical flux on the sample low, in order to avoid multiphoton processes.[3a]

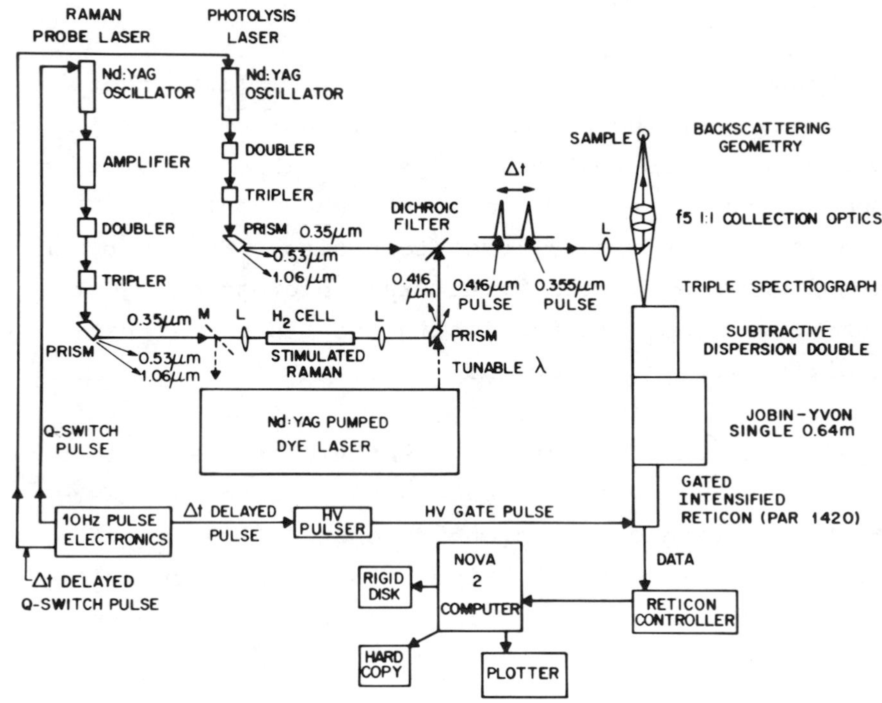

FIGURE 1. Apparatus schematic.

B. Microscopic Particulate Systems

Ionic micelles and semiconductor colloids represent microscopic analogues of biological membranes and semiconductor electrodes. At the molecular level, the same elementary surface processes occur in both "macroscopic" and "microscopic" systems. There are distinct experimental advantages in working with the microscopic analogues, when applying transient Raman scattering to surface (interface) problems. The effective surface area is extremely high, and the solution is optically clear, enabling accurate quantitative spectroscopic analysis. In the macroscopic systems one does have external potentiostatic control of redox potential. In the microscopic systems, redox potential can be established through appropriate choice of aqueous redox couples.

We describe our attempts to develop transient Raman scattering in these microscopically heterogeneous systems. The photochemistry of hydrophobic, often aromatic, species dissolved in micelles has been recently reviewed.[4] There are also two recent reviews of photoelectrochemistry occuring at semiconductor: electrolyte interfaces.[5]

II. MICELLAR SOLUTIONS

A. Detection Sensitivity

In systems without appreciable luminescence, the noise associated with the solvent Raman signal often limits transient Raman solute detection sensitivity. In hydrocarbon solvents this problem is especially severe. A substantial improvement in signal to noise can be achieved in micellar solutions while preserving the hydrocarbon-like environment of each solute molecule.[6] This effect occurs because the micellar solvent Raman signal (principally from water) is far weaker than that of pure hydrocarbons.

Figure 2a shows the 532 nm Raman spectrum of 10^{-3} M chrysene in hexane. Only hexane lines are observed, in contrast to 10^{-3} M in CTAB micelles (Figure 2c), where a distinct ground state (S_O) line is visible. If some chrysene molecules are excited to the lowest triplet state in hexane, then triplet resonance Raman lines occur on top of hexane lines (Figure 2b). In

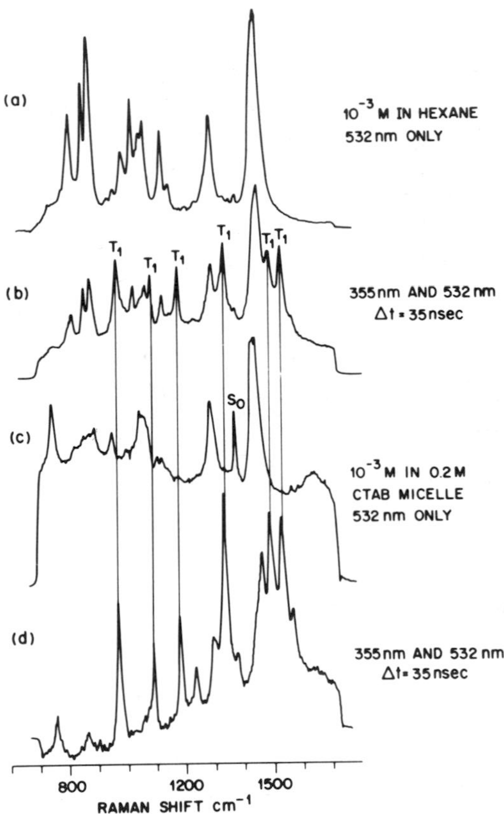

FIGURE 2. Comparison of chrysene Raman spectra in hexane and in CTAB micellar solutions.

the corresponding micelle experiment (Figure 2d), triplet lines are far stronger than solvent lines.

B. Interface Kinetics

Photoionization efficiencies of aromatic species inside micelles are distinctly dependent upon the sign of the micelle surface charge.[4] Negative surface charge (SDS micelles) increases ionization efficiency by stabilizing the resulting cation. Positive surface change (CTAB micelles) decreases ionization efficiency by destabilizing the cation.

We photoionized tetramethylbenzidine (TMB) in both SDS and CTAB micelles.[7] The time evolution of the TMB^+ resonance Raman spectrum at 416 nm was observed. In SDS micelles, the TMB^+ spectrum is present at our shortest delay times, and does not change with time. Bimolecular reaction products among TMB^+ molecules were not observed. In agreement with transient optical absorption studies,[8] it appears that TMB^+ remains stabilized inside the micelle.

In CTAB micelles (positive surface charge) a different and interesting reaction sequence occurs. The initial TMB^+ spectrum, due to TMB^+ inside the micelle, is the same (Figure 3b) as in SDS. This spectrum evolves in $\approx 10^{-6}$ sec into that of a perturbed TMB^+, labelled "I" for intermediate in Figure 3c. We suggest that "I" is TMB^+ on the micelle surface, with the positive charge localized on one end asymmetrically solvated by water. This spectrum evolves in $\approx 10^{-4}$ sec to that of TMB^{++} (Figure 3c), which appears to form by electron transfer as two micelles approach each other.

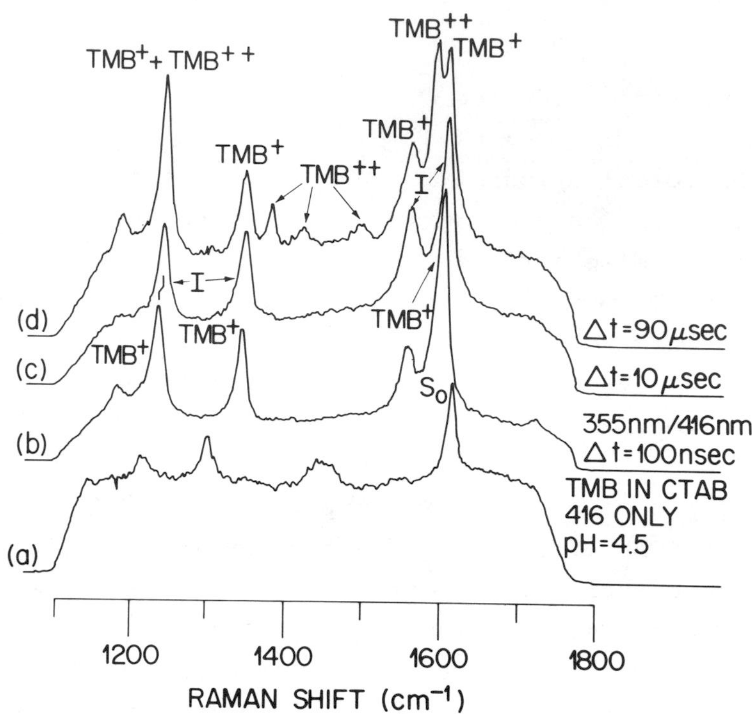

FIGURE 3. Time evolution of TMB^+ and TMB^{++} spectra following photoionization of TMB in CTAB micellar solution.

These experiments demonstrate the sensitivity of Raman spectra to solvation, and hence the possibility of detecting motion from one microenvironment to another.

III. SEMICONDUCTOR COLLOIDS

Semiconductor particles (~400 Å diameter) in aqueous colloids absorb photons at energies above the band gap. Absorption creates mobile electrons e^- and holes h^+ which may migrate to the particle surface and undergo redox processes with adsorbed chemical species. The small particles essentially act as photocatalysts; the carrier migration and surface reactions are not currently understood in detail.

Transient Raman scattering can provide a firm identification of initial reaction products, as well as detect possible surface structural modification and desorbtion kinetics. In an exploratory study, we have observed formation of reduced methyl viologen MV^+ (from MV^{++}) following ultraviolet excitation of colloidal TiO_2. The MV^+ spectra show a pH dependent risetime, as previously detected by transient absorption spectroscopy.[9] The earliest MV^+ resonance Raman spectra, at a delay of ≈5 nsec., are unchanged from those of aqueous MV^+. The surface reaction product MV^+ appears to be fully solvated by water at 5 nsec. It may be that MV^+ has desorbed and diffused away from the surface at this time. It may also be that MV^+ is created fully solvated in an e^- tunnelling process from the semiconductor surface. Further experiments at shorter times would be interesting.

REFERENCES

1. M. Bridoux and M. Delhaye, Adv. Infrared Raman Spectrosc. 2, Chpt. 4, (1976).
2. R. Wilbrandt, P. Pagsberg, K. B. Hansen, and K. V. Weisberg, Chem. Phys. Letters 36, 76 (1975).
3. Recent examples from our laboratory are: a) S. M. Beck and L. E. Brus, J. Chem. Phys. 75, 4934 (1981). b) S.M. Beck and L.E. Brus, J. Am. Chem. Soc. 104, 1805 (1982). c) S. M. Beck and L. E. Brus 104, 1103 (1982).

4. N. J. Turro, M. Gratzel, and A. M. Braun, Angew. Chem. Int. Ed. Engl. 19, 675 (1980).
5. a) A. Heller, Acc. Chem. Res. 14, 154 (1981).
b) A. J. Bard, J. Phys. Chem. 86, 172 (1982).
6. S. M. Beck and L. E. Brus, J. Chem. Phys. 75, 1031 (1981).
7. S. M. Beck and L. E. Brus, J. Am. Chem. Soc. (accepted).
8. S. A. Alkaitis and M. Gratzel, J. Am. Chem. Soc. 98, 3549 (1976).
9. D. Duonghong, J. Ramsden, and M. Gratzel, J. Am. Chem. Soc. 104, 2977 (1982).